New Age
Quantum Physics

New Age Quantum Physics

Al Schneider

Contents

Introduction

The goal of this book is to enable you, the reader, to understand quantum physics. If you read the first half of this book with a medium amount of diligence and some understanding of algebra, you should acquire a solid understanding of the subject. You will have earned the right to tell the person next to you on the bus, *I understand quantum physics.*

The first half of this book is devoted to a history of quantum physics. It travels back over 2000 years ago to begin with the early thoughts that became the core of quantum physics phenomena. An explanation of what quantum physics is begins about half way through this book. The author feels understanding the history of quantum physics is necessary for a complete understanding. In a sense, quantum physics is a way of thinking about the reality of the world around us. That way of thinking began with Aristotle and his contemporaries from many cultures of that time.

Is this Book Different?

How can this book succeed with such a complicated subject? Before an explanation of some phenomena is presented, we will first look at the people that discovered the phenomena. An effort is made to review the experiments that have been conducted down through the centuries. Instead of simply giving conclusions about the experiments, the attempt here is to show what was done and what was observed. Usually, someone interprets what was observed and the interpretation is reported to the world. Often what is reported is wrong, which we shall see as this book unfolds. A better way is to understand the events that led to such interpretations. This can appear to be a lesson in history rather than physics. The author feels this is necessary so when the moment comes to put the pieces together to grasp what quantum physics is, there will be enough understanding of each piece so they all fit together. It is a long road. Hopefully, the road is a straight line through this book.

Book Organization

Because of this approach, this book is organized somewhat historically and somewhat by discipline. The disciplines and the order of development in this book are:

- Light
- Electromagnetism
- Molecules and Atoms
- Electrons, Protons, and Neutrons
- Quantum Phenomena
- Matter Waves
- Future Speculation

Now, let's review the parts of this book. Normally, the task of an introduction is to describe what is in the book. This introduction adopts an additional task. This is to warn you of the level of math required to understand some of the material. Effort has been invested in keeping the math required to a minimum. As a result, many pictures have been included to enable the reader to visualize what is happening instead of relying on the math to describe the principles. When an equation is presented to explain some detail, a picture or pictures will be included to demonstrate the meaning of the equation.

The study of quantum physics begins with a study of light. If you attempt to determine where a discussion of light began, you find yourself reading about people almost 3000 years ago. Thus, we begin with the ancients.

The Ancients

This would be at the time of people like Aristotle and Epicurus. It was a time of developing basic thought and con-

templating the nature of the universe.. The basic words of
quantum physics emerged here. Quantum physics is about
light and matter waves. Some ancients compared light to the
waves of the ocean. Some thought light was little packets of
energy flying though space. These thoughts emerged from the
Orient, India, as well as from the Greek and Roman empires.

The Renaissance

The Dark Ages followed the fall of the Roman Empire. Not
much happened in Europe scientifically until the Renaissance. This was the time of Young, Newton and Huygens.
Who they are and what they did is discussed. The Renaissance included study of ancient philosophy and cod-
ification of the physical world around us. This included Newton's laws of motion. Those laws gave us power to build
machines and usher in the Industrial Age. Attempts are made
to describe how Renaissance men perceived the structure of
light. Several chapters are dedicated to describing light
waves as detailed by Young and Huygens. Some math is required to do this. A good deal of space is dedicated to these
ideas. Understanding light as described by these men is important, as these concepts are critical to how the scientific
community understands quantum physics. It is the beginning
point. Do not let the math stop you here. The pictures and associated descriptions should carry you through this area.

Electromagnetism

Secrets of electromagnetism are unraveled. The basis of radio
is established. The electric motor is born. Lights begin to
illuminate the world. These studies reveal more about the
structure of light and set the stage for seeing the true nature of
quantum physics. Water waves introduced us to the concept
of wave equations. The wave equation concept appears in
electromagnetism as well.

Molecules and Atoms

These chapters are about the discovery that the world around
us is made up of very small parts. The ancient concept that the
universe is composed of fire, earth, water and air is blown to

the wind.. A list of elements is created. The Periodic Table of
the Elements is constructed. The terms molecules and atoms
become everyday words. The scientific method becomes important, works and produces progress. Experimentation is
valued highly. However, there is another side to the scientific
method. That is organization and contemplation. The Periodic
Table of the Elements is an example that good organization
can yield very significant results. This simple ordered list was
instrumental in discovering protons and neutrons. These discoveries made a path leading toward an understanding of
quantum physics.

The Atomic Age
With the help of the Periodic Table of the Elements, science
unzips the atom and peers inside. The electron, proton and
neutron are found bouncing about. Mind you, we cannot see
these particles. This section shows how experiments revealed
what was inside the atom as well as the use of the Periodic
Table of the Elements. However, once we get our hands on
these basic sub-particles, they start to wiggle in weird ways.

A Second Revolution Emerges
Quantum weirdness emerges. A man named Max Planck
comes up with a constant that represents the birth of quantum
physics. The all-pervading ether suddenly vanishes. Einstein
popularizes the concept of special relativity. Just when the
world of physics was becoming a complete science, small
cracks grow into fissures.

A Paradigm Shift Occurs
The second revolution becomes a tidal wave. Common everyday particles like iron and salt are found to display wave characteristics just as light. The names of two men, Schrödinger and Heisenberg, become common names in the
world of science. Science fiction stories use these names and
their principles in many space opera plots.

More Quantum Weirdness

As this second revolution rolls on, more and more weirdness emerges. Some suggest time can go backwards, and that it is possible to travel faster than light. Some predict computers might fit inside wiggling particles.

Resolution and Speculation

At this point in the book, the present understanding of quantum physics is described. By now, you should have a sound understanding of what quantum physics is. If someone asks you what it is, you should be able to answer his or her question. Bear in mind, those that study this subject the most are the first to say no one understands it. However, you will know why this is true and will not shy away from answering such questions.

The book continues with the author's understanding of what quantum physics is.

What is a Photon?
The properties of photons are outlined and an explanation of photon structure is offered.

What is Subspace?
Einstein's special relativity is explained. This opens a door to understanding the foundations of space and time. As they come into view, we can see a structure that supports the existence of the universe. We develop a chance to go to the edge of existence and see from which we came. The goal is to answer the question, "What are space, time, mass and energy?" Bear in mind that to answer this question, one cannot use the words space, time, mass or energy, or anything similar. What then would the answer be? One interesting fact emerges: We are beings of light.

What is Space, Time, Mass, and Energy?
At this point in the book, each of these has been described to some degree. Then, we will take up each and attempt to

clearly, define the mathematical significance of each. Units
are clearly defined. The subspace concepts are connected to
measurable quantities in the real world. In addition, the properties of each of these are shown how they relate to Special
Relativity. In this area, each facet of special relativity is developed from the theory presented in this book. Then a complete description of kinetic and potential energy is offered.
There is some math here. The main text attempts to explain
this with pictures and discussion. For those readers that wish
a more detailed explanation, the mathematical developments
are placed in appendices at the end of the book so if you wish
to skip them, they will not be in the way.

Finally, What Is the Universe Made of?
If we consider string theory, we see that the universe is made
of vibrating strings. Part of this theory is that the vibration
determines the nature of the particle being studied. Here is a
major question: What is vibrating? Likewise, general relativity explains that space is warped and that is what causes
gravity. But again, what is warped? Some meta-physical
types and even physics types tell us that everything consists
of vibration. Again, what is vibrating? Then we are told that
all is energy. This does not make sense. Energy is the motion
of matter. When this author hears people say that all is energy, this implies to him that there is no matter. This is resolved in the last part of the book. The question, "What is
matter?" is answered.

Conclusion
The path presented seems to be a history of quantum physics.
Some of the items offered will not appear to be relevant to the
discussion at hand. The author is attempting to present a
broader perspective of what is going on. More than just a discussion of where quantum physics came from, there is an attempt to show the difficulties of the people that generated the
concepts. They had unique personalities. They often did not
know where they were going. They made blundering mis-
takes and produced valuable insights into the world around
us. An attempt is made to show that one person will consider

an insight from someone that has gone before. The insight is digested and found wanting. The person will correct the failures in the insight and produce a meaningful conclusion. Then, in turn, the next person in line will go through the same process. This long chain of events eventually brings us to the point we are today. If history repeats itself, these events will continue as we increase our understanding of the universe. The author hopes you, dear reader, will grasp this continuum of the whole process. Possibly, understanding the people behind the theories is as important as understanding the theories. Studying the history is justified. As a result, you will understand quantum physics as anyone else can.

Part 1

The Ancients

Light and wave theory play a significant role in quantum physics. If one looks into where the words used in quantum physics come from, one will find that a few originated thousands of years ago. Thus, it is fitting to review how light was understood by the ancients. In this chapter, we are going to look at concepts of waves that ultimately became intertwined with quantum physics.

We will trace a few other items as well. One is the ancient's thoughts of the structure of matter. Then another is a casual glance at how they arranged their thought patterns to approach an understanding of the world around themselves.

In this book, the study of quantum physics includes a philosophy of how the world is analyzed. That plays an important role in the explanation of understanding the subject. Understanding how a scientist perceives something tells us something about the conclusions he comes to. That extra bit of information can change our point of view.

Chapter 1

Common Ancient Beliefs

The source of some common ancient concepts is buried in antiquity. Did several cultures come up with these independently? Perhaps some wise person conceived of the ideas and traveled to distribute it or another traveled wide and far to get the information. Regardless, our goal here is to see what the ancients thought and how it affects our understanding today. As mentioned in the introduction, the first half of this book is about how man perceived reality over the ages. Half way through this book, we discover that those perceptions were in error. This discovery opens the door to understanding quantum physics. Thus, we must first grasp what those perceptions were to see how our understanding must change to appreciate quantum physics.

Before discussing some ancient personalities, we look at some common concepts from somewhere in our deep past. We begin with a concept of sight.

Ancient Vision Concept

One of the first items from antiquity is the way the ancients perceived sight.

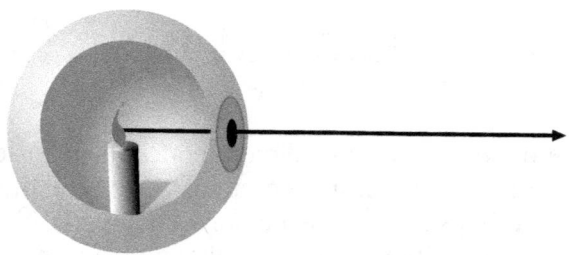

The ancient understanding was that something emanated from the human eye that enables man to see.

From where could this idea have come? Consider ancient man. They could see a tree. They could touch the tree with their fingertip. While both are not totally understood; touching the tree with one's finger probably seemed more real. One could ask another, *How did you know it was there?* The answer could be, *I reached out and touched it.* That, according to this theory is very understandable. One can touch a tree and see their finger do so. Apparently, it was not observed that sight was used to verify the sequence of actions. Then, one thought about how one could simply see the tree. If it were clear how one could sense the tree with a reach and a touch, then there must be something in the eye that could reach out and touch the tree and report that finding to the human mind. Therefore, it made "sense" that something in the eye could reach out to touch a tree. However, reaching out with your eye to touch something did not make sense. The ancients must have wondered how sight worked.

Now, if you felt that this was true, how would you explain it to someone else? What device could do something such as this? Well, it was observed that at night when there is no sun, a fire or a candle could enable one to see a tree. Therefore, a conclusion could be that a lit candle had some power to reach out and touch. Then, there must be something like a candle in the eye that had this power. This author believes that this kind of series of events would lead those that wanted to teach, would teach that there was something in the eye emanating light that traveled to that which was seen.

Air, Fire, Earth, and Water

The concept of the four classical elements in the Western tradition appears in Babylonian mythology. In an ancient text, Enuma Elis, a text written between the 18th and 16th centuries BC, describes four cosmic elements: the sea, earth, sky, and wind.

The concept of essentially the same four elements was similarly found in ancient India, where they formed a basis of analysis in both Hinduism and Buddhism. In Hinduism, particularly in an esoteric context, the four states-of-matter describe matter, and a fifth element describes that which was beyond the material world (non-matter). Similar lists existed in ancient China and Japan. In Buddhism the four great elements, to which two others are sometimes added, are not viewed as substances, but as categories of sensory experience.

The point here is that the origination of this concept seems to be buried in antiquity. Could this concept have originated even before these just mentioned? Does this simply emerge when uneducated man asks about what is around him? Ancient man had no cars or airplanes. They had what they found in the ground or what grew from the ground. They must have realized there was air around them. One could feel its presence when the wind hit ones face. However, it seemed different from the earth. One could see fire as something different from both. It was neither quiet like the earth nor soft like air. Then there was the water that seemed to be everywhere. It was soft but could not support itself. It would give way to everything.

Later, several ancient philosophers were said to conceive of this idea. Surely they taught it. The question is did they come up with this, did they hear it from some traveler, or did they get it from yet another source? This author has no resources to establish what happened. It can only be taken as a lesson of how the ancient egos fell to the pressures of happenstance and the drive for power that an educated person can have.

Light Particle Theory
Another item whose source seems to be buried in antiquity is the idea that light consists of particles shooting through space. This seems to be quite common until Aristotle or those around him proposed light is a wave.

Conclusion
A goal of this book is to trace the very early concepts that eventually evolved into quantum physics concepts. The point here is that if one searches for the information, one can only get a glimmer of what happened before the time of the Romans and Greeks. In studying that material, one gathers that those philosophers originated the material related to sight, the structure of light and the structure of matter. However, the glimmers from earlier times suggest that these concepts were found in many old societies. In a sense, this is a fitting beginning for a book such as this. We are after the roots of the subject. We are finding more. We not only get what was considered scientific fact but we see that the egos of people are intertwined with what they have recorded. The study of quantum physics is not only a study of mechanical structures but also a study of human frailty.

Chapter 2

The Ancients

The following personalities and groups popped up when doing a search for those that had anything to do with the history of quantum physics. This occurred while searching for ancient wisdom of light, particles, and so on. We begin with an ancient Indian Hindu school.

Vaisheshika Hindu School
(500BC), Indian Hindus.

They conceived of molecules and light particles.

The Vaisheshika school taught an atomic theory of the physical world on the non-atomic ground of ether, space and time. That is, they were not automatists in the sense the universe was built that way. They believed that the stuff other than ether, space and time was unitized. The basic atoms in this theory are earth (prthivl), water (apas), fire (tejas), and air (vayu). These are not to be confused with the ordinary meaning of these terms. In this theory, these atoms are taken to form binary molecules that combine further to form larger molecules. [1]

Light rays are taken to be a stream of high velocity fire (tejas) atoms. [1]

Two things are noteworthy here. One is that they conceived of universal building blocks that were combined to make what we call molecules. The other item of interest here is the idea that they thought of light as a form of some kind of particle flashing through space.

Leucippus
(First half of 5th century BC), Greek philosopher.

He was the first Greek to propose a theory of atomism.

Little is known about Leucippus. Leucippus was an Ionian Greek (Ionia, at present Turkey). Around 440 B. C. or 430 B. C. [2] Leucippus founded a school at Abdera. [3] What is known is that Aristotle and Theophrastus explicitly credit Leucippus with the invention of Atomism. [4] That is the idea that everything consists of small indivisible particles.

Democritus
(460 BC - 370 BC), Ancient Greek philosopher.

He developed the atomic theory to a higher level than those before him. [5]

Democritus was born in the city of Abdera in Thrace, an Ionian colony of Teos. [6] He was born in the 80th Olympiad (460-457 BC) His father was a very wealthy man. Democritus spent the inheritance which his father left him on travels into distant countries, to satisfy his thirst for knowledge. He traveled to Asia, and was even said to have reached India and Ethiopia. We know that he wrote on Babylon and Meroe; he apparently visited Egypt. After returning to his native land he traveled throughout Greece to acquire knowledge of its culture. [7]

Democritus died at the age of 90, which would put his death around 370 BC, but other writers have him living to 104, or even 109. [7]

His attitude toward life was described in two ways. Some said he was cheerful, and was always ready to see the comical side of life. Others said he was popularly known as the Laughing Philosopher for laughing at human follies as in scoffing and mockery. [7] Plato is said to have disliked him so much that he wished all his books burnt. [8] Many consider Democritus to be the "father of modern science". [9]

While many seemed to espouse the atomic theory, Democritus attempted to define the concept with detail. He proposed 6 points to his theory of atoms. [10] These were:

1. Matter consists of tiny, indivisible, invisible atom bits.

2. There is a void, which is empty space between atoms.

3. Atoms are completely solid.

4. Atoms are homogeneous, with no internal structure.

5. Atoms are different in their sizes, shapes, and weights.

6. Atoms are the building blocks of life.

Using analogies from sensual experiences, he gave a picture or an image of an atom that distinguished them from each other by their shape, their size, and the arrangement of their parts.

Moreover, connections were explained by material links in

which single atoms were supplied with attachments. Some were with hooks and eyes while others with balls and sockets. [11] Democritus reasoned that the solidness of the material corresponded to the shape of the atoms involved. Thus, iron atoms are solid and strong with hooks that lock them into a solid.

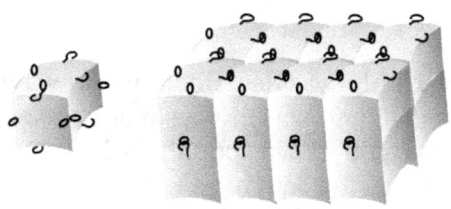

Iron, smooth and hard

On the other hand, water atoms are smooth and slippery and salt atoms, because of their taste, are sharp and pointed.

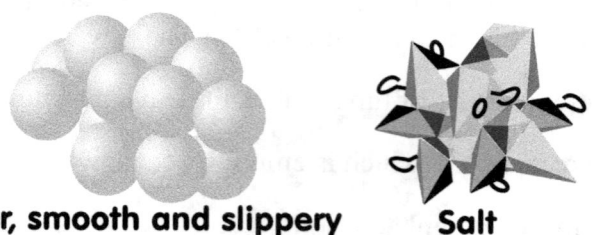

Water, smooth and slippery   Salt

Note that the Democritean atom is the smallest particle of some element. It was not a theory of atoms we are familiar with today that combine to make molecules.

Empedocles
(490-430 BC), Greek philosopher.

He challenged the concept of the eye emanating something that enabled eyesight and modified the concept adding that sunlight merged with that which emanated from the eye to enable sight.

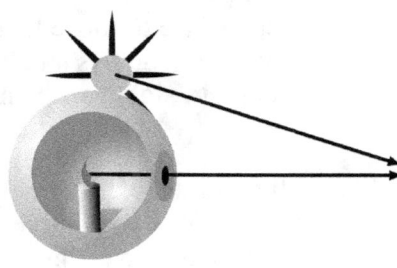

Empedocles was born in Agrigentum (Acragas) Sicily to a distinguished family. Very little is known about his life. His father Meto, seems to have been instrumental in overthrowing the tyrant of Agrigentum. That, presumably, was Thrasydaeus in 470 BC. Empedocles continued the democratic tradition of his house by helping to overthrow the succeeding oligarchic government. He is said to have been magnanimous in his support of the poor, severe in persecuting the overbearing conduct of the aristocrats, and he then declined the sovereignty of the city when it was offered to him. [12]

He traveled a great deal. He went to Athens and the newly-founded colony of Thurii, 446 BC. There are reports of him traveling far to the east to the lands of the Magi.[13]

According to Aristotle, he died at the age of sixty.

Empedocles' is best known for being the originator of the theory of the four classical elements ; fire, air, earth and water. This author doubts he originated that and suspects it was wisdom adopted from others before him or around him. [14] He

believed that Aphrodite made the human eye out of the four
elements and that she lit the fire in the eye, which shone out
from the eye making sight possible. He, however, observed a
problem with this scenario. If there were a fire in the eye to
make sight possible, then one could see during the night just
as well as during the day, so Empedocles postulated an interaction between rays from the eyes and rays from a light
source such as the sun. The assumption here is that right or
wrong did not matter. It made sense to those he preached to.
Apparently the message from Empedocles stuck for most
Europeans allowed this to explain sight for the next 2,000
years. [15]

It was written that his brilliant oratory, his penetrating knowledge of nature, and the reputation of his marvelous powers,
including the curing of diseases, and averting epidemics, produced many myths and stories surrounding his name. He was
said to have been a magician and controller of storms, and he
himself, in his famous poem *Purifications* seems to have
promised miraculous powers, including the destruction of
evil, the curing of old age, and the controlling of wind and
rain. [12]

There are two factors of interest here. One is that when a
person of authority speaks, that which is spoken of is believed. This seems to occur when no one knows the answer
and the speaker speaks with authority. The second is that
when attempting to explain an accepted theory that does not
make sense, man will add something that seems to make
sense. One tends to lack the initiative to simply challenge the
first non-acceptable concept and then adds something that
"fixes" it.

Epicurus
(341 BC - 270 BC), Greek philosopher.

Epicurus is a key figure in the development of science and the
scientific method because of his insistence that nothing

should be believed, except that which was tested through direct observation and logical deduction. [16]

His parents, Neocles and Chaerestrate, were both born in Athens. His father, a citizen, had emigrated to the Athenian settlement on the Aegean island of Samos about ten years before Epicurus's birth. As a boy, Epicurus studied philosophy for four years under the Platonist teacher Pamphilus. At the age of 18 he went to Athens for a two-year term of military service. [16]

After the death of Alexander the Great, Perdiccas expelled the Athenian settlers on Samos to Colophon. After the completion of his military service, Epicurus joined his family there. He studied under Nausiphanes, who followed the teachings of Democritus. In 311/310 BC Epicurus taught in Mytilene but caused strife and was forced to leave. He then founded a school in Lampsacus before returning to Athens in 306 BC. There he founded The Garden, a school named for the garden he owned about halfway between the Stoa and the Academy that served as the school's meeting place. [16]

Epicurus never married and had no known children. He suffered from kidney stones, [16] to which he finally succumbed in 270 BC at the age of 72.

He regularly admitted women and slaves into his school, introducing the new concept of fundamental human equality into Greek thought. He was one of the first Greeks to break from the god-fearing and god-worshiping tradition common at the time, even while affirming that religious activities are useful as a way to contemplate the gods and to use them as an example of the pleasant life. Epicurus participated in the activities of traditional Greek religion, but taught that one should avoid holding false opinions about the gods. Epicurus believed that the common person thought that the gods *send great evils to the wicked and great blessings to the righteous who model themselves after the gods.* Apparently Epicurus believed the gods did not concern themselves at all with human beings. [16]

Aristotle
(384 BC - 322 BC), Greek philosopher.

He is included in this brief history of the roots of quantum physics for he seems to be the first that claims light is a wave. Since then, light and wave theory became solidly intertwined.

Aristotle was born in Stageira, Chalcidice about thirty-four miles east of modern-day Thessaloniki. [17] His father, Nicomachus was the personal physician to King Amyntas of Macedon. Aristotle was trained and educated as a member of the aristocracy. At about the age of eighteen, he went to Athens to continue his education at Plato's Academy. Aristotle remained at the academy for nearly twenty years before leaving Athens in 348/47 BC. He then traveled with Xenocrates to the court of his friend Hermias of Atarneus in Asia Minor. Aristotle married Hermias's adopted daughter (or niece) Pythias. She bore him a daughter, whom they named Pythias. Soon after Hermias' death, Aristotle was invited by Philip II of Macedon to become the tutor to his son Alexander the Great in 343 BC.[18]

Aristotle was appointed as the head of the royal academy of Macedon. In his *Politics*, Aristotle states that only one thing could justify monarchy, and that was if the virtue of the king and his family were greater than the virtue of the rest of the citizens put together. Tactfully, he included the young prince and his father in that category. Aristotle encouraged Alexander toward eastern conquest, and his attitude towards Persia was unabashedly ethnocentric. In one famous example, he counsels Alexander to be, *a leader to the Greeks and a despot to the barbarians.* He said he should look after the former as after friends and relatives, and to deal with the latter as with beasts or plants.[19]

By 335 BC he had returned to Athens, establishing his own

school there known as the Lyceum. Aristotle conducted courses at the school for the next twelve years. While in Athens, his wife Pythias died and Aristotle became involved with Herpyllis of Stageira. She bore him a son whom he named after his father, Nicomachus.

When Alexander died, Aristotle fled the city to his mother's family estate in Chalcis. Some suspected Aristotle responsible for Alexander's death. Aristotle had publicly denounced Alexander's pretense of divinity. [20]

Why are we including Aristotle in this discussion? He concluded that light travels in a manner similar to waves in the ocean. [21] What is significant about this? This author believes that Aristotle was responsible for associating the term wave to light and having it stick all the way to our time. As you will see, the term wave is probably the most common technical term in this book.

What kind of observations could Aristotle and his fellow Greeks have done to conclude that light was a wave as the water waves that came in from the sea?

Consider what they might have experienced standing near a large bell. When struck, it would emanate sound that could be heard far away. If one were to touch the bell when making its sound, the fingers could sense the metal of the bell vibrating. Perhaps, if the bell were large enough, one could feel the vibration in the air as the sound moved through it.

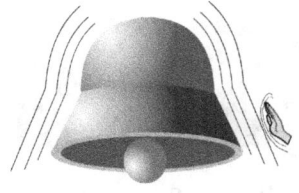

If one struck a small bell and then touched the surface of the

water in a bowl, one would see ripples like waves moving
away from the bell.

This would suggest a strong similarity between water waves
and sound waves.

Would it be a long mental jump to think Aristotle would see
water waves,

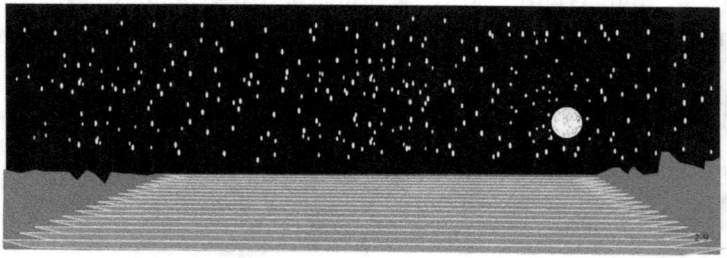

hear sound waves coming from a far hilltop and see the flash
of some silver thing in the distance; and conclude that light
was like the waves of the sea?

Certainly associative thinking had to play its role here. Aristotle studied many subjects and was accustomed to theorizing about all of them. He certainly had to have become accustomed to comparing a known thing to an unknown thing, arriving at some conclusion about the unknown thing even if to provide material to voice in his lectures. Regardless, that voice attached the word wave with that of light.

That stuck for over 2000 years.

Despite Aristotle's popularity, many today question if the mass of what he produced was Aristotle's own. [22] A point here is that Aristotle was a gatherer of information. It was offered, not as intelligence, but to make him look wise and knowledgeable. This person taught Alexander the Great. Unfortunately, this seems an example of those using scientific terminology to gain power over those that do not know what the terminology means.

Titus Lucretius Carus
(99 BC -55 BC), Roman poet and philosopher.

His only known work is *On the Nature of Things.* In this poem he presents a particle theory of light.

Little is known of his life. Any knowledge was derived from comments and notes in the works of others. [23] The reader should be aware that the Greeks made no distinction between a wise man, a thinker, a philosopher, or a scientist. Thus, a poet could easily be a theoretical physicist. Descriptions were expressed in verse. Amongst other scientific factors, the poem discussed atomic theory, conservation of matter and energy and a system whereby atoms could combine to form larger things. In this theory, atoms had small loops and hooks so atoms could link together. This appears to duplicate thoughts of Democritus.

While not accurate, it expresses the understanding that our world is a combination of some subset of things that form what we now call molecules.

Lucretiusis is recognized here for he was another ancient voice speaking up for a particle theory of light. The following suggests that even a candle is emitting particles into space.

Dignāga & Dharmakirti
(500 AD & 700 AD) Indian Buddhists.

They proposed an idea that the universe consisted of small packets of energy.

Indian Buddhists, such as Dignāga in the 5th century and Dharmakirti in the 7th century, developed a type of atomism that is a philosophy about reality being composed of atomic entities that are momentary flashes of light or energy. They viewed light as being an atomic entity equivalent to energy, similar to the modern concept of photons. They also viewed all matter as being composed of these light/energy particles. [24] The following picture attempts to show this. Not only does the candle emit light/energy particles but even we are composed of the particles as well.

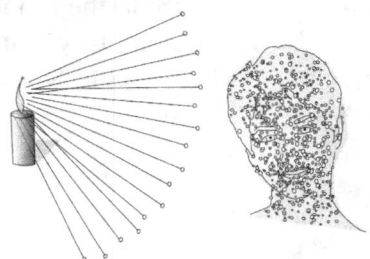

Ibn al-Haytham
(965 - 1039), Persian scientist.

Ibn al-Haytham's optical research utilized systemic and methodological experimental techniques and controlled testing. Combining classical physics with mathematics was far ahead of his time.

Ibn al-Haytham was one of the most advanced of all scientists before the Renaissance. His scientific method was very similar to the modern scientific method and consisted of the following procedures: [25]

1. Observation

2. Statement of problem

3. Formulation of hypothesis

4. Testing of hypothesis using experimentation

5. Analysis of experimental results

6. Interpret data and formulate conclusion

7. Publication of findings

He rejected the eye-beam idea that has stood with for over
2000 years. Ibn al-Haytham came to the modern conclusion
that light external to the eye fell on objects and was reflected
into the eye at right angles enabling our sight. [26] Of course,
his idea did not become popular.

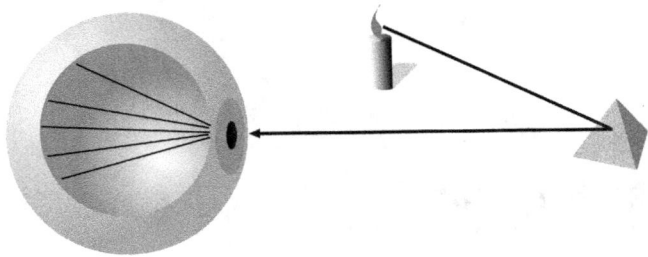

He was born in Basra, in the Iraq province of the Buyid Persian Empire. [27] He probably died in Cairo, Egypt. He was
educated in Baghdad. During a time in Buyid Iran, he worked as a civil servant and read many theological and scientific
books. [28]

One account of his career has him called to Egypt by Al-
Hakim bi-Amr Allah, ruler of the Fatimid Caliphate, to regulate the flooding of the Nile, a task requiring an early attempt
at building a dam at the present site of the Aswan Dam. [29]
His field work made him aware of the impracticality of a
scheme he developed, and fearing the caliph's anger, he
feigned madness. He was kept under house arrest from 1011
until al-Hakim's death in 1021. [30] During this time, he
wrote his influential *Book of Optics.*

He ventured to Egypt by 1038.[31] During his time in Cairo,
he became associated with Al-Azhar University, as well the
city's *House of Wisdom.* After his house arrest ended, he
wrote scores of other treatises on physics, astronomy and
mathematics.

Conclusion

Perhaps the most value gained here is to see that science is born of normal men. Men that are curious about their surroundings, anxious to use knowledge as power, and willing to get that power from other people without giving credit. The concept of the elements being air, fire, water and earth seem to emanate in many cultures. However, "learned" men often talked about it as if they conceived of it. Some commonly lectured on their perceptions as truth without establishing the truth of what they spoke. Then, there were the dedicated that developed the scientific method to make true progress.

Aristotle's ideas about light and waves are the highlight of this period and were carried from his time for over 2000 years and have not fallen to this day.

Before leaving this chapter, this author would like to introduce a concept that seems to exist in all of intellectual pursuit. It is taken from a concept called Occum's Razor. Occum was a monk that lived during the time of the Renaissance. He pursued scientific knowledge. Though a religious man, he apparently saw a great fault in using the concept of God to explain everyday phenomena. He proposed that using God to explain things was an unworkable path. Thus he proposed what became known as Occum's Razor. When studying some concept Occum's Razor suggests that one should cut (as in using a razor) away that which leads to complexity or non-understanding. Rather, one should examine a path of logic that could possibly lead to some intelligent conclusion. Today this is referred to as a simpler way. This concept is often interpreted to mean that one should select the simplest explanation. That is, if one is trying to solve a problem and one has many possible solutions in front of them, the solution is always the simple one. This is an inaccurate understanding. Occum's point was that if there were several solutions presented, select the simpler path to study. He believed that would lead to understanding faster.

Now, while this is interesting, why is this being presented
here? These comments are triggered by the attempt to explain
why some thought the light shining from the eye to see was
incomplete and attempted to explain it with something else
that was incomplete. The masters of knowledge noticed that
the light shining from the eye did not enable eyesight in a dark
room. They did observe that if a candle were in the room, one
could see. Their conclusion was that, whatever emanated
from the eye, the emanation mixed with external light sources
to enable sight. This represents the inability of man to let go
of some valued understanding and fix it with something just
as unreliable. This mechanism occurs in science repeatedly.
A term coined by this author to name this awkward process
is, Occum's Band-Aid. That is, when someone proposed that
external light mixed with the internal light emanated from the
eye to enable sight, that proposal was an Occum Band-Aid.

Part 2

The Renaissance

The Renaissance began in the 1700's. Europe had been ruled for centuries by the church. There was no development in what they called the natural sciences. However, the church suffered internal strife. The church split. A large section of it became the Lutheran church. As a result, those outside of the bickering began to look at concepts normally controlled by those in power. Part of this was a study of ancient wisdom, Greek and Roman philosophy. [1]

Chapter 3

The Renaissance Men

In science the Renaissance became a scientific revolution. [1]

The men representing change in the Renaissance were polymaths. They were doers of all things. They tended to be philosophers, theologians, scientists, astronomers and even civil servants. They were Renaissance men. [1]

Here, of course, we are considering those that had some effect on the development of quantum physics. Sometimes it is not what they did that counted but the thought process they brought to bear on their work.

We will begin this discussion with a bit about René Descartes [2] and Pierre Gassendi. [3] Their thinking led the way to understanding the makeup of matter. Next, we review the lives of Robert Hooke [4] and Sir Isaac Newton. [5] Although they were somewhat at odds with each other, together they formed our understanding that enabled those that followed to build the engines of modern society. Then we consider Francesco Maria Grimaldi [6], Christian Huygens [7], Thomas Young [8], Augustin-Jean Fresnel [9], and Jean Bernard Leon Foucault. [10] These men advanced the study of light and waves; the essence of quantum physics.

René Descartes
(31 March 1596 - 11 February 1650), French philosopher, mathematician, physicist, and writer.

René Descartes rejected the idea of many ancient philosophers but agreed with Aristotle about the wave theory of light.

Descartes was born in La Haye en Touraine (now Descartes), Indre-et-Loire, France. When he was one year old, his mother Jeanne Brochard died. His father Joachim was a member in the provincial parliament. Around the age of eleven, he entered the Jesuit Collčge Royal Henry-Le-Grand at La Flčche. After graduation, following his father's wishes that he should become a lawyer, he studied at the University of Poitiers, earning a Baccalauréat and Licence in law in 1616.

He subsequently turned his back on the pursuit of law and pursued philosophy and science. In 1618, Descartes joined the International College of War of Maurice of Nassau in the Dutch Republic. In 1622, he returned to France, and during the next few years spent time in Paris and other parts of Europe. He moved to La Haye in 1623. He also sold all of his property and invested in bonds. This provided Descartes with a comfortable income for the rest of his life. He returned to the Dutch Republic in 1628, where he lived until September 1649.

In Amsterdam, he had a relationship with a servant girl, Helena Jans van der Strom, with whom he had a daughter, Francine, who was born in 1635 in Deventer. Francine Descartes died in 1640 in Amersfoort, of Scarlet Fever.

In 1633, Galileo was condemned by the Roman Catholic Church, and Descartes abandoned plans to publish *Treatise on the World*, for fear of a similar treatment.

Descartes continued to publish works concerning both mathematics and philosophy for the rest of his life. In 1643, Cartesian philosophy was condemned at the University of Utrecht, and Descartes began his long correspondence with Princess Elisabeth of Bohemia. In 1647, he was awarded a pension by the King of France.

Rene Descartes died on 11 February 1650 in Stockholm,
Sweden, where he had been invited as a teacher for Queen
Christina of Sweden. The cause of death was said to be pneu-
monia. In a book, *The Mysterious Death of Rene Descartes*,
the German philosopher Theodor Ebert asserts that Descartes died from an arsenic-laced communion wafer given to
him by a Catholic priest. He believes that Jacques Viogué,
a missionary working in Stockholm, administered the poison
because he feared Descartes' radical theological ideas would
derail an expected conversion to Roman Catholicism by the
monarch of Protestant Lutheran Sweden.

In 1663, the Pope placed Descartes' works on the Index of
Prohibited Books.

He was buried in a graveyard mainly used for non-baptized
infants in Adolf Fredriks kyrkan in Stockholm. Later, his remains were taken to France and buried in the Abbey of Saint-
Germain-des-Prés in Paris.

Descartes' believed that matter was non-divisible. That is, regardless how much you magnified something, you would
never find a smaller part of it.

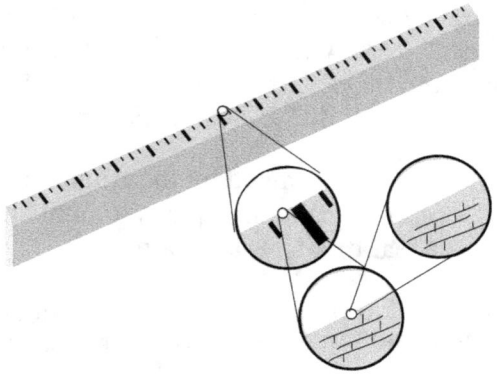

While Descartes' missed the mark on several issues, his mathematical theories provided the basis for the future development of calculus. Descartes also created analytic geometry, the Cartesian coordinate system and discovered an early form of the law of conservation of momentum.

Pierre Gassendi
(1592-1655), French philosopher, priest, scientist, astronomer, and mathematician.

He believed that the universe must consist of small things like quanta. This was in opposition to the idea of a continuous universe proposed by Descartes.

At a very early age, he attended the college at Digne, France, where he displayed a particular aptitude for languages and mathematics. Soon afterwards, he entered the University of Aix-en-Provence, to study philosophy. In 1612, the college of Digne invited him to lecture on theology. Four years later he received the degree of Doctor of Theology at Avignon. Then he assumed the chair of philosophy at Aix-en-Provence University. He gradually withdrew from theology.

In 1623 the Society of Jesus took over the University of Aix-en-Provence forcing him to return to Digne. Subsequently, he traveled to Grenoble. In 1642 he began to study the work of Rene Descartes. His objections to the fundamental propositions of Descartes appeared in print in 1642. In 1645, he accepted the chair of mathematics in the College Royal in Paris, and lectured for several years with great success.

In 1648, ill health compelled him to give up his lectures at the College Royal. Around this time, he reconciled his disagreements with Descartes.

In 1653, he returned to Paris and resumed his literary work. A lung problem led to his death in 1655.

In his philosophical writings, he tended to present problems of import with an apparent intention of explaining them.

However, his writings rarely had impact on the very problems he brought to light. It was said he wrote much about Aristotle's faults without knowing the work of the man.

Gassendi was also an active observational scientist, publishing the first data on the transit of Mercury in 1631. Some have said that Gassendi was one of the first thinkers to formulate the modern "scientific outlook" of moderated skepticism and empiricism.

Gassendi argued that there must be a bottom or "substance," to the physical universe. This was in opposition to the writings of Descartes. Gassendi expressed the idea that there must be some distance or extent to space that could not be divided. The following picture shows magnifying a ruler. If it is magnified again and again, eventually you will see some small part of its existence that can get no smaller. If that were divided there would be nothing.

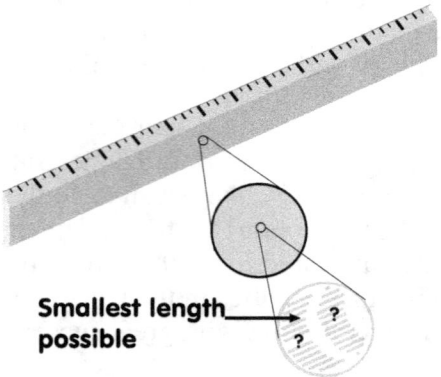

Smallest length possible

Do not confuse this with getting down to leptons, fermions and perhaps strings that we now know or have an idea about. Gassendi's theory suggests that those can be reduced to some smaller unit until there is no way to subdivide them.

He also proposed a particle theory of light which was published posthumously in the 1660s. Newton was apparently in-spired with this work.

Robert Hooke

(18 July 1635 - 3 March 1703), English philosopher, arch-itect and polymath.

Robert Hooke published a wave theory of light in 1660. This preceded Newton's corpuscular theory of light published in 1704.

Robert Hooke was born in Freshwater on the Isle of Wight to John Hooke and Mirena Blazer. Robert was the last of four children, two brothers and two sisters, and there was an age difference of seven years between him and the next youngest. Their father served the Church of England, specifically as the curate of Freshwater's Church of All Saints. His two brothers were also ministers. Robert Hooke was expected to follow and join the Church.

While growing up, Robert Hooke was fascinated by mech-anical devices. He dismantled a brass clock and built a wood-en replica that, by all accounts, worked "well enough". He learned to draw, making his own materials from coal, chalk and Iron ore.

On his father's death in 1648, Robert was left a sum of forty pounds that enabled him to buy an apprenticeship. With his poor health throughout his life but evident mechanical fa-cility, his father had it in mind that he might become a watch-maker. Shortly after that, however, he managed to enter Westminster School in London where he embarked on his life-long study of mechanics.

While playing organ at Christ Church, Oxford in 1653, Hooke met the natural philosopher Robert Boyle. Hooke could build machines and was an adept mathematician. Boyle had neither of these talents so hired Hooke. Some suggested that Hooke might well have developed the mathematics for Boyle's law.

Robert Hooke spent his life largely on the Isle of Wight, at Oxford, and in London. He never married. On 3 March 1703, Hooke died in London, having amassed a sizable sum of money, which was found in his room at Gresham College. He was buried at St Helen's Bishopsgate, but the precise location of his grave is unknown.

Hooke was often criticized personally. He was accused of plagiarizing other's material. However, time has revealed that many of these were attacks by various people that opposed his successes.

Hooke is known for his law of elasticity (Hooke's law). The following picture depicts Hooke's law.

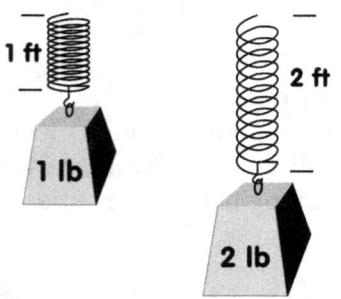

An elastic material will stretch proportinal to the force applied.

The picture shows that if a weight on a spring is doubled, the spring will stretch twice as long. This law is taught in every beginning class of physics that exists. He is also known for his book, *Micrographia*, and for first applying the word *cell* to describe the basic unit of life.

Even now, there is much less written about him than might be expected from the sheer industry of his life. He was at one time simultaneously the curator of experiments of the Royal Society and a member of its council, Gresham Professor of Geometry and a Surveyor to the City of London after the Great Fire of London.

In the capacity of surveyor, he appears to have performed more than half of all the surveys after the fire. He was also an important architect of his time, though few of his buildings now survive and some of those are generally misattributed. He was instrumental in devising a set of planning controls for London whose influence remains today.

His ideas about gravitation, and his claim of priority for the inverse square law, are part of a disagreement with others. He was granted a large number of patents for inventions and refinements in the fields of elasticity, optics, and barometry. The Royal Society's Hooke papers (recently discovered after disappearing when Newton took over) open up a modern reassessment.

Sir Isaac Newton

(4 January 1643 - 31 March 1727), English physicist, mathematician, astronomer, natural philosopher, alchemist, and theologian.

Isaac Newton was firmly opposed to the wave theory of light proposing that light consisted of small particles or something he called corpuscles. He presented this in 1704. The conflict between wave theory and particle theory would last well into the age of quantum physics.

He was born at Woolsthorpe Manor in Woolsthorpe-by-Colsterworth, a hamlet in the county of Lincolnshire. Newton was born three months after the death of his father, a prosperous farmer also named Isaac Newton. Born prematurely, he was a small child. His mother Hannah Ayscough reportedly said that he could have fit inside a quart mug. When Newton was three, his mother remarried and went to live with her new husband, the Reverend Barnabus Smith.

Hannah left her son in the care of his maternal grandmother, Margery Ayscough. The young Isaac disliked his stepfather and held some enmity towards his mother for marrying him. This was revealed by an entry in a list of sins committed up to the age of 19, *Threatening my father and mother Smith to burn them and the house over them.*

From the age of about twelve until he was seventeen, Newton was educated at The King's School, Grantham (where his signature can still be seen upon a library window sill). He was removed from school, and by October 1659, he was to be found at Woolsthorpe-by-Colsterworth, where his mother, widowed by now for a second time, attempted to make a farmer of him. He hated farming. Henry Stokes, master at the King's School, persuaded his mother to send him back to school so that he might complete his education.

In June 1661, he was admitted to Trinity College, Cambridge as a sizar, a sort of work-study role. In 1665, he discovered the generalized binomial theorem and began to develop a mathematical theory that would later become infinitesimal calculus. Soon after Newton had obtained his degree in August 1665, the university temporarily closed as a precaution against the Great Plague. Although he had been undistinguished as a Cambridge student, Newton's private studies at his home in Woolsthorpe over the subsequent two years saw the development of his theories on calculus, optics and the law of gravitation. In 1667, he returned to Cambridge as a fellow of Trinity.

While Newton was once engaged in his late teens to a Miss Storey, he never married, being highly engrossed in his studies and work.

Towards the end of his life, Newton took up residence at Cranbury Park, near Winchester with his niece and her husband, until his death in 1727. Newton died in his sleep in London on 31 March 1727, and was buried in Westminster Abbey. After his death, Newton's body was discovered to have had massive amounts of mercury in it, probably resulting from his alchemical pursuits. Mercury poisoning could explain Newton's eccentricity in late life.

Unfortunately, Newton had his squabbles with his contemporaries. In 1671, the Royal Society asked for a demonstration of his reflecting telescope. Their interest encouraged him to publish his notes *On Colour*, which he later expanded into his *Opticks*. When Robert Hooke criticized some of Newton's ideas, Newton was so offended that he withdrew from public debate. The two men remained generally on poor terms until Hooke's death.

Another popular dispute emerged with Leibniz over priority in the development of infinitesimal calculus. Most modern historians believe that Newton and Leibniz developed infinitesimal calculus independently. Starting in 1699, other members of the Royal Society (of which Newton was a member) accused Leibniz of plagiarism, and the dispute broke out in full force in 1711. The Royal Society proclaimed in a study that it was Newton who was the true discoverer and labeled Leibniz a fraud. This study was cast into doubt when it was later found that Newton himself wrote the study's concluding remarks on Leibniz. Thus began the bitter controversy which marred the lives of both Newton and Leibniz until the latter's death in 1716.

In later years, Newton moved into more social activities. He was a member of the Parliament of England from 1689 to 1690 and in 1701. He moved to London to take up the post of warden of the Royal Mint in 1696, a position that he had obtained through the patronage of Charles Montague, 1st Earl of Halifax, then Chancellor of the Exchequer. Newton took charge of England's great mint. He became perhaps the best-known Master of the Mint, a position Newton held until his death. This appointment was intended as a gift without an intention to occupy the position seriously, but Newton took it seriously. He retired from his Cambridge duties in 1701, and exercised his power to reform the currency and punish clippers and counterfeiters.

Newton is considered by many scholars and members of the general public to be one of the most influential people in human history. His 1687 publication of the *Philosophie Naturalis Principia Mathematica* is considered to be among the most influential books in the history of science. In this work, Newton described universal gravitation and the three laws of motion that dominated the scientific view of the physical universe for the next three centuries. His work significantly advanced the Scientific Revolution.

Many people are aware of the image of Newton sitting under an apple tree and being hit on the head by a falling apple. This, apparently, is a story of how he discovered gravity though it is a fabrication. He often told the story that he was inspired to formulate his theory of gravitation by watching the fall of an apple from a tree. John Conduitt, Newton's assistant at the Royal Mint and husband of Newton's niece, described the apple event when he wrote about Newton's life. *In the year 1666 he retired again from Cambridge to his mother in Lincolnshire. Whilst he was pensively meandering in a garden it came into his thought that the power of gravity (which brought an apple from a tree to the ground) was not limited to a certain distance from earth, but that this power must extend much further than was usually thought. Why not as high as the Moon said he to himself & if so, that must influence her motion & perhaps retain her in her orbit, whereupon he fell a calculating what would be the effect of that supposition.*

The question was not whether gravity existed, but whether it extended so far from Earth that it could also be the force holding the moon in its orbit. Newton showed that if the force decreased as the inverse square of the distance, one could indeed calculate the Moon's orbital period, and get good agreement. He guessed the same force was responsible for other orbital motions, and hence named it "universal gravitation".

Curiously, Newton's postulate of an invisible force able to act over vast distances led to him being criticized for introducing "occult agencies" into science.

Francesco Maria Grimaldi
(April 2, 1618 - December 28, 1663), Italian Jesuit priest, mathematician and physicist.

He was the first to make accurate observations of the diffraction of light and coined the word "diffraction." The study of diffraction is a pivot point of quantum physics today.

Francesco grew up in a wealthy 17th century family in Bo-

logna, where his father made a great deal of money selling
silk. When Francesco was old enough for school, he decided
to leave his family and live a life of God and science, as a Jesuit. Jesuits are men who are part of a Roman Catholic religious order. They are something like monks but they also feel a
need to teach and help others and not live in solitude. In the
year of 1632, Grimaldi went to study in the Jesuit college at
Novellara. Grimaldi eventually took on teaching jobs. He
taught rhetoric, humanities, astronomy, mathematics, and optics. Grimaldi died in Bologna on Dec.28, 1663 of natural
causes.

Grimaldi's spent much of his own time working on optics. He
made many discoveries of fundamental importance. They
were ahead of the theory of the time and their significance
was not recognized until over a century later.

Our interest here is his observation of sunlight. In a dark
room, Grimaldi allowed sunlight to pass through a small hole.
He recorded that the circle of light formed on a screen by the
rays passing through a very small perforation in a plate of lead
was greater than it would be if its magnitude depended solely
on the divergence of the rays.

He arrived at the conclusion that the rays of light change direction in passing near the edges of objects. He called this effect "diffraction."

That name is used for this phenomenon today.

Christiaan Huygens
(14 April 1629 - 8 July 1695), Dutch mathematician, astronomer, physicist, horologist, and writer of early science fiction.

He developed a wave theory of light in 1678, and published it in his *Treatise on Light* in 1690. This was overshadowed by Newton's corpuscular theory of light transmission that was presented in 1704.. However, by the 19th century, Newton's corpuscular theory was rejected by the scientific community and Huygens's ideas became the theory of the day.

Huygens was born in April 1629 at The Hague, the second son of Constantijn Huygens, and of Suzanna van Baerle, whom Constantijn had married on 6 April 1627.

Huygens studied law and mathematics at the University of Leiden and the College of Orange in Breda. After a stint as a diplomat, Huygens turned to science. The Royal Society elected Huygens a member in 1663. In the year 1666 Huygens moved to Paris where he held a position at the French Academy of Sciences under the patronage of Louis XIV. In 1681, Huygens returned to The Hague in Holland. Although ill health was the immediate cause, additional personal and religious pressures combined to make permanent his return to his native country. He had never married, and his later years were characterized by considerable solitariness. In his correspondence, he often lamented the absence of anyone with whom to discuss scientific topics. He did, however, maintain his extensive correspondence, and although his mathematical and abstract studies suffered a marked diminution after 1680, the general pattern of his life remained little changed until his death on July 8, 1695.

Unlike many men of science in the 17th century, Huygens

never occupied himself to any significant extent with either philosophy or theology. He devoted his efforts entirely to the pursuit of science, and his contributions to astronomy, dynamics, and optics were of fundamental importance.

Huygens experimented and developed techniques in a wide range of areas. He formulated what is now known as the second law of motion of Isaac Newton in a quadratic form. He also worked on the design of accurate clocks, suitable for naval navigation. His invention of the pendulum clock, patented in 1657, was a breakthrough in timekeeping.

Thomas Young
(13 June 1773 - 10 May 1829), English scientist.

Young presented a wave theory of light in 1801 that was accepted some 50 years later. This won out over Newton's theory that was presented about 100 years earlier.

Young was born in 1773 in Milverton, Somerset to a Quaker family. He was the eldest of ten children. At the age of fourteen, Young had learned Greek and Latin and was acquainted with over ten other languages.

He began to study medicine in London in 1792, moved to Edinburgh in 1794, and a year later went to Goettingen, Lower Saxony, Germany where he obtained the degree of doctor of physics in 1796. In 1797, he entered Emmanuel College, Cambridge. That year he also inherited the estate of his granduncle, Richard Brocklesby, which made him financially independent. In 1799, he established himself as a physician at 48 Welbeck Street, London. From 1811 to the time of his death, he served as a physician at St. George's Hospital. His main medical interest, though, was not in treating patients but in doing research. Human vision and the mechanism of the eye held a special fascination for him.

Young published many of his first academic articles anonymously to protect his reputation as a physician.

In 1801, Young was appointed professor of natural philosophy (mainly physics) at the Royal Institution. In two years, he delivered 91 lectures. In 1802, he was appointed foreign secretary of the Royal Society, of which he had been elected a fellow in 1794. He resigned his professorship in 1803, fearing that its duties would interfere with his medical practice. His lectures were published in 1807 in the *Course of Lectures on Natural Philosophy* and contain a number of anticipations of later theories.

Thomas Young died in London on 10 May 1829, and was buried in the cemetery of St. Giles Church in Farnborough, Kent, England.

In Young's own judgment, of his many achievements, the most important was to establish the wave theory of light. To do so, he had to overcome Isaac Newton's view that light is a particle. Nevertheless, in the early 1800's, Young put forth a number of theoretical reasons supporting the wave theory of light, and presented demonstrations to support this viewpoint. One was the double-slit experiment, which demonstrated interference with water waves that could be duplicated with light.

Normally, Young would have gotten a lot of attention for this work. However, even though Newton had been dead eighty years when Young officially published his findings on interference in 1807, the godlike status of the great man in Britain meant that Young's compelling results were ignored.

Augustin-Jean Fresnel
(10 May 1788 - 14 July 1827), French physicist.

In 1817, Frenchman, Augustin Fresnel, presented conclusive experiments that light was a series of waves and not a movement of minuscule particles.

Fresnel was the son of an architect, born at Broglie (Eure).

His early progress in learning was slow, and he still could not
read when he was eight years old. At thirteen, he entered the
Ecole Centrale in Caen, and at sixteen and a half the Ecole
Polytechnique, where he acquitted himself with distinction.
From there he went to the Ecole des Ponts et Chaussees. He
served as an engineer successively in the departments of Ven-
dee, Drome and Ille-et-Vilaine; but having supported the
Bourbons in 1814 he lost his appointment on Napoleon's return to power.

On the second restoration of the monarchy, he obtained a post
as engineer in Paris, where he spent much of his life from that
time onwards. He appears to have begun his research in optics around 1814 when he prepared a paper on the aberration
of light, although it was never published. In 1818, he wrote a
memoir on diffraction for which he received the prize of the
Academie des Sciences at Paris in the ensuing year. He was
the first to construct a special type of lens, now called a Fresnel lens, as a substitute for mirrors in lighthouses. In 1819, he
was nominated to be a commissioner of lighthouses. In 1823,
he was unanimously elected a member of the academy, and in
1825 he became a member of the Royal Society of London. In
1827, the time of his last illness, the Royal Society of London
awarded him the Rumford Medal.

Fresnel died of tuberculosis at Ville-d'Avray, near Paris.

He received only scant public recognition during his lifetime
for his labors in the cause of optical science. Some of his papers were not printed by the Academie des Sciences until
many years after his death.

His discoveries and mathematical deductions, building on
experimental work by Thomas Young, extended the wave
theory of light to a large class of optical phenomena. In 1817,
Young had proposed a small transverse component to light,
while yet retaining a far larger longitudinal component. Fresnel, by the year 1821, was able to show via mathematical
methods that polarization could be explained only if light was

transverse, with no longitudinal vibration whatsoever. This went a long way to validating Young's wave theory of light. Thus, polarization and its understanding played a pivotal role in establishing the wave theory of light and remain important to quantum physics today.

Jean Bernard Leon Foucault
(18 September 1819 - 11 February 1868), French scientist.

Foucault performed critical experiments with the speed of light that supported Young's wave theory.

Foucault was the son of a publisher, Jean Leon Fortune Foucault. He was born in Paris on September 18, 1819. Leon was a very frail child. He had one eye which was short-sighted and the other long-sighted. It gave him a rather awkward appearance and this was made worse by the fact that Leon became self-conscious about his appearance. He tended to prefer being by himself. His mother sent him to College Stanislas but he did poorly. He only made progress when his mother hired tutors and watched over him. As a teenager, he loved to construct toys and machines, some of which were highly sophisticated such as a steam engine and a telegraph. His dexterity suggested to his mother than he would make a superb surgeon and so, having obtained his high school diploma, he entered medical school in Paris in 1839. A problem arose when he saw some blood and fainted. Alfred Donne, his professor, decided to employ him as an assistant.

Foucault had attended talks by Daguerre on his photographic methods. Foucault's friend Fizeau had been with him and the two experimented with the photographic process. Foucault combined his new photographic skills with his work for Donne and devised a method of taking photographs through a microscope. In 1845, Foucault and Donne published *A Course of Microscopy*, which contained 80 photographs of objects under a microscope. Foucault eventually replaced Donne when he retired as the scientific editor of the *Journal des Debats*.

Bertrand writes about this task which Foucault carried out with remarkable success. *At the age of 25, not having learnt anything at school nor from book, enthusiastic about science but not about study, Leon Foucault took on the task of making the work of scientists understandable to the public and of passing judgment on the value to the work of leading men of science. From the start, he showed great subtlety, good judgment based on more prudence than would be expected. His first articles were remarkable; they were spiritual. He took his duties seriously. Launched, without any experience, into the highest level of science with all its confusion and problems, he was assured carrying out a role in which mediocrity would mean failure, with complete success . . . Always polite, yet seeking the truth, Foucault applied carefully considered judgments. Previously an unknown, this young man with no scientific publications nor known scientific discoveries, displayed a quiet authority and frankness which irritated many leading scientists.*

The Academy of Sciences approached Foucault with a task of taking photographs of the sun. That being successful, the academy asked if he could measure the speed of light in water. In April 1850, he showed that light travels slower in water than in air. This was in accordance with what the wave theory of light predicted, but contradicted what the corpuscular theory of Newton predicted. This was viewed as *driving the last nail in the coffin* of Newton's corpuscle theory of light.

Foucault died of what was probably a rapidly developing case of multiple sclerosis on February 11, 1868 in Paris and was buried in the Cimetiere de Montmartre.

Conclusion

The ancients have passed. The Renaissance began with some vague philosophies of the laws of the universe. By the end of this period, the Scientific Revolution is in full swing. Newton's laws and others have been established. Man now has the math to make machines and tools to penetrate deeper into the secrets of the universe. However, the all knowing philosopher is gone. The polymath, the individual master of all has passed. Individuals that are dedicated to a single purpose replace them.

Chapter 4

The Meaning of Light

Insofar as quantum physics is concerned, the primary developments around the time of the Renaissance were about light. Three men and their ideas stand out here. They were Huygens, Newton, and Young. Huygens presented his theory in 1678. Newton presented his in 1704. Young followed in 1801. Huygens presented a wave theory that was for the most part ignored. It did appear 26 years before Newton's. [1] Then, during Newton's time, due to his stature in the physics community, Huygens' theory was totally discarded. Newton proposed what he called a corpuscular theory of light. It described light as particles. Newton's theory reigned for some 150 years. It fell from favor, as it did not explain diffraction and interference. [2] Young came along with his experiment about 100 years later.[3] It captured attention but fell short due to a few small problems and the stature of the Newton image. Some of the problems with Young's theory were resolved about 50 years later and Young's theory reigned until the emergence of the quantum age.

Even then, Young's theory was not rejected. Rather, it became part of the mix of quantum weirdness. Both Huygens and Young's experiments remained important factors in the quantum age. Although Newton's theories of light were closer to some concepts of the new age of quantum physics, his ideas were not close enough until we entered that age.

Wave phenomena is important in the study of quantum physics. Both Huygens and Young's experiments remained as tests to demonstrate the wave nature of light and the wave nature of matter waves. In this chapter, we briefly examine the concepts of both Huygens and Young about the structure of light. This chapter quickly describes the experiments demonstrating light is a wave. These concepts remain today as models of the structure of our existence.

Huygens' One Slit Experiment

In 1678, Huygens observed diffraction in water waves and experiments with light. To him, both phenomena appeared the same.

The following depicts the diffraction of water waves.

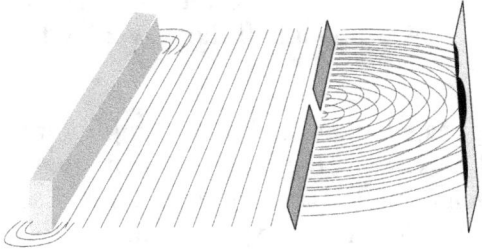

The wood block on the left of the picture moves. The motion creates water waves that move to the right. There is a barrier with a small opening in it, which only allows a small section of water waves to slip through. That generates circular waves that interact with each other. The dark areas on the barrier on the right have depicted where the water waves converge and splash water higher. [4]

The following depicts light waves doing the same thing. Here, however, we cannot see the light waves moving through the air. The light is forced to pass through a similar opening as in the previous picture. Here, the light waves form a pattern similar to the water waves on the barrier on the right.

The white areas represent where the light strikes on this barrier. [4]

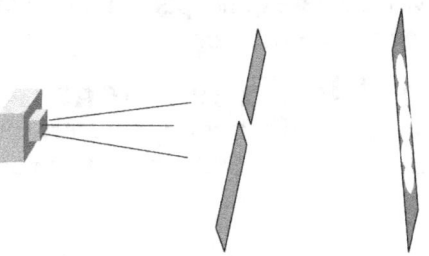

Both experiments produce the same result. The size of the opening in the barriers was about half the wavelength of the waves.

Light, Huygens suggested, consisted of the longitudinal vibrations of an all pervasive ether composed of small, hard, elastic particles, each of which transmitted the impulses it received to all contiguous particles. [5] According to this theory, the particles did not move but remained in a given position.

This presented a bit of a barrier at the time as this back and forth motion would not support polarization. Water waves do not display an analogy to polarization. Nor do sound waves. Yet, light is polarized. However, the real barrier to the acceptance of Huygens' idea was the popularity of Isaac Newton and his concepts. Both would fall away as time passed.

The Young Double Slit Experiment

Young's double slit experiment appears very similar to Huygens' experiment. It has another slit for the waves to go through. Before considering why this is important, here is how the experiment appears.

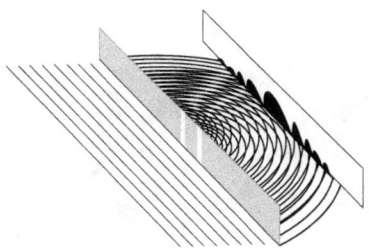

The double slits are close together. As the water waves move through the slits, they form two circular waves moving away from the slit barrier. They interfere to produce the interference pattern on the barrier on the right. Some water waves splash higher on the barrier to from the pattern shown in black. [6] The primary thing to notice is that the interference pattern is much higher than a pattern with a single slit.

To see the importance of this experiment, block off one of the slits.

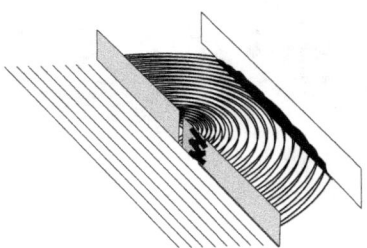

The heavy interference pattern suddenly becomes the single

slit pattern, which is considerably smaller. Try blocking off
the other slit.

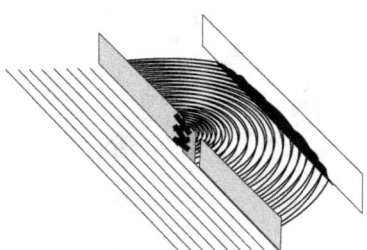

The result is the same. This illustrates an important fact about
waves in this experiment. When the waves are going through
both slits, the waves interfere with each other. Young observed the same behavior with light. [6] Consider a similar
experiment with light.

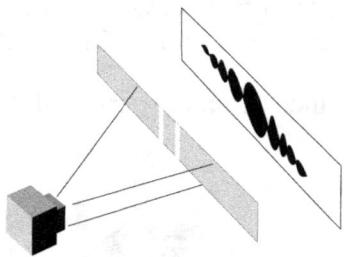

A pattern similar to water appears on the far barrier when light shines through two slits. Now, cover one of the slits.

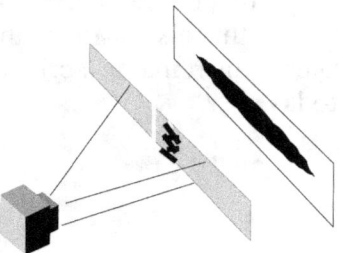

Here, the pattern changes just as in the water experiment. Try the other slit.

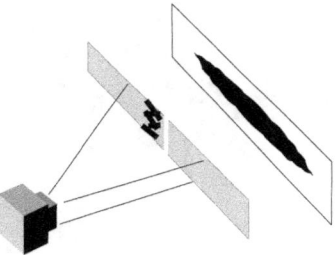

The result is the same.

As mentioned, there were a few problems with this theory of light. However, Fresnel and Foucalt resolved them about 50 years after Young introduced his theory. [7]

The next chapter looks at Huygens' experiment in some detail.

Conclusion

There are a few fascinating observations here. One is the realization that some experiments performed in 1678 have a profound effect on our understanding today. Another is that we get to see why light is considered a wave. We, the public, are told that light is a wave. Looking at these experiments, we see that science has come to this conclusion because a light pattern on a wall produces a similar pattern as water waves when they splash against a wall. As time passed, the ether was shown to be something that did not exist. Non-the-less, the key element to get from this discussion is the similarity between light and water. This remains a critical component of quantum physics today.

Chapter 5

Wave Theory

In this chapter, we take a closer look at Huygens' wave experiment.

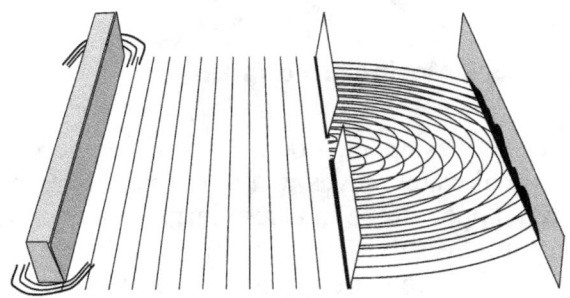

In the previous chapter, we reviewed Young and Huygens' experiments with water waves and light. Here, we focus on Huygens' experiments and his conclusions. The goal here is to get a deeper understanding of why the scientific community attached so much importance to water waves in a tank and how it was interpreted to be a description of the light around us. This is a deeper mathematical analysis of the mechanics of wave motions. You need not dwell on the math. Concentrate on how two waves can add or subtract from each other. Also, focus on the aspects that water waves and light waves have in common. The water waves make a pattern on the splash wall that is similar to the pattern light makes when traveling through slits and illuminates a wall. The important thing to realize is that these similarities form the basis to claim light is a wave. As you will see, as this book unfolds, waves are the beginning and the ending of what quantum physics is about. The waves can be made of water, light and even matter.

The Water Wave

The following picture depicts a wooden block moving up and down in a large tray of water.

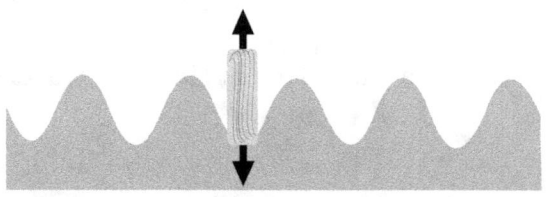

Wooden Block Moving Up and Down

If the wooden block moves up and down at the right speed some very symmetrical water waves will be formed. Notice that the water waves appear to be in a very large container so we are not looking at waves that might bounce off the edges of the container and interfere with waves we wish to study.

Before going further, we need to clearly identify what we call the parts of the water wave.

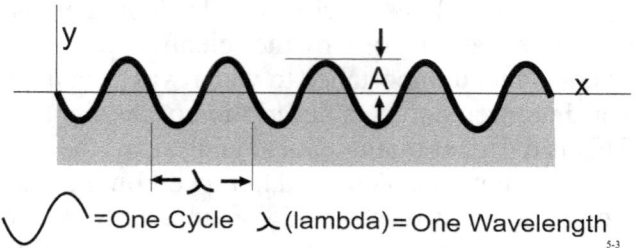

The black line on top of the shaded area represents the wave. Notice that y and x axes have been superimposed on the black line that represents the wave. Notice that the x axis goes through the middle of the wave. The amplitude of the wave is A. That is the height of the wave above the center of the wave indicated by the x axis. The wave will move above and below the center by this amount. One up and down motion is considered one cycle. This is commonly called one wavelength or lambda. Note that in this picture a wave cycle begins when the black line crosses the middle line or the x axis. The crest of a wave is the part that is high above the x axis. The valley of a wave is the part that is lowest. If this nomenclature is used, one can talk clearly about a wave.

Superposition

Another factor in all of this is superposition. This term refers to what happens when two waves fall on top of each other. The amplitude (The A just introduced.) of each wave is added. The following picture depicts the crests of two waves that line up when falling on top of each other. In this case the amplitude of each wave is added. This is called, "in phase." Consider the following picture,

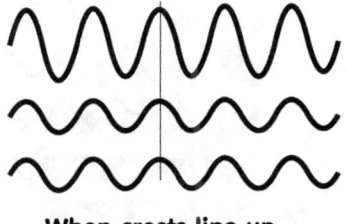

When crests line up.

The upper wave is the addition of the lower two.

The following picture demonstrates two waves are on top of

each other but the crests are not lined up. This is called, "out of phase."

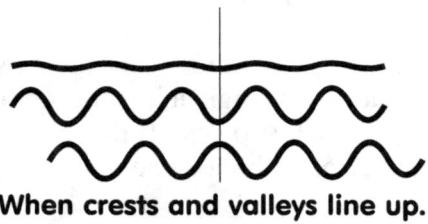

When crests and valleys line up.

Then, the lower two waves add to cancel each other out as demonstrated by the top wave.

Waves Can Pass Each Other
Part of wave phenomena is that waves can move by each other without permanently changing the character of each other. The following picture demonstrates this.

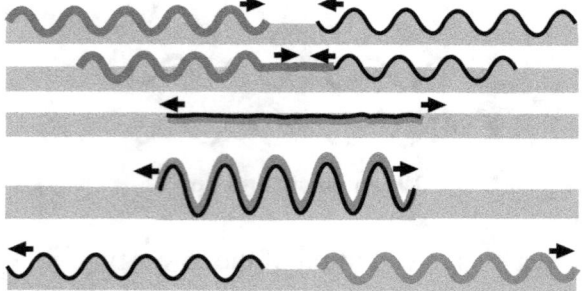

The five lines in this picture demonstrate a wave moving to the right and a wave moving to the left. The waves move past each other. When crests and valleys line up, each wave cancels the other out. When the crests line up, they add to each other. After they pass, both waves appear as they did before the interaction.

The Water Wave Diffraction Experiment

Now, let's take a look at Huygens' water diffraction experiment. The following picture shows a large tray of water with a wooden paddle, a barrier with an opening in it and a barrier that the waves moving through the opening splash against. We are viewing this from above looking downward.

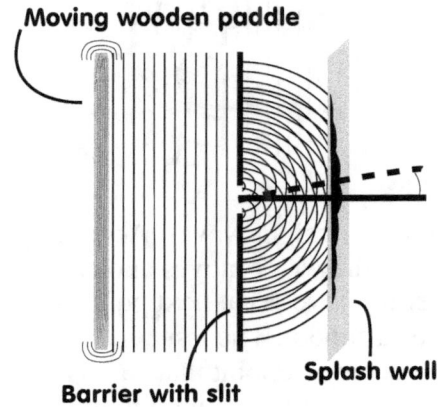

The wooden paddle is moving in the water. This generates straight waves that move to the right. The straight waves come against the slit. Only a small part of the straight waves go through the opening in the barrier. As the waves go through the barrier, they move in circular patterns to the right. The waves interact; creating valley lines and crest lines. When the crest lines hit the splash wall, the water hits a bit higher against the wall as depicted by the darker areas on the splash wall.

For a clearer understanding, let's examine some parts of this carefully.

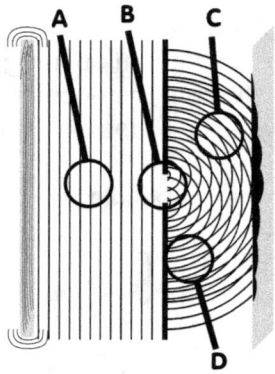

Area A focuses on the straight waves. According to Huygens' theory, each point in the water moves and causes adjacent points in the water to move. That next point is then the driver to move the points around it. The issue here is that each point in the water then becomes a point that cases a wave to move away from it in a circle. Thus, each point in a straight wave is generating a circular wave. However, all the circular waves add up to produce another straight wave. The following picture of area A depicts that a straight line is generating circular waves. However, the circular waves add together to form another straight wave.

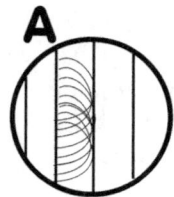

Area B demonstrates that the bit of wave moving through the opening or slit generate circular waves from the edges of the opening. In the following picture, the dots show the center of the two circular waves moving away from the opening edge.

As the two circular waves move away from each other, they are superimposed. The picture of area C focuses on where the crests of the waves fall on each other. At the points where the lines cross in this picture, the crests add and the wave is high.

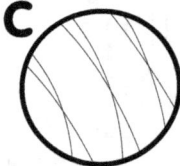

The following picture attempts to show that the crests of each wave are far apart and the valleys and crests of the two waves fall on each other. Here the waves cancel each other out and the water is flat at those points.

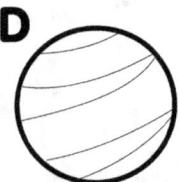

Let's take another look at the entire experiment. In the following picture, the shaded stripes mark the areas where the crests of the waves line up and they add to be higher than surrounding waves. Because the two waves are adding together, they will splash higher on the splash board than the waves on either side. This is depicted by the shaded areas on the splash-board.

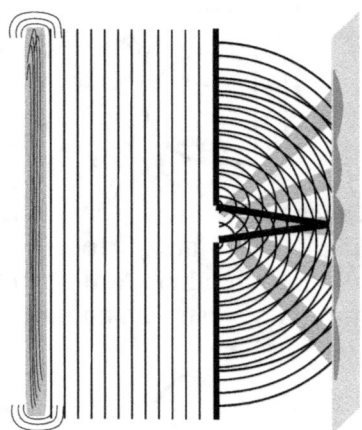

The two black lines in the middle of the picture show the path of the two waves where the crests line up in the middle of the circular waves. Notice that these two lines are of equal length. In the experiment depicted, this allows four wave-lengths to travel this length.

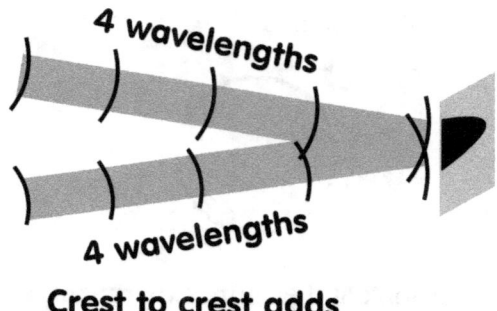

Because one crest is on top of the other when the waves hit the splash wall, the splash is high on the wall.

Now consider the waves that do not fall on each other crest to crest. Instead, consider the path of the waves where the crest falls on the valley of the other waves. The two black lines here represent the path of that situation.

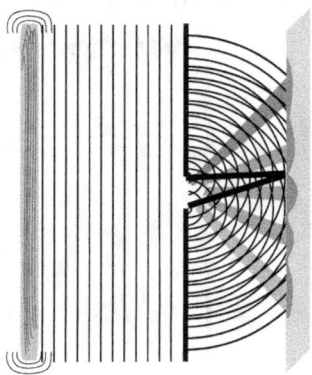

Notice that one path is longer than the other. The lower is longer by 1/2 a wavelength. Thus the crest/valley combination hits the wall and each wave cancels the other out. Here, there is no splash against the wall.

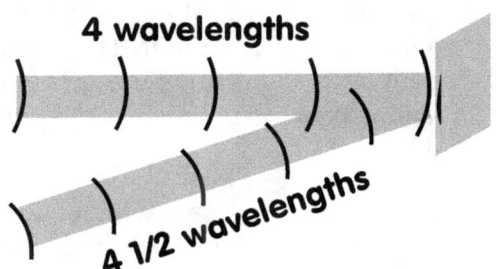

Crest to valley cancels

This occurs with all the waves all along the wall on the right producing the interference pattern we are accustomed to.

Now let's take a brief look at the experiment with light.

The Light Wave Diffraction Experiment

Part of Huygens experiment was to duplicate the water wave experiment with a monochromatic light source.

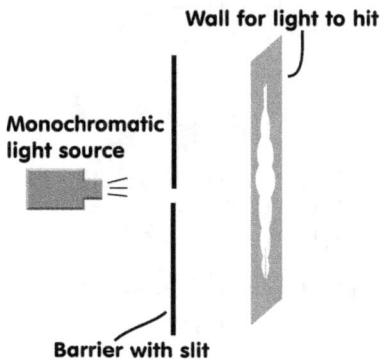

The experiment is very similar but there is no water. Instead of a wooden bar on the left making water waves, there is a light source. The light source is monochromatic so it projects light of a single wavelength. This would be similar to the wooden paddle producing waves of the same length. The light shines on a barrier with a slit in it. The slit is about one half the wavelength of the light being used. The light passes through the slit and makes a diffraction pattern on the wall or board on the right.

You can actually run this experiment without all this equipment. Hold your first finger and second finger close together.

Put both near to your eye and make a thin slit between the two fingers.

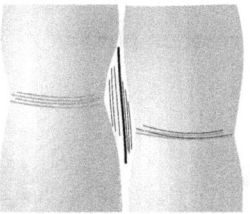

Hold your hand in front of some light source like a window or a light turned on so you can see the light while looking through this slit. Adjust the width of this slit by moving your fingers until you see lines that appear as in the picture above. These are lines of diffraction.

Coming to a Conclusion

Now, let's look at the items in common between these two experiments.

- The setup of the experiment is essentially the same. They both have something sending waves, a barrier with a slit or opening and a wall to display the resulting wave interference.

- The transmission of the waves is inserted into the experiment from the left.

- The transmission hits a wall with a slit in it.

- The slit or opening is one half the wavelength of the transmission.

- A similar pattern is formed on the opposing wall.

Perhaps the most solid observation for this theory is the angle of the crest to crest path to the angle of the crest to valley path. That angle depends upon the wavelength of the impinging transmission.

Consider that with a representation for water waves.

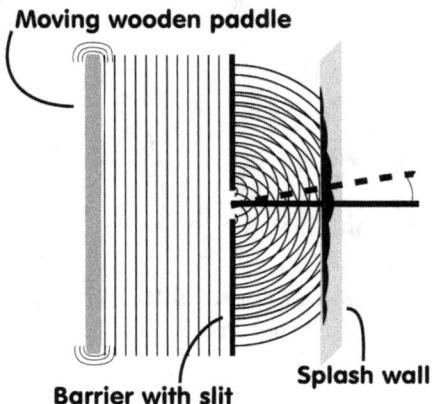

The dash line represents the crest to valley path. The black horizontal line represents the crest to crest path. The angle between these depends on the water wavelength.

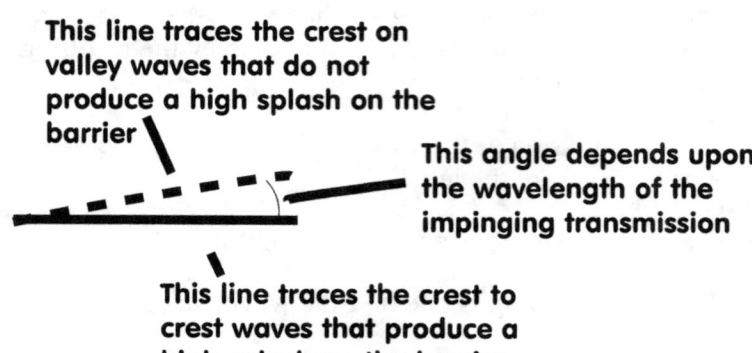

Here is a representation for light waves.

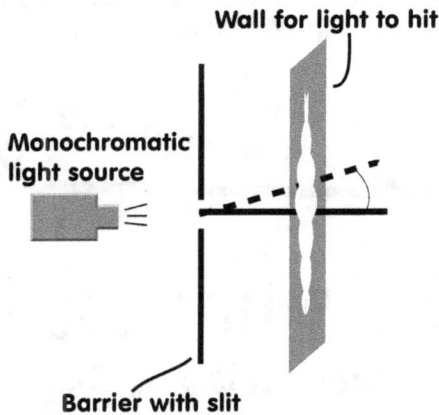

The dash line represents the crest to valley path. The black horizontal line represents the crest to crest path. The angle between these depends on the wavelength of the impinging light.

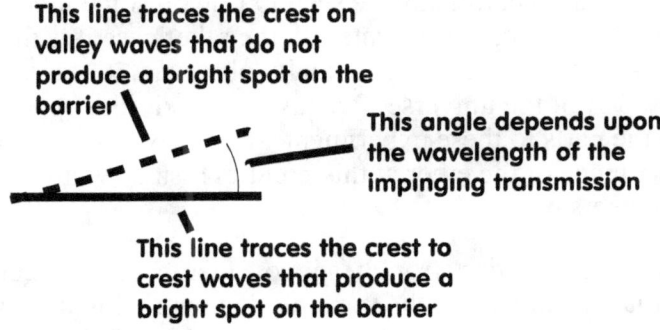

This angle between the line, here called theta, can be calculated using the wavelength lambda. The dash line represents the path of the crest over valley interaction. The black horizontal line is the path of the crest on crest interaction.

theta θ
lambda λ

$$\theta = \cos\left(\frac{n\lambda}{n\lambda + 1/2\,\lambda}\right)$$

Where n is 1, 2, 3 ...

The point here is that in both experiments, the pattern displayed on the barrier opposite the wave source is the same and depends upon the wavelength of the waves generated.

Finally, both experiments are seen to ride on some kind of medium. In one case it is water. In the other case it is the ether. While the ether is not actually observed, it is an accepted part of the universe. Because of the similar appearance of all the parts to these experiments, it clearly exists. To doubt the existence of the ether at this point in history would be scientific hearsay.

So, these similarities strongly suggest that there is medium in both that is similar and the transmission is moving through each medium as a wave. This came to stand as fact until the quantum age.

Conclusion
If you followed all of this at any level, you have a good idea of what a wave is. You can also see why the experiments demonstrate that light is a wave. There are several significant similarities. However, studying this experiment is also instructive about good science. The conclusion of the ex-

periment is that light is a wave. Huygens would have you believe that. If however, you were a critical observer and could
look into the performance of the experiment and be allowed to
ask some critical questions, you might question the accepted
idea of the ether that light depends upon.

Chapter 6

Polarization

Light manifests polarization. Water waves do not nor do sound waves. Water waves move up and down as they move through the slits in the experiments. They also move forward and backward along the direction of motion. Water waves do not move side to side as they pass through the slits. Thus, polarization does not appear in water wave experiments. Sound waves are longitudinal. That is, the wave oscillates along the direction of motion. They do not display polarization either as they pass through a slit in some kind of barrier. Any explanation of the nature of light will need to explain the polarization effect. Thus, we need to study it here.

Initial Discoveries

Early understanding of polarization involved calcite crystals. These crystals were initially found in Iceland but Mexico is now a common source for them. If one had an optically clear crystal and looked through the crystal at something, one would see a double image of it.

If one attempted to look at the double image with another

crystal, the effect would not seem to work.

One would see only the two images.

Those that studied this phenomenon discovered that the two images were caused by polarization. One image was polarized in one direction while the other was polarized ninety degrees in another direction.

What is Polarization?

The theory is that light is a transverse wave, as the wave oscillation moves either up and down, or sideways; perpendicular to the direction of motion of the wave. [1] The following picture describes two waves.

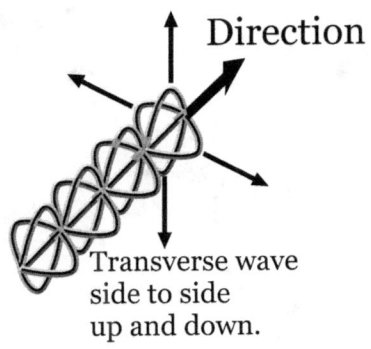

Direction

Transverse wave
side to side
up and down.

One is going up and down and one is side to side. Polarizing devices are capable of allowing one of these waves to go though a barrier while stopping the other wave. A common device to do this consists of plastic material with very long molecular chains. You can see these lines in the following image. In an actual filter, you would not see the spaces between the lines. The surface of the filter would simply appear dark.

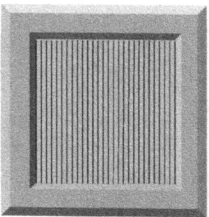

The vertical molecular strands allow the vertical portion of a beam of light to go through. It stops the horizontal portion from going through. In the following picture, a candle emits not-polarized light toward a polarizing device. The device allows the vertical component of the beam to travel to the eye, which perceives it.

To detect non-polarized light, one would hold such a device and look at the light while holding the barrier at all angles.

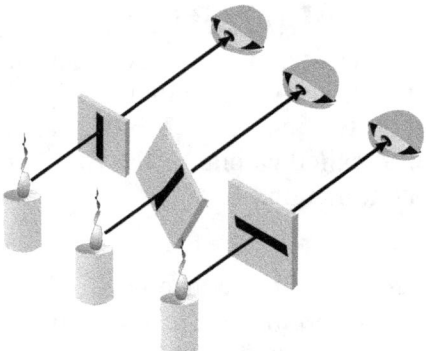

The image observed through the slit will not change. If however, the light travels through a substance that polarizes the light, the position of the device will change the image observed. The following picture depicts this. The light from the candle passes through some kind of a crystal. The light is polarized. The light will pass through a horizontal slit. Some passes through a slit held at an angle. No light passes through a slit held vertically.

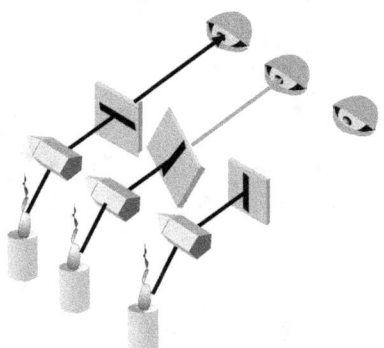

The understanding was that the polarizer somehow stopped

light that oscillated in one direction and allowed light that
moved in another direction to pass.

A History of Polarization

The first observation of polarization is one of those discoveries buried in antiquity. More than likely, the first observations occurred on Iceland about 700 AD. Iceland Spar, a
crystalline material called calcite, is the first known material
to display this phenomenon.

First Discovery

Bartholinus is credited with the discovery of polarization. [2]
In 1669, Erasmus Bartholinus performed some experiments
with the crystal that consisted of optical quality calcite,
$CaCO3$. He was a Danish mathematician at the University of
Copenhagen. He may be the one that gets credit for its discovery for he was apparently the first to document its
properties.

Reflected Polorization

Another discovery was made while Etienne Louis Malus was
toying with a piece of the calcite crystal. He noticed how the
intensity of reflected light varied when he rotated the crystal.
He conducted some experiments showing light could be

polarized when reflected. [2] Here, the candle light is reflected by glass.

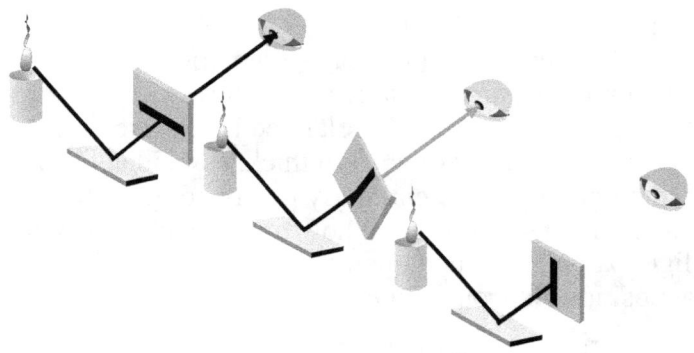

When the slit is horizontal, one can see the candle clearly. As
the slit rotates, the image disappears.

Brewster's Angle
In 1812, Brewster enhanced the experiments of Malus and
made a few more discoveries. He discovered a critical angle
called Brewster's angle. [2]

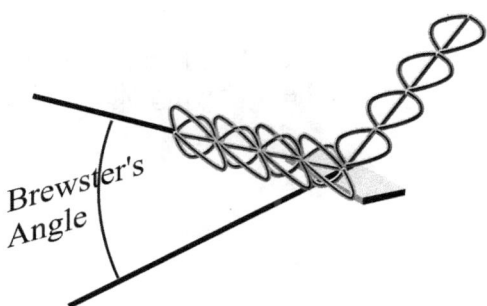

At this precise angle, all parts of the light that are vertical are
absorbed during the reflection of the beam. The reflected part

consists only of the horizontal part. This effectively polarizes
the beam of light.

Circular Polarized Light

In 1811, The Frenchmen Francois Arago noticed circular
polarization by placing a quartz crystal between a glass re-
flector and a calcite prism. If polarized light were to
fall on a sheet of quartz of a given thickness and at a given
angle, the light would be circularly polarized. [2] The following picture depicts this and how to detect circularly polarized light. A polorizer and quartz plate are held at the necessary positions in front of the source of light.

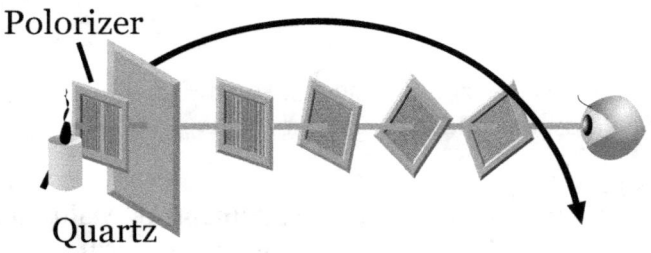

A polarizing device is held in front of the quartz sheet and rotated until the source of the light is seen. Then the polarizing
device is moved away from the quartz and rotated, the image
of the light source remains visible. In this scenario, the electric field is rotating as in the following picture. This picture

depicts the electric vector rotating as the wave moves to the right.

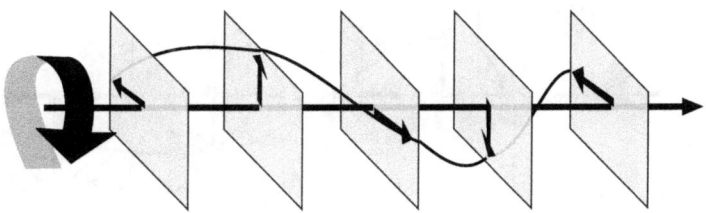

The light is moving according to the right hand rule. That is, if your thumb points in the direction of motion of the wave, your fingers will curl and point in the direction of rotation.

Light can be polarized in the other direction. The following shows the light moving through the initial polarizer and quartz but rotating in the other direction. Again, the detector-polarizing device moves away from the quartz at various positions and distances so the image of the source remains in view.

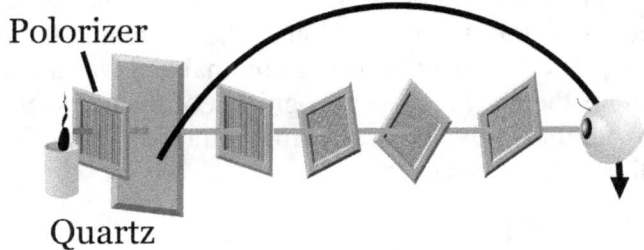

Here, the light is moving according to the left hand rule. That is, if your left thumb points in the direction of the motion of

the wave, your left curled fingers point along the direction of
rotation of the electric vectors of that wave.

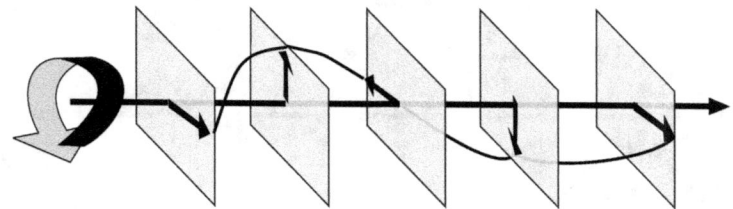

Fresnel's Observation

Fresnel worked on a wave theory of light around 1820. Non-
polarized light confused him. Its behavior suggested that light
had a longitudinal component that would prevent the total
interference of light in double split experiments. Around
1821, he realized that non-polarized light was made of transverse waves changing direction very fast. With that, he com-
pleted his own wave theory of light. [3]

Young had already captured the recognition of the community
for identifying the structure of light. The thought at the time
was that part of the wave was longitudinal and was a sticking
point in his theory. Fresnel's recognition of the transverse behavior of light resolved that sticking point. [3] Thus, Young's
wave theory became accepted.

Conclusion

Polarization is very significant part of light phenomena. Any
complete description of light must address what polarization
is without the artwork of displaying the electric fields we cannot see.

Chapter 7

Wave Functions 101

The purpose of this chapter is to present the importance of the wave function in quantum physics.

To this point the discussion of waves has been about something we can see. In quantum physics, there is nothing to see. Instead, we depend upon the wave function to determine what is going on. Thus, we must wean ourselves from looking at water waves and judge our environment from what we see in a mathematical equation, the wave function.

It may help if you understand why this is being introduced now in this book. The wave function is an important device in quantum physics. Normally, it is introduced when a discussion of quantum physics is begun. At that point, it seems to come out of nowhere and the student can be confused as to what it actually means. The goal here is to let the student know that the concept has been around since day one. It emerged with the water wave. Thus, the water wave is introduced here, not only as a wave but also as a wave function. If this concept is ignored early on, later the student will be confused when the discussion shifts to wave functions. The problem is that, after water waves are discussed, there is not a wave to be seen. Electromagnetic radiation is considered a wave function. Likewise, quantum physical phenomena is considered from the point of view of wave functions. The student will look for waves and become confused because there are none. Wave functions are mathematical beasts. While they are used to produce real world answers that can be observed, the waves are not. Contrast this with water waves that can be seen and have a wave function that predicts sim-

ilar behaviors. In essence, the student must get comfortable
with the idea there will be no mental image of a wave to work
with. It is pure mathematics.

Now, let us look at water wave functions.

(In the following discussion, the terms sine and cosine are
used. For a description of these, see Appendix A "Sine and
Cosine Functions.")

Wave Phenomena

The mathematics of both phenomena, water and light, appear
the same. Both experiments with water waves and light focused the waves on some kind of small opening in a barrier.
The size of the opening was about half the wavelength or half
of lambda used in the equations of each. (In discussions of
light, lambda is often used as the wavelength of the wave.)
The experiments produced something that was observable
from which frequencies and wavelengths of each phenomenon could be physically measured.

We can derive a mathematical equation for the water waves.
It appears as follows.

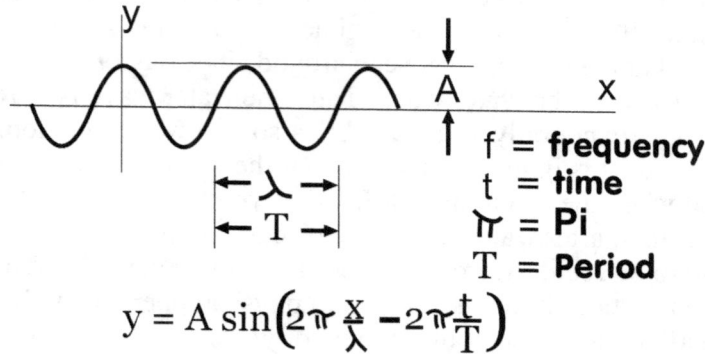

$$y = A \sin\left(2\pi \frac{x}{\lambda} - 2\pi \frac{t}{T}\right)$$

The bottom of this display is a water wave equation. In this
equation, y is a function of x and t. The small letter x is the

distance from some arbitrary location to plot the equation. The small letter t is time. Note that we have selected two lines from which to plot this equation. The zero position on x, where the y-axis crosses the x-axis, was selected as the beginning of one wave. A is the maximum height of the wave from the x-axis. Lambda is the length of a single wave. Frequency (f) is the number of times the wave goes up and down in one second. It is given in cycles per second.

Note that a position of the wave is given by both x, some distance, and t, some time. The t tells us the wave is moving to the right. The x tells us which part of the wave we are looking at, at time t.

For continuity with what follows, one may call the above equation a wave function for a water wave. Due to the similarity of the water wave experiment and light wave experiment, the above equation is considered to be useful for light as well. A goal of this chapter is to clarify the importance of using this equation for light.

A water wave function can be easily seen giving it an advantage over other kinds of wave functions we will discuss as this book unfolds. There are several other aspects of a water wave that might be interesting. For example, we might find it interesting to observe a particular point on the water wave. Let's put an imaginary tiny bead that floats in the water right on top of the wave. Our task might be to determine the equation of motion of this imaginary bead. This would describe the motion of the bead as it rides the wave. To get this equation we operate on the water wave function.

To make this operation easy, let us assume the bead is located
where x is zero.

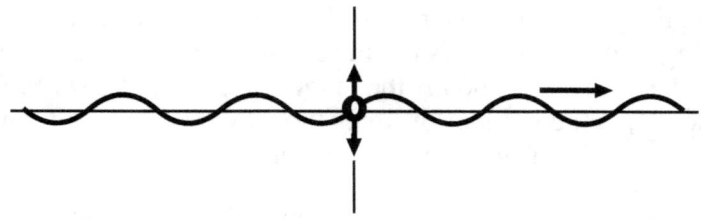

As we are interested in the up and down motion of the bead
only, we assume the bead does not move back and forth, as
the wave passes by. In reality it does just a bit. We ignore that
here. Thus, x in the equation is zero.

$$y = A \sin\left(2\pi \frac{x}{\lambda} - 2\pi \frac{t}{T}\right)$$

With x = 0,

$$y = A \sin\left(2\pi \frac{0}{\lambda} - 2\pi \frac{t}{T}\right)$$

$$y = A \sin\left(2\pi \frac{t}{T}\right)$$

The above equations show the development of the equation of
motion of a bead floating on the water. We began with the
water wave function. We set x to zero. The bottom equation
of the above image is an equation representing the motion of
the bead. This equation shows, mathematically, the bead is
oscillating up and down as time increases.

Look at what happened here. We began with a water wave
function. We wanted to know what happened to a particular

point in the water wave. We manipulated the water wave function and got some other point of view about the water wave.

We could have used other operations to determine other factors of the water wave.

For example,

What is the speed of a point on the water wave as it moves up and down?

How much energy does the wave transfer as it moves from left to right?

When a point in a wave goes up and down, how fast does it accelerate and decelerate?

To get an answer for each example, we start with the water wave function and manipulate it in some way to get the desired answer. Later, this will be extensively used with light and matter waves. All the commotion about quantum physics is about these manipulations. The matter wave function is manipulated with extremely complicated mathematics to get answers about the world around us. One such device is called the Schrödinger Wave Equation. Therefore, getting a grasp on the concept of wave functions is important to the understanding of quantum physics.

To complete a discussion of why the analogy of water waves and light waves is so strong, consider putting something analogous to the bead in a water wave into a light wave. Instead of a bead, place an electron there. Like the bead, an electron will oscillate up and down.

What does a Wave Function Mean?

Many believe that a wave function in quantum physics represents the structure of matter. They think it is like a picture of what the heart of our universe looks like. Beware that, right now, this is a mathematical picture. Water waves are real and we can see them. Other wave functions that we introduce in this book cannot be seen by humans. Thinking that the wave function explains what our universe is, is like saying the weight of an apple explains what an apple is. To say an apple weights two ounces does not describe the apple. While our goal is to understand what the universe is, our first goal is to get answers we can observe. If we wish to determine the weight of an apple, two ounces is an appropriate answer. The wave function is a mathematical device that we can manipulate to get answers.

The most common use of the wave function is simply to determine where the wave is at any given time. Consider how his would apply to water waves. At some particular x, we insert a value for time. Then solve the equation for those two values and you get a value indicating the height of the wave at that particular time and that particular position. If you run an experiment and take pictures of a water wave as it moves and look at the picture for the time entered into the function, you will see solving for those two values produced a correct answer.

Conclusion

This chapter is not intended to present some deep thought. It is only to caution that as this book moves forward, the reader will have a desire to "see" an electromagnet wave or "see" a photon wave. If one realizes there is nothing to see, the temptation to grasp at some physical meaning of the space carrying the radiation will be eased. Most importantly, one must understand that the similarities between water waves and light waves have convinced us that they are similar phenomena. Because of this, a light wave function is considered the same as a water wave function.

Be aware that making this assumption may lead you down a wrong path.

The important issue to carry away here is that the concept

came from observing water waves. The equations for water waves and light waves are the same. However, because there is some physical manifestation for water wave equations, do not assume there is some physical manifestation of light waves. When specific values are placed into the light wave equations, physical phenomena will be described. The desire to see light wave equations spawned the idea of ether that pervades all of space. This concept was shown to be wrong. This demonstrates that if you attempt to see something like a water wave in an electromagnetic wave, you may be disappointed.

Again, the goal of this book is to enable you to understand quantum physics. This is the first significant step.

Part 3

The Scientific Revolution

The Renaissance began. Shortly after, the Scientific Revolution began. It rolls forward into the 20th century. In this part of the book, we look at some of the men that pushed the ball of progress forward. They worked with little but wires and magnets to unravel the secrets of electromagnetism. The words molecule and atom become household words. They even began prying open the lid on atoms and get a first glimpse of what is going on inside. Explanations are offered that lead to more complex questions. Their work brings us to the edge of the quantum age.

The Scientific Revolution is marked with real experimentation in addition to logical thought. It began with electromagnetism. Batteries come into existence and wires are wrapped around bobbins to study current traveling around a wooden shaft. Experimentation leads to the understanding that the old elements of air, earth, water and fire; are no longer useful concepts. Electrons are the first subatomic particle to be recognized. The elements are codified and the Periodic Table of the Elements emerges. This tool essentially becomes a microscope enabling man to see inside the atom. He sees how electrons fit into the atom and finds protons and neutrons there also.

This is valuable in our quest to understand quantum physics. How do we study that which we cannot see? Here, we get a good idea of how the problem of seeing the small was solved. We see the importance of the periodic table and how it aided in discovering protons and neutrons.

Chapter 8

Electromagnetic Personalities

Many aspects of light have been addressed. The squabble over light as a particle or a wave has been settled for now. Its important properties as a wave: diffraction, interference, and polarization are well documented. Now, we change our attention to electromagnetism. In this chapter, we introduce some men that were important in the study of electromagnetism and electromagnetic radiation. Alessandro Volta [1] is the first man we consider. He is significant for he developed a way for the researchers that followed him to produce electricity for their experiments. Charles Coulomb [2] was one of those individuals using Volta's batteries to perform experiments. Coulomb mapped out the forces between electrically charged particles. Michael Faraday [3] mapped fields of the forces that Coulomb found. This set the stage for James Clerk Maxwell [4] to codify Coulomb's, Faraday's and the works of others to produce Maxwell's equations. The significance here is that these works eventually produced a wave function for light that eventually became the cornerstone for quantum physics.

Alessandro Volta
(18 February 1745 – 5 March 1827), Italian physicist.

Invented battries in 1800.

Volta was born in Como, Italy, and taught in the public schools there. In 1774, he became a professor of physics at the Royal School in Como. In 1779, he became professor of experimental physics at the University of Pavia, a chair he occupied for almost 25 years. In 1794, Volta married Teresa

Peregrini, with whom he raised three sons, Giovanni, Flaminio and Zanino.

In 1800, he invented the voltaic pile, an early electric battery, which produced a steady electric current. Volta had determined that the most effective pair of dissimilar metals to produce electricity was zinc and silver. Initially he experimented with individual cells in series, each cell being a wine goblet filled with salt water into which the two dissimilar electrodes were dipped.

The voltaic pile replaced the goblets with cardboard soaked in salt water.

In honor of his work, Volta was made a count by Napoleon in 1810. Volta retired in 1819 to his estate in Camnago, now called Camnago Volta after him, where he died on March 5, 1827. He is buried in Camnago Volta.

Charles Augustin de Coulomb
(June 14, 1736 - August 23, 1806), French physicist.

He developed Coulomb's law which defined electrostatic forces of attraction and repulsion.

Coulomb was born in Angoulźeme, France, to a wealthy family. His father, Henri Coulomb, was inspector of the Royal Fields in Montpellier. His mother, Catherine Bajet, came from a wealthy family in the wool trade. When Coulomb was a boy, the family moved to Paris, and there Coulomb studied at the Collčge des Quatre-Nations. While taking courses in mathematics, he decided to pursue mathematics and similar subjects as a career. From 1757 to 1759, he joined his father's family in Montpellier. Coulomb returned to Paris in 1759 where he attended a military school in Mézičres.

After he left the school in 1761, he performed a variety of civil engineering tasks including eight years directing the construction of Fort Bourbon in Martinique.

Upon his return, with the rank of Captain, he was employed at La Rochelle, the Isle of Aix and Cherbourg. He discovered an inverse relationship of the force between electric charges and the square of its distance, later named after him as Coulomb's law.

In 1781, he was stationed permanently at Paris. On the outbreak of the Revolution in 1789, he resigned and retired to a small estate that he possessed at Blois. He was recalled to Paris for a time in order to take part in the development of weights and measures, which had been decreed by the Revolutionary government. He was one of the first members of the National Institute and appointed inspector of public instruction in 1802. His health was already very feeble, and four years later, he died in Paris, France.

Coulomb explained the laws of attraction and repulsion between electric charges and magnetic poles, although he did not find any relationship between the two phenomena. He thought that the attraction and repulsion were due to different kinds of fluids.

Michael Faraday
(22 September 1791 – 25 August 1867), English chemist and physicist.

He developed the concepts of electric fields that eventually led to the electromagnetic wave function.

Faraday was born in Newington Butts, now part of the London Borough of Southwark. His family was not well off. James Faraday, Michael's father, moved his wife and two children to London during the winter of 1790 -1 from Outhgill in Westmorland, where he had been an apprentice to the village blacksmith. Michael was born the autumn of that year. The young Michael Faraday, the third of four children, received little education. At fourteen, he began an apprenticeship with a local bookbinder and bookseller George Riebau in Blandford St. There, he read many books, essentially educating himself.

At the age of twenty, in 1812, at the end of his apprenticeship, Faraday attended lectures by the eminent English chemist Humphry Davy of the Royal Institution and Royal Society, and John Tatum, founder of the City Philosophical Society. Afterwards, Faraday sent Davy a three hundred page book based on notes taken during the lectures. Davy's reply was immediate, kind, and favorable. When Davy damaged his eyesight in an accident with nitrogen trichloride, he decided to employ Faraday as a secretary. When John Payne, one of the Royal Institution's assistants, was relieved of his position, Sir Humphry Davy was asked to find a replacement. He appointed Faraday as Chemical Assistant at the Royal Institution on 1 March 1813.

Faraday was a devout Christian and met Sarah Barnard. They met through their families at the Sandemanian church. They married on 12 June 1821. They never had children.

In 1812, Faraday was successful at producing the first working electric motor. After some personal conflict on the job, he was assigned other projects for about ten years. He eventually continued his laboratory work with electromagnetism. In 1824, Faraday briefly set up a circuit to study whether a magnetic field could regulate the flow of a current in an adjacent wire, but could find no such relationship. Two years after the death of Davy, in 1831, he began a series of experiments in which he discovered electromagnetic induction.

Faraday's breakthrough came when he wrapped two insulated coils of wire around an iron ring, and found that, upon passing a current through one coil, a momentary current was induced in the other coil. His demonstrations established that a changing magnetic field produces an electric field. James Clerk Maxwell expressed this effect mathematically. It became Faraday's law, one of four Maxwell equations.

Faraday later used the principle to construct the electric dynamo, the ancestor of modern power generators.

Near the end of his career, Faraday proposed that electromagnetic forces extended into the empty space around the conductor. His fellow scientists rejected this idea. This idea was accepted after his death. Faraday's concept of lines of flux emanating from charged bodies and magnets provided a way to visualize electric and magnetic fields. That mental model was crucial to the successful development of electromechanical devices that dominated engineering and industry for the remainder of the 19th century.

Faraday was an excellent experimentalist who conveyed his ideas in clear and simple language. However, his mathematical abilities were limited. He only knew basic algebra. Maxwell took the work of Faraday, and others, and consolidated it with a set of equations that the modern theories of electromagnetic phenomena are based upon.

Faraday had many interests. He spent extensive amounts of time on projects such as the construction and operation of light houses and protecting the bottoms of ships from corrosion. Faraday was also active in what would now be called environmental science. He investigated industrial pollution at Swansea and was consulted on air pollution at the Royal Mint. In July 1855, Faraday wrote a letter to The Times on the subject of the foul condition of the River Thames. When the British government asked for his help to produce chemical weapons for use in the Crimean War (1853 – 1856), Faraday refused to participate citing ethical reasons.

In 1848, Michael Faraday was awarded a house in Hampton Court, Surrey free of all expenses or upkeep. This was the Master Mason's House, later called Faraday House, and now No.37 Hampton Court Road. In 1858, Faraday retired to live there. Faraday died in this house on 25 August 1867.

James Clerk Maxwell
(13 June 1831 – 5 November 1879), Theoretical physicist and mathematician.

He developed a wave function for electromagnetic transmissions, which is identical to the wave function developed for water waves.

James Clerk Maxwell was born at 14 India Street, Edinburgh, to John Clerk Maxwell, a lawyer, and Frances Maxwell . Maxwell's father was a man of comfortable means. He had been born John Clerk, adding the surname Maxwell to his own after he inherited a country estate in Middlebie, Kirkcudbrightshire from connections to the Maxwell family. Maxwell's parents did not meet and marry until they were well into their thirties, unusual for the times, and Frances Maxwell was nearly 40 when James was born. They had had one earlier child, a daughter, Elizabeth, who died in infancy.

The family moved when Maxwell was young to a house his parents had built on the 1500 acre Middlebie estate. His mother Frances took responsibility for James' early education. She was however taken ill with abdominal cancer and died in December 1839 when Maxwell was only eight. James eventually went to the prestigious Edinburgh Academy.

The ten-year old Maxwell, raised in isolation on his father's countryside estate, did not fit in well at school. Any social isolation at the Academy however ended when he met Lewis

Campbell and Peter Guthrie Tait, two boys of a similar age. They also would become notable scholars. They would remain lifetime friends.

In his eighteenth year, Maxwell contributed two papers for the Transactions of the Royal Society of Edinburgh. In October 1850, already an accomplished mathematician, Maxwell left Scotland for Cambridge University.

In 1854, Maxwell graduated from Trinity with a degree in mathematics. Maxwell decided to remain at Trinity after graduating and applied for a fellowship, a process that could take a couple of years. Maxwell was made a fellow of Trinity on 10 October 1855, sooner than was the norm. However, he accepted a professorship at Aberdeen, leaving Cambridge in November 1856.

The twenty-five year old Maxwell was a decade and a half younger than any other professor at Marischal in Aberdeen. He aggressively assumed his new responsibilities as head of department, devising the syllabus and preparing the lectures. He committed himself to lecturing 15 hours a week, including a weekly pro bono lecture to the local working men's college. He lived in Aberdeen during the six months of the academic year, and would spend the summers at his family home, which he had inherited from his father.

Maxwell married Katherine Mary Dewar in Aberdeen on 2 June 1859. Comparatively little is known of Katherine, seven years Maxwell's senior. In 1860, he lost his position at Aberdeen and managed to get the Chair of Natural Philosophy at King's College London. After recovering from a near-fatal bout of smallpox in the summer of 1860, Maxwell headed south to London with his wife. Maxwell's time at King's was probably the most productive of his career. In 1865, Maxwell resigned the chair at King's College London and returned to his family home with Katherine. In 1871, he became the first Cavendish Professor of Physics at Cambridge. He died in

Cambridge of abdominal cancer on 5 November 1879 at the age of 48.

Maxwell is considered by many physicists to be the 19th-century scientist with the greatest influence on 20th-century physics. His contributions to the science are considered by many to be of the same magnitude as those of Isaac Newton and Albert Einstein.

Conclusion

Though the material they had to work with was very crude, the advancements were very significant. Volta created a battery with wine glasses and salt water. Then there was Maxwell. He and those that worked with him at the time discovered much with very crude instruments on a wooden bench. They came up with values of the speed of light and the mathematical structure of radio waves.

Interestingly, discovering something significant did not automatically get you a place in scientific history. While Faraday discovered induction, others did before him. Faraday apparently spent a great deal of time communicating with others throughout Europe. In his spare time from this optics work, Faraday continued publishing his experimental work (some of which related to electromagnetism) and conducted foreign correspondence with scientists (also working on electromagnetism) he previously met on his journeys about Europe with Davy. [5] Joseph Henry likely discovered self-induction a few months earlier and both may have been anticipated by the work of Francesco Zantedeschi in Italy in 1829 and 1830. [6] Apparently, these individuals were not as popular as Faraday.

Chapter 9

Electromagnetic Theory

The primary goal in this chapter is to present the electromagnetic wave function developed by Maxwell in 1861. If
you can keep in mind an understanding of wave functions and
a basic understanding of the diffraction and interference phenomena, you are in a good position to understand quantum
physics.

This chapter will discuss electromagnetism from a historical
point of view. We are not going to start with the ancients but
somewhere along the way when electromagnetism became
significant. Here, we will take for granted some of the pione-
ering work that was done. This includes the work from people
like Ben Franklin, Ebenezer Kinnersley, André-Marie
Ampčre, Georg Simon Ohm, Robert Boyle, and Cavendish to name a few. We are of course, focusing on a path that
leads to an understanding of quantum physics. Essentially,
we pick up at a time when batteries were being used to power
many experiments. Up until this time, static electricity machines rubbed or rolled things together to generate electricity
used in experiments.

In this chapter, light is shown to be an electromagnetic phenomenon. This is a significant step away from water waves
and a step toward the stuff that travels through space.

Batteries

We need to talk a bit about batteries. Electricity is so common
today we have trouble understanding that getting it was difficult in the recent past. In the experiments discussed here,
one would think that the experimenter plugged a cord into a

wall to power the experiment. In general, these people had to
be expert at building a battery, then called a pile, to power the
experiment.

It began with frog legs. Back in 1790, Prof. Luigi Alyisio
Galvani was inspired when he saw frog legs twitch when
subjected to an electric current. [1] While conducting experiments on this "animal electricity," he noticed that muscles of
a frog, which was linked through its dorsal column in a circuit
with iron and copper, convulsed without electricity.

Upon presenting this phenomena and ideas about it to the
community, the scientists of that time began heated discussions of what was going on. One of the most prominent
was Alexander Volta. He made numerous experiments to
support a theory of his and ultimately developed the pile or
battery. [2] In 1800 he constructed the first device to produce
a large electric current, later known as the electric battery.
Thus, batteries came into being.

The first experiment that he recorded utilized the construction
of a voltaic pile with seven English coins (copper), stacked
together with seven disks of sheet zinc, and six pieces of paper moistened with salt water.

Lines of Electromagnetic Force
Michael Faraday began his epoch-making research about
electric induction in 1831. He constructed what is now and
was then termed an induction coil, the primary and secondary

wires of which were wound on an iron ring, side by side, and insulated from one another. In the circuit of the primary wire, he placed a battery of approximately 100 cells. In the secondary wire, he inserted a galvanometer (a device that detects the flow of electricity). He observed that if the switch connecting the primary wire to the battery was turned off and on, the galvanometner attached to the secondary wire would indicate a current flowing through the wire. This was the first he observed an alternating current causing an electromotive force by electromagnetic induction. [3]

From this little step, he went on to discover, by experiment, virtually all the laws and facts now known concerning electromagnetic induction and magneto-electric induction.

Magnetic Lines of Force

During other investigations, he took note of the peculiar manner in which iron filings arrange themselves on a cardboard or glass in proximity to the poles of a magnet. Faraday conceived the idea of magnetic "lines of force" extending from pole to pole of the magnet following the lines of iron filings.

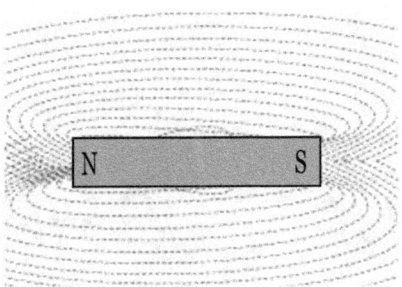

Hans Christian Oersted had discovered that electricity produced a magnetic field. [4] Faraday discovered that a current in a wire produced the same effect as a magnet. He used a compass to trace the magnetic field around a wire carrying a current. [5] Faraday eventually came to the idea that a charged particle somehow changed the space around the particle.

Then another charged particle immersed in that space would
experience a force on it due to the condition immediately surrounding the particle. [6]

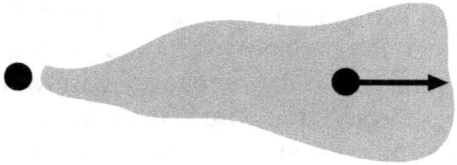

He eventually interpreted this as many lines representing the
force on another charged particle. Faraday published his ideas
of lines of force in *On Physical Lines of Force*, March 1861.

Electric Lines of Force

Part of this theory is that an electron or charged particle generates an Electric Field around the space it occupies. Our task
now, is to see where these electric lines of force or electric
field comes from.

It begins with Coulomb's law. Here is a picture demonstrating
this law. [7]

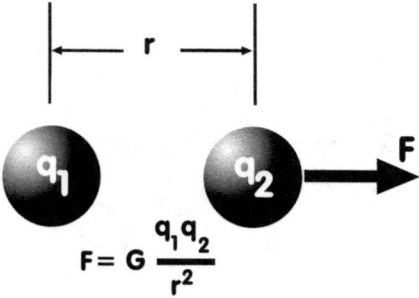

The picture lays out the elements in this law. There are two
electric charges depicted. q sub 1 on the left and q sub 2 on
the right. These could be two electrons. Or, they could be two
metal spheres covered with electrons. They are two charges

with the amount of charge specified in coulombs. Negative charges repel each other and q sub 1 and q sub 2 are negative. The arrow F represents a force pushing q sub 2 to the right. There is a similar force pushing the left charge to the left. However, we are only concerned with the charge on the right. The small r is the distance between the two charges. Note that r is the distance between the centers of the charges.

To get the electric field in this simple example, begin with Coulomb's Law.

$$F = G \frac{q_1 q_2}{r^2}$$

Then get an equation for E by removing q sub 2 from the equation.

$$E = G \frac{q_1}{r^2}$$

Now, in our scenario above, the charge on the ball never changes. We can, however, generate a number for E at each point around the ball. Just plug in the distance from the center of the ball and we have it. Now, we can conveniently determine the force on any electron that might jump out before us. When this happens, we need to find how far from q sub 1 the electron is. That is the value for r. Put that into our E equation and multiply that by the amount of charge that jumped in front of us.

$$F = E q_2$$

The attempt here is to show you the source of the E field. The E field is the subject of much study. The E field generates the lines of force that science greats: Faraday, Gauss, Lenz and Ampere studied so intently. [8]

Before we go further, perhaps we should be sure we know what a field is and what lines of force are.

Fields and Vector Fields

What is a vector? It is a number. Most numbers we are accustomed to are called scalars. They mostly are used for counting things such as apples. One would say, *There are 25 apples in the basket.* It could also be used for miles per hour as in, *He is going 20 miles per hour.* Scalar numbers such as, 20 miles per hour or 25 apples are good for counting things. A vector number also has a direction. For example, *He is going 50 miles per hour due south.* When specifying a vector number, one must also specify the direction of the number some how. Often this is done by drawing an arrow to represent the direction of the number. Often, the length of the arrow represents the scalar value (miles per hour) of the number. Here is an arrow representing a vector number. It could represent a bug in the text traveling at one inch per second moving directly to the right.

Now let's look at the term field. We can begin with a field of corn.

Now, let's see what a field of points look like.

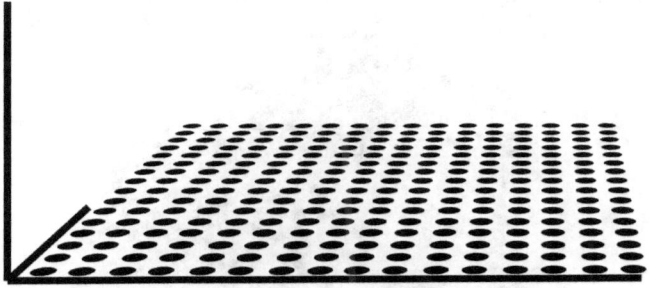

Unfortunately, we do not live in a two dimensional world. Most things around us are three dimensional. So, let's see what a three dimensional field of points looks like.

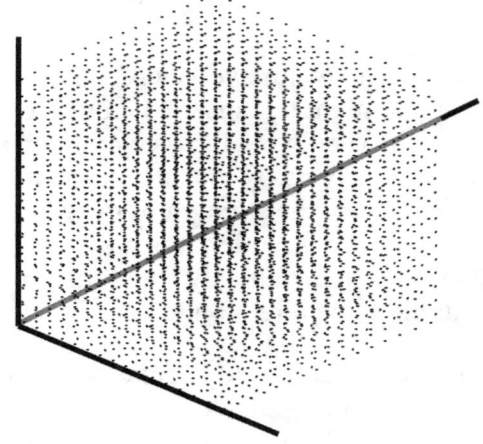

Most of the time, we will be talking about vectors. So we
need to get an idea of what a vector field looks like.

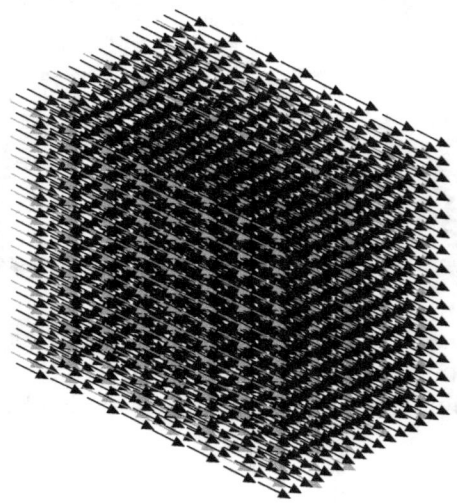

If you look close at the picture you can see it is made up of
small arrows. What might this mean? It could represent the
molecules in the air when wind blows by you on a windy day.
All the molecules are flying by you at a constant speed. However, the arrows indicate the strength of the wind at each
point in the field and the direction it is moving. This sample
implies the wind is moving quickly for all the molecules seem
to be moving in the same direction.

Normally, drawing a vector field is quite difficult so a three
dimensional image is not used. Normally, a two dimensional
image is used with the understanding that the image represents something in three dimensions. The following picture

would then be a representation of the previous, three dimensional, vector field.

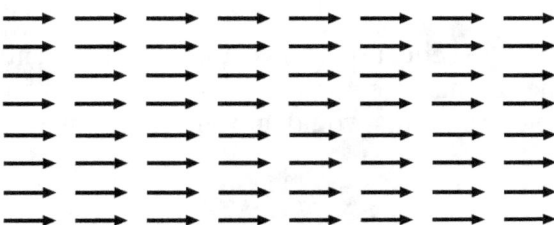

Notice that this picture is still very complicated. Also, notice
that many of the arrows seem to fall on the same line.

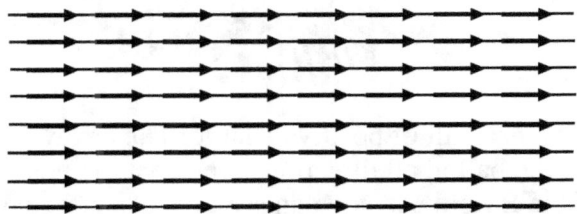

The arrows are then replaced with lines of force that are to
describe the position and direction of many vectors along
those lines.

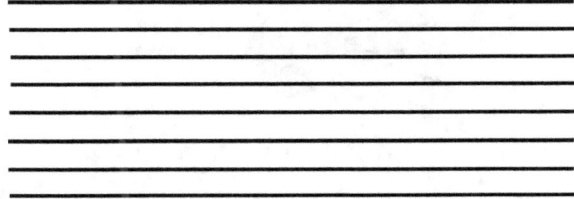

Let's consider the vector field and lines of force for the E field we just discovered. Here is the equation of the E field.

$$E = G \frac{q_1}{r^2}$$

Now, the E in this equation is a vector field. It essentially says that there is an E vector at all points around a ball of charge. A picture of this would appear something like this.

This represents a ball of positive charge. The arrows pointing away from the ball in all directions represent the positive direction of the E field. As you can see this gets quite complicated. So, let's try a two dimensional image and be sure to tell everyone it represents a three dimensional image.

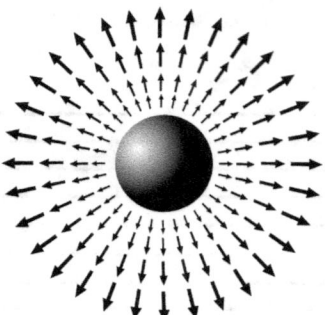

This is still fairly complicated. The science guys don't like to spend a lot of time drawing pictures like this. So, again, notice that a lot of the vectors line up in what could be simple lines. With that in mind, a drawing might appear as follows.

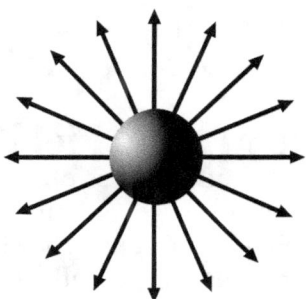

The lines that end in arrow heads represent the E field of this equation.

$$E = G \, \frac{q_1}{r^2}$$

They are called lines of force. Hopefully, this description offers a good explanation of vectors and fields.

Getting more Complex
Note that the charge above is in the shape of a single sphere. Two such charges look quite a bit different as is shown below.

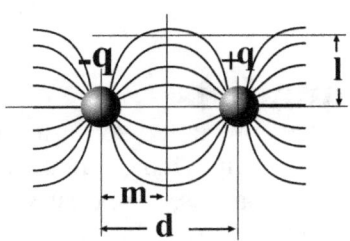

The math that represents this E field is much more complex

than the one that applied Coulomb's Law to a simple situation. The following equation represents the above E field.

$$E = \frac{q}{2\pi \epsilon_0 \left(d^2 + 2\, l^2\right)} \left((d-m)^2 \vec{x} + 2 l \vec{y}^2\right)$$

What is a Line in an E Field?

In all of this, the E field and the lines of the E Field were studied carefully. But let us not forget an important factor. What do the lines of the E Field represent? The size and direction of a line of E Field determines the strength and direction of the force the field exerts on a charged particle on that line.

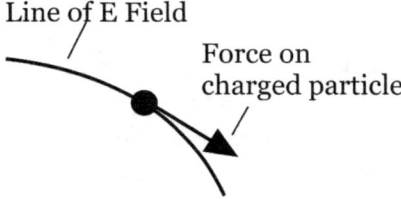

The black dot in this picture represents some kind of charged particle. In this picture, the charged particle was dropped into a line of the E field. An E field equation describes the fate of such a particle.

Maxwell's Equations

The examples of charges presented so far are static charges. They are not moving. Moving charges makes things much more complex. In static charges, the negative force tends to drive another negative force away in a straight line. When a charge is moving, this is not true. The E fields and lines of force get very complicated. Likewise, the mathematics gets complicated as well.

Faraday studied these lines of force extensively. In addition, Gauss, Lenz and Ampere as well as many others, studied the lines. Maxwell took some of Faraday's work and developed equations to express the concepts. Maxwell also gathered a number of equations from those three men. These mathematical equations were gathered together in a package and popularized as Maxwell's Equations. They were published in a book called *On Physical Lines of Force*. [9]

This package is very complicated and beyond the math used in this book. But we need to study one aspect of this package that has a great impact on the wave functions being studied here and ultimately on our study of quantum physics. This aspect can be introduced with an experiment using two wires. Consider the following.

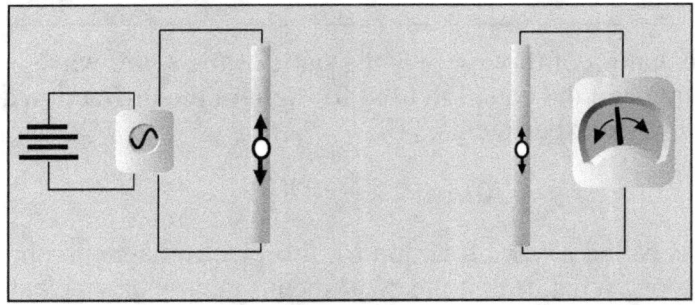

On the left a battery is connected to an oscillator. The oscillator outputs an alternating current to a wire. The electrons in the wire first go up. Then the electrons in the wire go down. They continue this up and down motion. Here the wire on the left is enlarged to show an electron in the wire oscillating up and down. That is the dot in the middle of the left wire. On the right of the picture is another enlarged wire to show an electron in the wire. That wire is attached to a meter that can display the current in the wire. The needle is moving back and forth. There is no connection between the left wire and the right wire. However, the electron in the left wire moving up and down; causes the electron to move up and down in the right wire. This causes the needle in the meter to go back and forth. This demonstrates that an electromagnetic emission is traveling from the left to the right. There is nothing to see. This represents how radio and TV transmissions work.

When Maxwell's equations are applied to this problem, the E Field representing this field would appear as this.

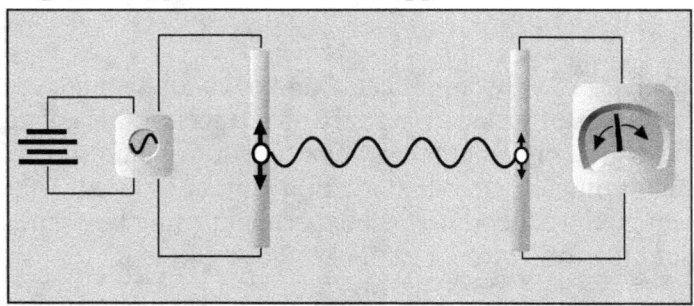

The equation of this curve is the same as the water wave equation and the equation used for light in the diffraction and interference experiments. [10]

$$y = A \sin\left(2\pi\frac{x}{\lambda} - 2\pi\frac{t}{T}\right)$$

This is called a wave function for this electromagnetic phenomenon. What does it mean? It means that when an electron is oscillating regularly like this, it somehow changes the environment around it. The curved line in the picture represents the change in the environment. It represents an E field or electric field. The various positions represent the strength of the E field in that space. Where the curve is high represents the E field is strong. Where the field is low indicates the E field is weak. So why do we want to know what the E field is? Of what use is it? Go back and look at the effect of an E field on a charged particle. When a charged particle is immersed in an E field, there is a force exerted on the charged particle. That force causes the charged particle to move, as depicted in these pictures. However, the E field is oscillating. Therefore, the force on the particle is oscillating and a charged particle immersed in it will move up and down.

The similarities between water waves and electromagnetic waves are staggering. The following picture puts them side by side. The upper image represents a water wave with wooden blocks at each end of the wave. The lower image represents an electromagnetic wave with wires at each end of the wave.

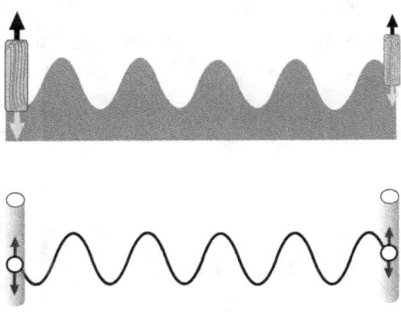

In the water wave, on the left, we have a block of wood moving up and down. In the electromagnetic wave, we have an electron moving up and down in a wire at the left of the screen. Then each has a wave that moves to the right. Both are wave functions and have the same function or equation form.

$$y = A \sin\left(2\pi\frac{x}{\lambda} - 2\pi\frac{t}{T}\right)$$

For water, the wave function affects the wooden block on the right. The electromagnetic wave function affects an electron in the wire at the right side of the picture.

Let's look at some differences between water waves and electromagnetic waves. First, the water wave equation represents the position of the surface of the water as the waves move from left to right. Second, we can see the water wave. The equation represents something in our physical world we can sense without other instruments.

This is not so with the electromagnetic wave function. The

wave function represents some change in space we cannot see until we drop something into its field of influence. Let's look at this for a moment. Although the picture of the wave is a wiggly line moving up and down, the space we are concerned with is a straight line. The following shows the space we are concerned with and a charge placed in that space. We are concerned what happens along a straight line from the electron in the wire and some external electron.

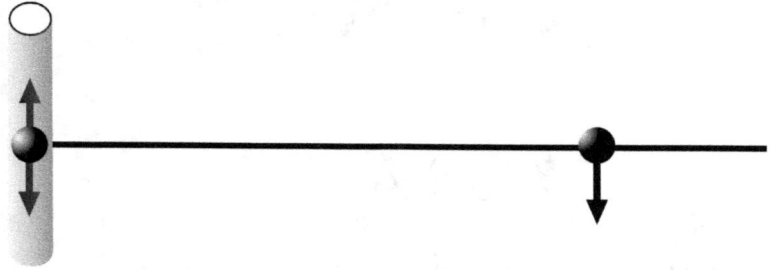

Because the electron on the left is moving, it generates the wave function in space along that line. The wave function is a field of vectors as depicted in this image.

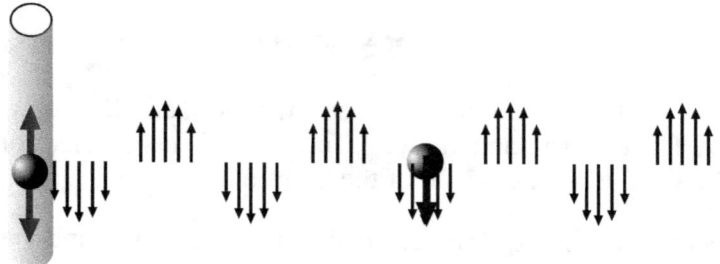

Note, there are more vectors in the space around the line than depicted here. This picture is illustrating those in the x-y plane of the space. That is, the picture is showing the E field vectors that occur in the plane of the page you are looking at.

Note the length of those arrows are not the length of something. They represent the direction and strength of the E field at those positions.

The electromagnetic wave function or equation appears as follows to indicate to engineers the intensity of the E field vectors.

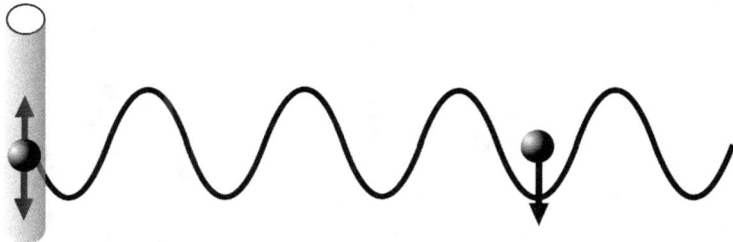

Thus, the water wave function is of the surface of the water. The electromagnetic wave function is of the E field vectors resulting from the oscillating electron.

Conclusion

In this chapter, you should have gotten some idea of what a wave function is for an electromagnetic wave. With this, you should understand that the wave function is a mathematical equation that represents how space is altered to affect a charge placed in this field. This process represents a description of space, the E field. It has no meaning until something is put there it can affect. Then we observe the effect.

There is one more little thing you should take away from this. Observe what determines the equation of the E field. The shape and form of the charged boundaries of the space determine what the E field looks like. In the examples here, a single sphere of charge resulted on one equation. Two spheres produced a more complex, different equation. The electrons moving up and down produced another equation. Although Maxwell's equations explain the change in space around charged objects, one must know the shape of the charged objects to determine the equation. The shape of the charges is referred to as the boundary conditions.

Chapter 10

The Men of Molecules and Atoms

Around the early 1800's, human thought moved away from superstition and into an era of investigation leading to understanding. Here we are introduced to some men that decided that the world around them was built from small tiny things. They conceived of molecules and atoms. These entities had to be recognized to enable us to probe deeper into the meaning of existence to get close to quantum physics.

Antoine Lavoisier [1] led the way with an experimental approach. Instead of contemplating the meaning of life, he executed experiments and formed opinions based on observation. He was the first that performed a number of tests and gathered a large amount of data that marked the end of the air, water, fire and earth beliefs. Dmitri Mendeleev [2] popularized the first Periodic Table of the Elements. A tool that is significant when observing the structure of nature. John Dalton [3] put forth a formalized description of what atoms or molecules could be. Jean Perrin [4] extended that idea and proved that substances consisted of molecules. Avogadro [5] came up with a way to count the number of molecules in a given amount of substance. These men and those they worked with, studied matter, and prepared us to look inside the atom.

Antoine-Laurent de Lavoisier
(26 August 1743 – 8 May 1794), French chemist and biologist.

He was one of the noteworthy individuals that altered the view of fire, earth, air, and water. That view was replaced with the modern view of elements and their combinations to make compounds. He compiled the first extensive list of elements.

Born to a wealthy family in Paris, Antoine Laurent Lavoisier inherited a large fortune at the age of five when his mother died. He attended the Collĕge Mazarin in 1754 to 1761, studying chemistry, botany, astronomy, and mathematics. His first chemical publication appeared in 1764. At the age of 25, he was elected a member of the French Academy of Sciences, France's most elite scientific society. This was a result of an essay on street lighting, and in recognition for his earlier research.

In 1771, at the age of 28, Lavoisier married 13-year-old Marie-Anne Pierrette Paulze. She became his lab assistant in all his experiments.

While an outstanding scientist, Lavoisier discovered no new substances, devised no really novel apparatus, and worked out no improved methods of preparation. He was essentially a theorist. He picked up experiments from others and brought them to completion. He finished the work of Black, Priestley and Cavendish, and gave a correct explanation of their experiments. Often, however, he did not give credit where credit was due.

He stated the first version of the law of conservation of mass, recognized and named oxygen (1778) and hydrogen (1783), helped construct the metric system, wrote the first extensive list of elements, and helped to reform chemical nomenclature.

Lavoisier was branded a traitor during the Reign of Terror by French Revolutionists in 1794. Lavoisier was tried, convicted, and guillotined on 8 May in Paris, at the age of 50.

One and a half years following his death, Lavoisier was exonerated by the French government. When his private belongings were delivered to his widow, a brief note was included reading, *To the widow of Lavoisier, who was falsely convicted.*

Overall, his contributions are considered the most important in advancing chemistry to the level reached in physics and mathematics during the 18th century.

Dmitri Ivanovich Mendeleev
(8 February 1834 – 2 February 1907), Russian chemist and inventor.

He is noted for the creation of first significant version of the Periodic Table of the Elements. This became a powerful tool for chemists and discovery of protons and neutrons.

Mendeleev was born in Verhnie Aremzyani village, near Tobolsk, to Ivan Pavlovich Mendeleev and Maria Dmitrievna Mendeleeva. Mendeleev is thought to be the youngest of 14 siblings, but the exact number differs among sources. At the age of 13, after the passing of his father and the destruction of his mother's factory by fire, Mendeleev attended the Gymnasium in Tobolsk.

In 1849, the now poor Mendeleev family relocated to Saint Petersburg, where he entered the Main Pedagogical Institute in 1850. After graduation, tuberculosis caused him to move to the Crimean Peninsula on the northern coast of the Black Sea in 1855. While there he became a science master of the Simferopol gymnasium Number 1. He returned with fully restored health to Saint Petersburg in 1857.

On 4 April 1862 he had got engaged to Feozva Nikitichna Leshcheva, and they married on 27 April 1862 at Nikolaev Engineering College's church in Saint Petersburg. In 1876, he became obsessed with Anna Ivanova Popova and began courting her; in 1881 he proposed to her and threatened sui-

cide if she refused. His divorce from Leshcheva was finalized
one month after he had married Popova on 2 April 1882. Even
after the divorce, Mendeleev was technically a bigamist. The
Russian Orthodox Church required at least 7 years before
lawful re-marriage. His divorce and the surrounding controversy contributed to his failure to be admitted to the Russian Academy of Sciences. His daughter from his second
marriage, Lyubov, became the wife of the famous Russian
poet Alexander Blok. His other children were son Vladimir
and daughter Olga, from his first marriage to Feozva, and son
Ivan and a pair of twins from Anna.

In 1907, Mendeleev died at the age of 72 in Saint Petersburg
from influenza.

After becoming a teacher, Mendeleev wrote the definitive
two-volume textbook at that time: *Principles of Chemistry*
(1868–1870). As he attempted to classify the elements
according to their chemical properties, he noticed patterns
that led him to postulate his "Periodic Table." Mendeleev was
unaware of the other work on periodic tables going on in the
1860s.

John Dalton
(September 5, 1766 - July 27, 1844), Mathematics and physics tutor.

Dalton is best known for the establishment of the concept of
an atom. He is known as the father of modern atomic theory.

Born in Eaglesfield, England of Quaker parents, he was
strongly influenced by his Quaker faith. Dalton grew up in
modest surroundings and continued to live simply throughout
his life. He attended the local school at Eaglesfield until the
age of 11. A year later, he returned as a teacher at the same
school. As a scientist, he was self-taught. At the age of 15,
Dalton became an assistant at the Quaker school in Kendal.
About four years later, he was appointed principal of the
school, a post he held for eight more years. While at Kendal,

Dalton also offered a series of public lectures on natural philosophy. The series was unsuccessful, as Dalton was shy before an audience.

He was a tutor in mathematics and natural philosophy at the New College in Manchester for a few years. He left there and became a private tutor in mathematics and natural philosophy.

Dalton's interest in weather caused him to think about the nature and composition of air. He eventually concluded, as had a few scholars before him, that air consists of tiny, individual particles. But Dalton went beyond the musings of his predecessors and hypothesized that all forms of matter consisted of these tiny particles.

He first developed a formal theory about these particles between 1803 and 1805. The atoms that Dalton had in mind were tiny, indivisible particles that constituted all chemical elements. Dalton's theory also dealt with the composition of compounds. The smallest particle of any compound, he said, was a compound atom. Thus, he taught that water was composed of compound water atoms.

Dalton is called the father of modern atomic theory partly because of his clear statement of that theory and partly because of his emphasis on atomic weights. No proponent of the concept of atoms had previously made clear the fact that atoms must have weights that can be determined experimentally. Dalton did. He said that finding the weights of atoms was a relatively straight forward task that any chemist could accomplish.

Dalton's discussion of atomic weights gave chemists a concrete plan for exploring at least this aspect of the atomic theory. Within a few years, Dalton had prepared a list of atomic weights for the known elements.

Dalton's theory was quickly accepted by the vast majority of chemists. One reason for its rapid success was that it explained certain experimental results that had recently been announced, most notably Joseph Proust's law of constant composition which stated that a chemical compound always contains the same constituents in the same proportions.

He died quietly at his home in Manchester on July 27, 1844.

Lorenzo Avogadro
(9 August 1776 – 9 July 1856), Italian physicist and mathematician.

Avogadro is hailed as a founder of the atomic-molecular theory. Notable for establishing that a specific mass of material consists of a specific number of molecules

Amedeo Avogadro was born in Turin to a noble family of Piedmont, Italy. He graduated in ecclesiastical law at the early age of 20 and began to practice. Soon after, he focused on physics and mathematics. In 1809 he started teaching them at a high school in Vercelli, where his family had property.

Little is known about Avogadro's private life, which appears to have been sober and religious. He married Felicita Mazzé and had six children. Two sons rose to positions of distinction: Luigi, who became general of the Italian army, and Felici, who became president of the Court of Appeal.

In 1820, he became professor of physics at the University of Turin. After the downfall of Napoleon in 1815, northern Italy came under control of the king. He was active in the revolutionary movements of 1821 against the king. As a result, he lost his chair in 1823

Later, Avogadro was recalled to the university in Turin in 1833, where he taught for another twenty years. Avogadro held posts dealing with statistics, meteorology, and weights and measures and was a member of the Royal Superior Council on Public Instruction.

Avogadro was the first to recognize that all matter was composed of molecules and atoms; and that molecules were composed of atoms. This was a significant advancement. John Dalton had not considered this possibility.

Avogadro became the first to show that water is composed of molecules of two atoms of hydrogen and one of oxygen. The related concept, called Avogadro's number, is that when the mass of a compound in grams is equal to the molecular weight, the total number of molecules is always the same, equal to 1 mole.

Unfortunately, the simplicity and clarity of Avogadro's views were not compelling to the majority of chemists. This was due to the fame of John Dalton and the interest in the performance of experiments. Logical inference was not fashionable

Avogadro's hypothesis was ignored. However, by the 1880s it was universally accepted, thanks to Stanislao Cannizzaro, who created a table of atomic weights based on Avogadro's work. Unfortunately, this was four years after Avogadro's death.

Modest and retiring, he was indifferent to honors and scrupulously avoided those public struggles for prestige, which were a characteristic of continental scientific society in the mid-19th century.

Jean Baptiste Perrin
(30 September 1870 - 17 April 1942), French physicist.

His experiments proved the physical reality of molecules.

Born in Lille, France, Perrin attended the École Normale Supérieure, the elite grande école in Paris. He became an assistant at the school during the period of 1894-97 when he began the study of cathode rays and X-rays. He was awarded a PhD in 1897. In the same year he was appointed as a lecturer in physical chemistry at the Sorbonne, Paris. He became a professor at the University in 1910, holding this post until the German occupation of France during World War II.

Perrin was an officer in the engineer corps during World War I. When the Germans invaded France in 1940, he escaped to the U.S.A.

Jean Perrin showed that cathode rays were made of corpuscles with negative electric charge. He computed Avogadro's number through several methods. He explained solar energy by the thermonuclear reactions of hydrogen. After Albert Einstein published his theoretical explanation of Brownian motion in terms of atoms, Perrin did the experimental work to test and verify Einstein's predictions, thereby settling the century-long dispute about John Dalton's atomic theory.

While in the USA he died in New York City in 1942. After the War, his remains were transported back to France by the battleship Jeanne d'Arc and buried in the Panthéon.

Conclusion

The degree that scientists work together is striking here. The degree to which acceptance is social and not scientific is striking. Avogadro's discovery is very remarkable. Yet, it was ignored. John Dalton was not a professor at some prestigious university. Yet he apparently communicated widely with many in the scientific community. That community seemed to hold his word dear. That was very good for a man that was shy in front of a group of people. Unfortunately, Avogadro had died before a member of that community discovered and publicized the importance of his discovery.

Chapter 11

Molecules and Atoms

Now, we are aware of what atoms and molecules are. This was not true in the 1800's. Furthermore, it is difficult to believe that we did not know conclusively until 1926 that molecules existed. In this chapter, we discuss the predominate players that proposed and found how small we could divide the matter that is around us.

Why are we looking at this? Up until the mid-eighteen hundreds, the common person viewed the universe a place of analog design. Water flows and a steel bar is a continuous piece of matter. This chapter is about a step forward in which matter is found to be made of small particles called molecules and atoms. This must be mastered before we can look deeper. There is another important issue here. It is valuable to observe how we learned to look and see that which cannot be seen. Seeing a single molecule or atom is difficult. How can one prove that they even exist?

The Real Elements Step Forward

Around the time of Lavoisier, people still believed in the Hellenistic concept of fire, air, earth and water. Even those that began to unwind this belief used those concepts while studying what the universe is made of. Early people in this period considered their findings some subset of that Hellenistic point of view. Lavoisier's experiments began to deviate from this custom and simply looked at what was there. Here is a description of a couple of his experiments.

Lavoisier did an experiment demonstrating the existence of oxygen. [1] He heated metallic tin in a sealed flask. A grayish

ash appeared on the surface of the melting tin, which Lavoisier heated until no more ash formed. After the flask cooled, he
inverted it and opened it underwater.

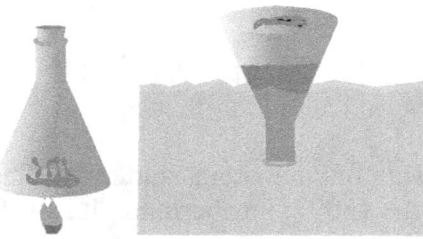

He discovered the water rose one-fifth of the way into the
glass, leading Lavoisier to conclude that air itself is a mixture,
with one-fifth of it having combined with the tin, yet the other
four-fifths did not. He concluded that this demonstrated that
air was not an element. Lavoisier repeated the experiment
again, substituting mercury for tin, and found that the same
happened. Yet after heating gently, found that the ash released the air, showing that the experiment could be reversed. He
concluded that the ash was a compound of the metal and oxygen, which he proved by weighing the metal and the ash, and
showing that it was heavier than the original metal. Lavoisier
then recorded that combustion was not an element, but instead was a chemical reaction of a fuel and oxygen.

Lavoisier continued his experiments and derived a list of elements. He named oxygen and hydrogen. His list was the first
of its kind. However, the Hellenistic belief of air, earth, water
and fire had not completely died but its end was not far away.

The important issue here is that, through the efforts of people
like Lavoisier, the scientific community began to realize that
the matter around us was made up of elements. This author
considers this the first significant step in establishing the true
structure of the universe. The concept of an element leads
directly into an understanding of atoms and molecules.

In addition, opposed to the ancient philosophers, the understanding was based on physical experiments not mind experiments.

The Atom Concept is Formalized

Atomic theory was born with Dalton when he published his theories in 1803. [2] His theory consists of a five parts:

- Elements are composed of tiny particles called atoms.
- All atoms of a given element are identical.
- The atoms of a given element are different from those of any other element; the atoms of different elements can be distinguished from one another by their respective relative weights.
- Atoms of one element can combine with atoms of other elements to form chemical compounds; a given compound always has the same relative numbers of types of atoms.
- Atoms cannot be created, divided into smaller particles, nor destroyed in the chemical process; a chemical reaction simply changes the way atoms are grouped together.

The atom appeared to be the smallest thing from which the universe was constructed. It still appeared to be a small round ball of something. Usually it was considered to be some kind of jelly like substance. One could guess that was because we observed the material around us in a variety of textures being sometimes hard and sometimes pliable. Dalton proposed that atoms combined in ratios. [3] For example, he suggested that the elements oxygen and hydrogen consist of atoms. He further suggested that oxygen gas combines with hydrogen gas to form water. He believed that one atom of oxygen combines with one atom of hydrogen to form one compound atom of water, $H + O = HO$. He erred a bit here. However, the concept of atoms combining in definite ratios added to the acceptance of his theories.

Avogadro's Number Emerges

Avogadro came up with the concept that one molar mass contained a specific number of molecules. [4] To clarify what this means, consider helium. Its nucleus has two protons and two neutrons for an atomic weight of four.

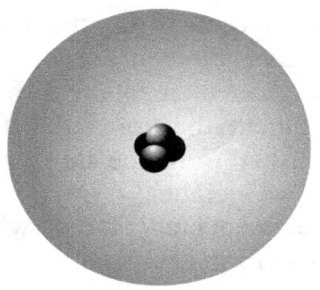

This means its molar mass is 4 grams. Avogadro's number is 6.0221415 X 1023 .

Atomic weight of helium is 4

Molar Mass = 4 grams

The number of molecules in 1 molar mass of helium is

240885660000000000000000

During Avogadro's development of his theory, he had to differentiate between molecules and atoms. Up until then, the words molecule and atom were interchangeable. For Avogadro to complete his theory, he had to differentiate between the two although he did not specifically name them as such. He cleared up one of the points that Dalton slipped up on. Avogadro clearly and correctly pointed out that two hydrogen atoms bonded with one oxygen atom to make water. [4] Avogadro was also correct in identifying the structure of several other molecules.

Avogadro's number is important for it enabled those that studied the elements to determine an element's atomic weight. That is, with the number, one could determine how many protons and neutrons were in the neucleus.

The Periodic Table of the Elements

Dmitri Mendeleev presented the Periodic Table of Elements in 1869. [5] The original table was created without knowledge of the inner structure of atoms. Instead, physical and chemical properties of the elements were correlated with atomic mass. [6] If the elements are ordered by atomic mass then a certain periodicity, or regular repetition, of physical and chemical properties can be observed. For example, the inert elements formed a column in the table.

On 6 March 1869, Mendeleev made a formal presentation to the Russian Chemical Society, entitled *The Dependence between the Properties of the Atomic Weights of the Elements,* which described elements according to both atomic weight and valence. This presentation stated that the elements, if arranged according to their atomic weight, exhibit an apparent periodicity of properties. [6] Similar properties could be gathered in specific groups based on their atomic weights. The atomic weights would appear in a specific order of these weights.

The point here is that simply gathering, observing properties and organizing like things together led the way to understanding electrons, protons and neutrons.

We, as nonprofessionals in science, hear the scientific community say, *There is a nucleus in the center of the atom that is made up of protons and neutrons.* Taken at face value, we believe one of these great minds grabbed an atom, unzipped it, and looked inside as if lifting the hood of a car and finding the engine that powered the car. This is not so. The concept of the proton grew over a long time. Many guesses were made about its existence. It worked out that many of the guesses indicated the same thing. Therefore, some head guesser announced to the world that there was a proton in the middle of the atom.

The Periodic Table of the Elements played a key roll in this. A number of people worked on it. However, historically, it came to focus on one individual, Dmitri Mendeleev. He put the properties of the then known elements on small cards. He shuffled them around attempting to find some correlation between them. He ultimately came up with the organization now called the Periodic Table of the Elements.

With his crude table, he predicted the existence of elements similar to aluminum and silicon. These were soon discovered and named gallium and selenium. This discovery confirmed the usefulness of the table and a lot of scientific effort was invested into expanding the table and explaining its structure.

One striking effect of the periodic table was the weights of the elements. They were all an integer multiple of the weight of the hydrogen atom. This strongly suggested that all elements were made up of some kind of basic building block. This eventually became the proton. The assumption was that all elements consist of protons and the number of protons was somehow responsible for the characteristics of each element. As the number of electrons in an atom were determined, and found to determine element characteristics, the periodic table indicated that there was something else in the atom about the same size as protons. Eventually, the neutron was found. [7]

As theories of atoms, molecules, electrons, protons and neutrons emerged; the concepts were taken and matched against the Periodic Table of the Elements. As a result, the concepts gained acceptance or lived a short life. In addition, the new concepts changed the periodic table. It has been constantly changing up to today.

The importance of this is to answer the question, *How do scientists know what electrons, protons and neutrons are?* Part of the answer is that the elements were organized into the Periodic Table of the Elements. This categorization of properties of elements revealed an inner structure of the elements even though the inner structure could not be observed directly. This categorization became a guiding light for those trying to unzip the atom and peer inside.

Molecules Identified

The year 1873, by many accounts, was a seminal point in the history of the development of the concept of the "molecule". In this year, the renowned Scottish physicist James Clerk Maxwell published his famous thirteen page article *Molecules?* in the September issue of *Nature*. [8] In the opening section to this article, Maxwell clearly states: An atom is a body which cannot be cut in two; a molecule is the smallest possible portion of a particular substance.

After speaking about the atomic theory of Democritus, Maxwell goes on to tell us that the word "molecule" is a modern word. He states, *It does not occur in 'Johnson's Dictionary.' The ideas it embodies are those belonging to modern chemistry.* We are told that an "atom" is a material point, invested and surrounded by "potential forces" and that when "flying molecules" strike against a solid body in constant succession it causes what is called pressure of air and other gases. At this point, however, Maxwell notes that no one has ever seen or handled a molecule.

Moreover, at this point, the pudding picture of an atom persists.

Molecule Existence Proven

In 1926, French physicist Jean Perrin received the Nobel
Prize in physics for proving, conclusively, the existence of
molecules. [9] He did this by calculating Avogadro's number
using three different methods. 1926 is not that long ago.

Conclusion

This pattern of behavior is of interest in the development of
quarks. During research of sub-atomic particles, high energy
accelerators forced small bits of matter to smash against each
other. Many unusual particles flew away from these collisions. In the process, nuclear physicists measured the properties of these little bits. As in the case of the elements, the
table of little bits was organized in some periodic manner.
Like the elements that found the mass of hydrogen to be a
common factor in the elements, the quark was found to be a
common factor in the bits that formed this little bit table. The
point being made is that scientists cannot see into the smallest
things of our universe but compare things of similar nature
and draw conclusions based on these similarities and
differences.

Chapter 12

Atomic Men

Up to this point, the atom remained a closed door. There was an idea that everything around us was made of very small things. The idea that the small things consisted of smaller things had not occurred to anyone. Even the ancients of the past did not seem to have any idea of this. The men presented here grab that unseen door and open it to see what is inside of the little balls of jelly that made the universe tick. J. J. Thomson [1] led the way by studying and establishing the existence of the electron. Ernest Rutherford [2] pursued an experiment suggested by Thomson and revealed the hard core of atoms. He went on to identify the proton as one of the constituents of the nucleus. Henry Moseley [3] came up with a way to count electrons in an atom. This expanded the Periodic Table of the Elements considerably adding to its power in science. James Chadwick [4] recognized the neutron as a mysterious component in the nucleus.

Sir Joseph John "J. J." Thomson
(18 December 1856 – 30 August 1940), British physicist.

He is credited with the discovery of the electron.

Joseph John Thomson was born in 1866 in Cheetham Hill, Manchester, England. His mother, Emma Swindells, came from a local textile family. His father, Joseph James Thomson, ran an antique bookshop founded by his grandfather from Scotland.

J J's early education took place in small private schools where he demonstrated great talent and interest in science. In 1880, he was admitted to Owens College though he was only 12 years old. His parents planned to enroll him as an apprentice engineer to Sharp-Stewart & Co., a locomotive manufacturer, but these plans were cut short when his father died in 1873. He moved on to Trinity College, Cambridge in 1876. In 1880, he obtained his BA in Mathematics and MA in 1883. In 1884, he became Cavendish Professor of Physics. One of his students was Ernest Rutherford, who would later succeed him in the post. In 1890, J J married Rose Elisabeth Paget, daughter of Sir George Edward Paget, a physician and then Regius Professor of Physic at Cambridge. He and Rose had one son, George Paget Thomson, and one daughter, Joan Paget Thomson.

Thomson's most significant work was his investigation of cathode rays in evacuated glass tubes.

This led to his discovery of the electron.

In 1906, Thomson demonstrated that hydrogen had only a single electron per atom. Previous theories allowed various numbers of electrons.

He was knighted in 1908 and appointed to the Order of Merit in 1912. In 1914 he gave the Romanes Lecture in Oxford on *The Atomic Theory*. In 1918 he became Master of Trinity College, Cambridge, where he remained until his death. He died on August 30, 1940 and was buried in Westminster Abbey, close to Sir Isaac Newton.

Ernest Rutherford
(August 30th 1871 - 19th of October 1937), Physicist.

He formed the idea that the atom consists of a small nucleus with electrons zipping about it like planets around a sun.

Ernest Rutherford was born at Spring Grove, Nelson on August 30th 1871, the second son and fourth child of twelve born to James and Martha Rutherford. James Rutherford migrated with his family from Scotland to New Zealand in 1843 when four years old. He became a wheelwright and engineer, and later a flax-miller. Martha Rutherford was born in England and arrived in New Plymouth in 1855 when thirteen years old. She was moved to Nelson in 1860 during the Taranaki Land War. Martha became a teacher at the Spring Grove School.

Rutherford led the life typical of a child growing up in rural New Zealand. This included milking cows and gathering firewood. In 1887, He won a Scholarship to Nelson College. For the next three years, he boarded at Nelson College. In 1889, he won one of the ten scholarships to the University of New Zealand. From 1890 to 1894, he attended Canterbury College in Christchurch. In 1892, he got a degree in Pure Mathematics, Latin, Applied Mathematics, English, French and Physics.

His mathematical ability won him the one Senior Scholarship in Mathematics available in New Zealand. This allowed him to return for a further year during which he took both mathematics and physics. In 1893, he obtained a Master of Arts degree in Mathematics and Mathematical Physics and in Physical Science (Electricity and Magnetism).

However, having failed to obtain a permanent job as a schoolteacher, Rutherford had few other options for a career. He seemed to be limited to tutoring. Yet, he continued research in electrical science. At this time, the Royal Commissioners for the Exhibition of 1851 had just initiated scholarships to allow graduates of universities in the British Empire to go anywhere in the world and work on research of importance to their home country's industries. In response to this, Ernst returned to Canterbury College to pursue a BS degree in geology and chemistry which qualified him for one of these scholarships. Through a few twists of fate, he got the scholarship and went off to Cambridge University in London. He elected to work with Professor J J Thomson of Cambridge University's Cavendish Laboratory. J J Thomson quickly realized that Rutherford was a researcher of exceptional ability. Thomson invited him to join in a study of the electrical conduction of gases. Thus, Ernst aided Thomson during the discovery of the electron.

Cambridge University rules prevented Rutherford's advancement at the university. In 1898, he accepted a professorship at McGill University in Montreal, Canada. The laboratories at McGill were very well equipped. As Rutherford wrote to his future wife, *I am expected to do a lot of work and to form a research school in order to knock the shine out of the Yankees!*

Rutherford returned to New Zealand in 1900 to marry Mary Georgina Newton, the daughter of his landlady in Christchurch. They were to have one child, Eileen.

In 1919, Rutherford became the Director of Cambridge University's Cavendish Laboratory. He was responsible to set up a research team for the university.

His daughter, Eileen, married Ralph Fowler, a mathematical physicist at the Cavendish Laboratory. They had four children: Peter who became a distinguished cosmic ray physicist, Elizabeth a doctor, Patrick an electrical engineer monitoring safety at nuclear power plants and Ruth a research physiologist. Sadly, Eileen died of an embolism at age 29, nine days after the birth of her fourth child and just two days before Christmas of 1930.

Rutherford's life was marked with many significant achievements. Three are particularly striking. He explained radioactivity as the spontaneous disintegration of atoms. He was, in a sense, the world's first successful alchemist. During his experiments, he converted nitrogen into oxygen. These experiments were responsible for the discovery of the proton. He is most famous for his explanation of the structure of the atom. It is a very common icon in our society and displayed prominently in many locations.

Ernest Rutherford died at the age of 66 on 19 October 1937, the result of delays in operating on his partially strangulated umbilical hernia. His ashes were interred in London's Westminster Abbey. When JJ Thomson died in 1940, he was interred next to Rutherford. Newton presides above them and they are surrounded by other greats of British science.

Henry Gwyn-Jeffreys Moseley
(23 November 1887 - 10 August 1915), English physicist.

He developed Moseley's Law, which had a profound effect on the organization of the Periodic Table of the Elements. His law essentially provided a way to count the number of electrons in an atom.

Moseley was born in Weymouth, Dorset, on the southwestern coast of England in 1887. His father Henry Nottidge Moseley, who died when Henry Moseley was quite young, was a biologist and a professor of anatomy and physiology at the

University of Oxford. Moseley's mother was Amabel Gwyn-Jeffreys Moseley, who was the daughter of the biologist John Gwyn Jeffreys.

Henry Moseley performed well in school and was awarded a King's scholarship to attend Eton College. In 1906, Moseley entered the Trinity College of the University of Oxford, where he earned his bachelor's degree. Immediately after graduation from Oxford in 1910, Moseley entered the graduate school at the University of Manchester to study and work under the supervision of professors such as Sir Ernest Rutherford. During Moseley's first year at Manchester, he had a teaching load as a graduate teaching assistant, but following that first year, he was reassigned from his teaching duties to work as a graduate research assistant.

Here he used X-ray crystallography to reveal the structure of atoms and answer some questions that plagued chemists for a number of years. Moseley predicted the existence of element 61 that was previously unsuspected. Quite a few years later, the element was created artificially in nuclear reactors and was named promethium.

Sometime in the first half of 1914, Moseley resigned from his position at Manchester, with plans to return to Oxford and continue his physics research there. However, World War I broke out in August 1914, and Moseley turned down this job offer to enlist in the Royal Engineers of the British Army. Moseley served as a technical officer in communications during the Battle of Gallipoli, in Turkey, beginning in April 1915, where he was killed in action on August 10, 1915. Moseley was shot through the head by a Turkish sniper while in the act of telephoning a military order.

Moseley could in many scientists' opinions have contributed a lot to the knowledge of atomic structure had he survived. As Niels Bohr once said in 1962, *You see actually the Rutherford work [the nuclear atom] was not taken seriously. We cannot understand today, but it was not taken seriously at all.*

There was no mention of it any place. The great change came from Moseley.

Sir James Chadwick
(20 October 1891 – 24 July 1974), English physicist.

He is noted for the discovery of the neutron.

Chadwick was born in Bollington, Cheshire to John Joseph Chadwick and Anne Mary Knowles. He went to Bollington Cross C of E Primary School, attended the Central Grammar School for Boys in Manchester, and then studied at the Universities of Manchester and Cambridge.

In 1913, Chadwick went and worked with Hans Geiger at the Technical University of Berlin. He also worked with Ernest Rutherford. He was in Germany at the start of World War I and was interned in Ruhleben P.O.W. Camp just outside Berlin. While he was interned, he had the freedom to set up a laboratory in the stables. With the help of Charles Ellis, he worked on the ionization of phosphorus and on the photochemical reaction of carbon monoxide and chlorine. He spent most of the war years in Ruhleben until Geiger's laboratory interceded for his release.

In 1932, Chadwick discovered the previously unknown particle now called neutron in the atomic nucleus. This discovery enabled the creation of atom bombs. He was awarded the Hughes Medal of the Royal Society in 1932 and the Nobel Prize for Physics in 1935.

His discovery particularly inspired Enrico Fermi, Italian physicist, to discover nuclear reactions brought by slowed neutrons, and led to the revolutionary discovery of "nuclear fission."

Chadwick became professor of physics at Liverpool University in 1935. He visited North America in 1940 to collaborate with the Americans and Canadians on nuclear research.

Returning to England in November 1940, he concluded that
nothing would emerge from this research until after the war.
In December 1940, it was reported that isotope uranium-235
could be extracted. James Chadwick later wrote that it was at
that time he, *realized that a nuclear bomb was not only possible, it was inevitable. I had to then take sleeping pills. It was
the only remedy.*

Shortly afterward, he joined the Manhattan Project in the United States, which developed the atomic bombs dropped on
Hiroshima and Nagasaki. Chadwick was knighted in 1945.

He died on 24 July 1974 at the age of 82 in Cambridge,
England

Conclusion
These men worked in a remarkable time that began without
knowing what was in the atom. Traditional wisdom suggested
that the atom was naught but a bit of jelly. Eventually they
saw the atom cracked and the production of the atom bomb.
They unzipped the atom and peered inside. This set the stage
to observe the innards behaving in weird ways that became
famous for very small particles.

Chapter 13

Atomic Theory

Relative to our quest for knowledge of quantum physics, this is a very important chapter. Up to now, man perceived the universe as built from small billiard balls. Dalton advanced this in 1803. Since then, man has peered inside the billiard balls and discovered electrons, protons, and neutrons. However, no one has actually seen any of these particles. So, how can we be so confident that they are there? Today, the words electron, proton, and neutron are used as if everyone knew about them forever. The neutron was only discovered in 1932. That is less than 100 years ago. This is not ancient knowledge much less very old. It is easily in the lifetime of our fathers or grandfathers. This chapter attempts to show the experiments and the thoughts of the men that told us about them. In addition, the mastery of concepts about things the eye cannot see demonstrates the use of the Periodic Table of the Elements. This chapter is important because it demonstrates man's ability to organize data, execute experiments, and use logic to discern, that which cannot be seen.

It is important to see how the electron was detected. In the future quantum physics would paint a different picture of the electron. Observing the kind of picture Thompson and Rutherford painted for us is interesting.

Discovery of the Electron

In the year 1889, the British physicist J.J. Thompson discovered the electron. [1] Thompson conducted a number of experiments using cathode rays. Cathode rays are produced in a glass envelope. This is constructed by sealing two electrodes in a glass tube and removing most of the air from it.

When the electrodes are attached to high voltage, a beam of radiation is emitted from the negative electrode. These beams are called cathode rays.

In that day, it was popular for some enterprising individuals to make these tubes with traces of various kinds of gasses in the tubes. Then, the tubes would glow with various colors when a high voltage current was passed through it.

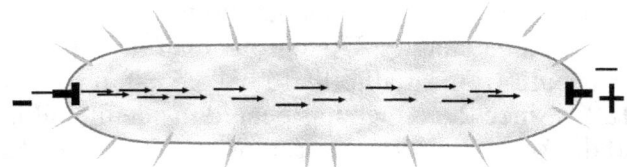

These enterprising individuals would put on demonstrations showing this technology. These demonstrations eventually became the neon lights we are familiar with today.

Thompson performed some experiments with these tubes in an effort to determine what the beam consisted of. The first of these experiments attempted to see if the charge could be separated from whatever was flowing though the tube. [2] In the experiment the beam was aimed at some device to measure charge.

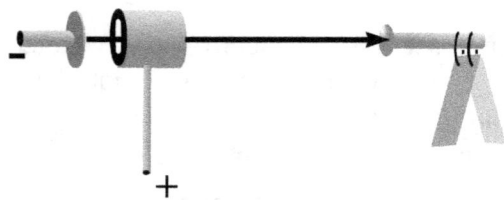

The electrons entered from the left and were attracted by a cylinder attached to the plus side of a battery. The electrons accelerated through the plus charged cylinder and traveled to

a conductor at the right end of the tube. This, in turn, was
attached to a device to detect charge. In this picture this is
depicted as two thin foils of metal. When charged, the two
thin metal plates would open up.

Then, Thompson set up an experiment to see if the charge
could be separated from the beam. [3] He positioned a magnet
near the beam to see if charge would still go to the measuring
device. It did not so he concluded that they could not be separated.

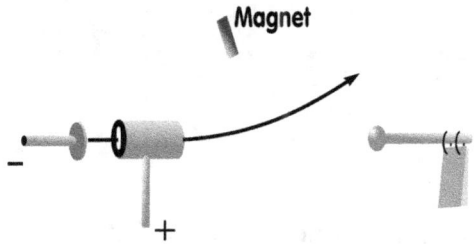

The next experiment was about bending the beam with an
electric field. At that time, no one had shown an electric field
had any effect on the beam. He suspected that some gas left in
the tube was interfering with such experiments. He built an
experiment that took care to be sure all gas was removed from
the test chamber. Then he turned on the beam and saw it bend
demonstrating that, in fact, the beam responded to an electric
field. [3]

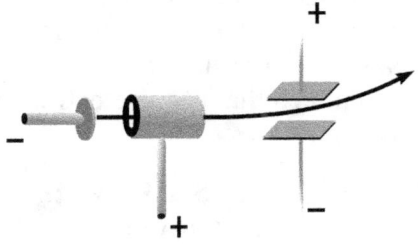

He had assumed that the beam consisted of some kind of particle. His next test was to see if he could determine the mass of those particles. He constructed an experiment that passed the beam through a magnetic field. The experiment was designed to measure the force causing the beam to turn and thus determine the mass of such particles. He set up the experiment so a magnet would cause the beam to curve in a known way. He placed a graduated scale at one end of the beam so he could measure its curvature.

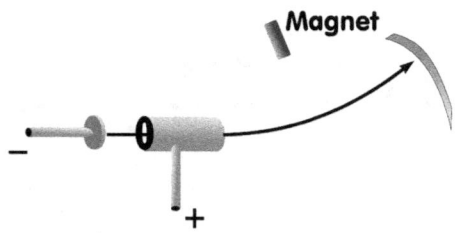

He discovered that the particle was 1/10000 the mass of a hydrogen atom. [3]

Consider what he might think. He understood that the little particles came from the atoms in the metal disk. The scientists of the day accepted that atoms were little balls of jelly. They were normally neutral or not charged. He guessed that the little particles were some part of the little ball of jelly. But they were very small and negatively charged. Is it a wonder that he concluded that most of the jelly ball was made of positive material with a few spots of negative material. To him it suggested a "plum pudding" model. In this model, bits of "plum" were the electrons which were floating around in a "pudding" of positive charge to match that of the electrons and make an electrically neutral atom.

Atomic Theory Leaps Forward

JJ Thompson recommended to one of his students, Ernest Rutherford, an experiment that would examine the jelly model of the atom. Thus, Rutherford designed an experiment that probed the depth of that idea. [4] The experiment used a high

energy source of alpha radiation that was aimed at a small
hole in a lead plate. We know today that alpha radiation is
two protons and two neutrons clumped together. This is the
nucleus of a helium atom. Some radioactive substances constantly emitted these particles as the radioactive material
decayed. This material was put into a box with a little hole in
it to allow a stream of these particles to shoot out. This produced a narrow beam of alpha particles that was aimed at a
thin foil of gold. On the other side of the gold foil was a
screen of material that gave off flashes of light when struck
with an alpha particle. When the experiment was executed,
alpha particles zipped through the lead hole and zipped
through the gold as if it were not there.

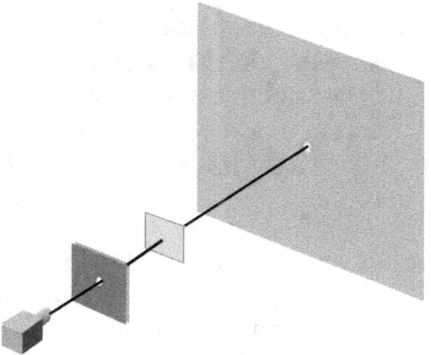

The screen glowed mostly with a spot on it that resembled the
size of the hole in the lead plate. As time passed, flashes were

occurring on the screen far away from the bright spot on the
screen.

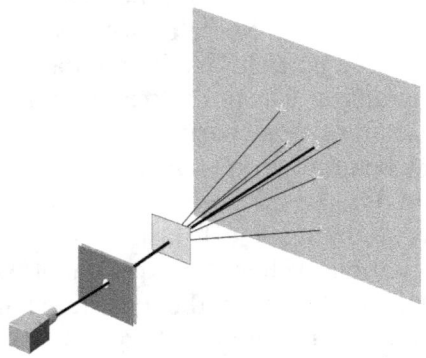

The assumption was that something hard in the middle of the
gold atom was causing the alpha particles to change direction
considerably as they passed through a gold atom. A few alpha
particles were deflected back the way they came. Rutherford
was heard to say, *It was as if you fired a 15-inch shell at a
piece of tissue paper and it came back and hit you.* [5]

Counting the flashes on the screen and using probability, they
could estimate the size of the thing deflecting the alpha particles as they moved through the gold atom. As the alpha particles were positively charged, the assumption was that the

object was positive. The probability they measured indicated that the object was very small. [6]

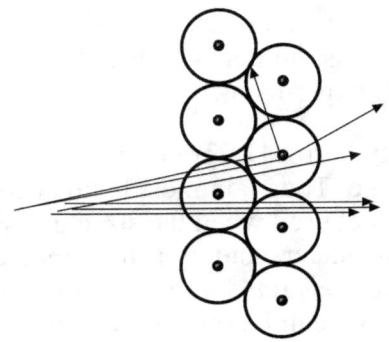

It is simply interesting to note that a beam of alpha particles striking the screen would give us some idea about the structure of the atom. We cannot see the nucleus. But, back then, the experiment was as if you could unzip the atom and see what was going on. All this was perceived because the alpha particles appeared as a simple dot with small flashes appearing some distance from that dot. There were more small flashes nearer the dot than farther away. From this they drew the conclusion the atom must be a hard core nucleus with very small electrons flying around it.

However, could there be any other conclusion to draw? They could have decided that the jelly was very thin and light and small spots on the surface of the jelly were homes for the electrons. One's imagination could have come up with other scenarios. As it turns out, the resulting conclusion was inaccurate. But quantum physics was needed to clarify what was going on.

Notice that Rutherford was affected by Thompson's idea of the plum pudding model. Thompson's experiments suggested that the electron was considerably smaller than the positive portion of the atom. This thought persisted when Rutherford saw the electron as some tiny bit of stuff that was a hardball thing flying around the nucleus. This shows again that some old idea persisted and was modified with some new information. Sometimes the understanding of what goes into an experiment is more useful than reviewing the experimenter's conclusions of the experiment.

The conclusion Rutherford offered follows and was presented to the world at large. Today, in most layman environments, this is still the model used when the structure of the atom is discussed. This misunderstanding is highlighted here for this model has been demonstrated to be very inaccurate with the study and impact of quantum physics on our society. [7] Yet the following model persists from one edge of our society to the other. The atom is perceived to be a solid ball in the middle with electrons zipping around the nucleus at some distance.

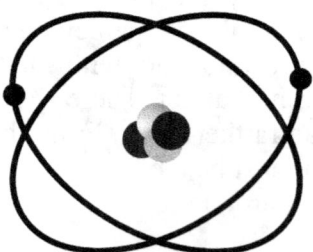

However, as this model was established, the door was opened to discern the structure of the hard ball in the middle of the electron bullets. Thus, further investigation began and we continued to advance.

Henry Moseley Counts Electrons
In 1913, Henry Moseley developed and used X-ray spectroscopy to count the number of electrons in an atom. [8] This enabled the number of electrons of each element in the Periodic Table of the Elements to be determined. [9]

We need to understand a bit about electron theory to get a grasp on how this works. The following picture shows an atom with one electron in it. In this picture the electron, the small ball, has jumped to a higher orbit. This can happen several ways. It can happen when someone heats the material. Then, the heat knocking the atoms around can bump an electron into a higher orbit. Normally, electrons do not stay in a higher orbit. They quickly drop back into the orbit they came from. When they do that, they give off a photon of light. This is how a light bulb works. Electricity runs through a wire. The current or the heat causes electrons to jump to a higher orbit. Then, when they fall back, light is radiated and you can read the book you are holding.

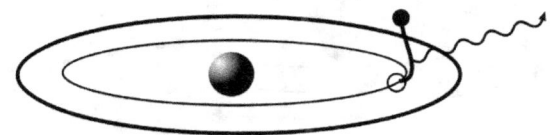

This technique can be used on very complex molecules. The following atom has four electrons.

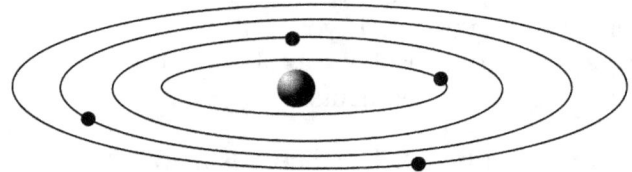

However, normal heat rarely causes the lower electrons to move. So, to meet our ends, we need to get at the inner electrons.

This can be accomplished by pointing an X-ray machine at the atoms. [10] The X-rays are very small and powerful. They can zing inside complex atoms and hit an inner electron. The

X-rays carry enough energy to knock one of the electrons out of its orbit.

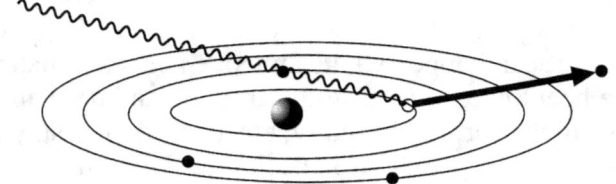

That orbit, then, is empty. So immediately, a nearby electron will quickly fall into that orbit. That electron changing that quickly will give off a photon of light.

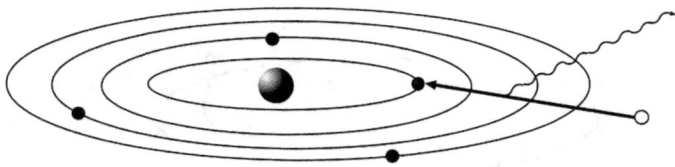

Now, here is the interesting thing about this. All electrons give off a different wavelength or color of light when they fall into a lower orbit. What this means is that, if you hit any material with X-rays, you can differentiate between the electrons. [11] Each has a fingerprint, if you will, of what happens when hit by X-rays. Henry Moseley was able to do this and counted the electrons in several elements.

This was very significant. The theory is that, for each electron in an atom, there should be a positively charged proton in the nucleus. This would balance out the charge in the atom. Now, at this time the significance of the neutron was not known. The result was that Moseley's finding answered a great number of questions. However, more questions emerged. In spite of that, the ability to know how many electrons were in each element advanced physics and chemistry significantly.

Before we leave this, let us look at how Moseley's process worked. The following picture outlines the equipment needed.

A special flask is on the left. The material to be tested is in the flask. It is heated to get some of the material vaporized. While the vapor of the material is suspended in the flask, an X-ray generator is turned on hitting the vapor with X-rays. This causes the above mentioned dance of the electrons and photons or electromagnetic waves are allowed to exit the flask. The waves hit a barrier with a slit in it so the waves will form a line later. They enter a prism or some device to break apart the waves and the various waves can be seen through a device that magnifies them. The circle on the right shows what the observer might see. The observed lines represent the energy in an orbit around the atom.

Measuring this contribution by Henry Moseley is very difficult. This enhancement to the Periodic Table of the Elements is a very significant step forward in chemistry and physics.

Discovery of the Proton

Rutherford noticed that when alpha particles (helium nuclei or two proton, two neutron particles) were shot into nitrogen gas, his scintillation detectors showed the signatures of hydrogen nuclei. Rutherford determined that this hydrogen could only have come from the nitrogen, and therefore nitrogen must contain hydrogen nuclei. He named it proton. [12]

The following depicts something of this experiment. A box is filled with nitrogen. A small box in the chamber contains some radioactive material that emits alpha particles or helium ions that consist of two protons and two neutrons. There is a hole in that small box so a beam of these particles, shown as white, shoots out into the nitrogen gas. Some of the alpha particles hit nitrogen atoms and knock protons out of those atoms. The protons that have been hit fly free to hit the scintillation screen to the right in the big box. There they cause flashes of light. Apparently, Rutherford could gage by the size of the flash they were the size of hydrogen nuclei.

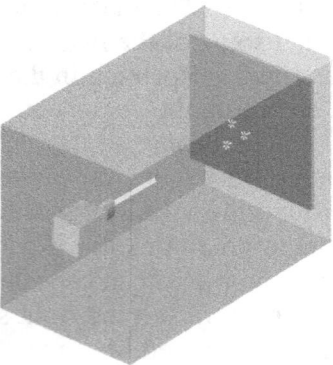

Rutherford and other physicists performed many experiments transmuting one atom into another. In every case, hydrogen nuclei were emitted. It was apparent that the hydrogen nucleus played a fundamental role in atomic structure. By the late 1920's physicists were regularly referring to hydrogen nuclei as "protons." Thus, the word proton and its place in the universe became established.

As the hydrogen atom has an atomic weight of 1 and all other elements were some multiple of that weight, the nuclei of all elements were considered some combination of protons.

As a side note, we once again see what power the Periodic Table of the Elements had. Without this table, the photon's place would probably take longer to discern. With it, however, there was a place where new concepts could be tested to see if they "fit in."

Discovery of the Neutron

Before the discovery of the neutron, part of the periodic table did not make sense. For example, Oxygen had an atomic weight of 16. But it only had 8 electrons around the nucleus. With the understanding that there was an electron for each proton in the nucleus balancing the electric charge, what were the eight extra protons doing in the nucleus? An idea was devised to solve this problem. This idea suggested that there were nucleonic electrons in the nucleus that balanced out 8 of the protons in the nucleus. With some elements, this did not work out. The discovery of the neutron inside the nucleus solved this problem. [13]

In 1931, Walther Bothe and Herbert Becker in Germany focused fast moving alpha particles from radioactive material on beryllium, boron, or lithium. [14] A very high energy radiation was emitted from those materials. That is, they thought the radiation was high energy because it seemed to penetrate everything quite easily. At first, this radiation was thought to be gamma rays which are a high energy electromagnetic radiation. However, the associated energy was so far above normal gamma radiation that it did not appear to be a good conclusion to draw.

The next important contribution was reported in 1932 by Irène Joliot-Curie and Frédéric Joliot in Paris. They showed that if this unknown radiation fell on paraffin, or any other hydrogen-containing compound, it ejected protons of very high energy. The unknown radiation remained a mystery. [15]

The following depicts this experiment. The radioactive alpha source is on the left. It shoots high energy alpha particles into the plate of beryllium. There, the unknown particles travel to the right. There is nothing there to show the particles, as they appear to be neutral. However, when the invisible particles hit the paraffin, protons are apparently hit and knocked into the scintillation screen on the right where a flash occurs indicating the presence of a proton.

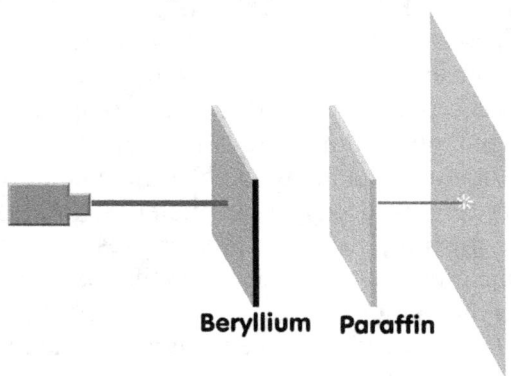

Beryllium   Paraffin

In 1932, James Chadwick performed a series of experiments at the University of Cambridge and also came up with what looked like gamma radiation. He did not accept that conclusion. Instead he suggested that the new radiation consisted of uncharged particles of approximately the mass of the proton. He performed a series of experiments verifying this idea. [16] He called these uncharged particles, neutrons. However, this idea was not accepted at first.

When the place of the odd particles in the Periodic Table of the Elements was filled with the new neutron and its characteristics worked in with the characteristics of all the elements in the table, everything fit well together. Then, the neutron was accepted as a basic structural unit of atomic nuclei.

Once again, it is interesting to see the roll the Periodic Table of the Elements played in understanding what unseen things could be.

Conclusion

For this author, seeing how each of the particles was found and defined is very comforting. The processes clearly demonstrate that scientists do not have supreme power to see everything. They are human like the rest of us and must make intelligent decisions (guesses?) to make progress.

How does this affect our goal of explaining quantum physics? As the details of quantum physics are not clear to those that work with it, we can only understand it if we understand that from which the concepts come. Traditionally, as stated elsewhere, we are offered conclusions about the experiments run but usually do not see from whence they came. This chapter clearly lays out some conclusions that were off the mark and were corrected. We need to see this to understand that what we are told does not necessarily match the reality of the situation. We see that one man's work clarifies a previous man's work. And we see that one man will produce something knowing full well that his work will be treated likewise. The hope here is that when you get to the end of this chain of effort, you will understand quantum physics as well as anyone.

Part 4

The Second Scientific Revolution

In 1874, when Max Planck entered school to study physics he is told that all of physics has been solved. He was told there were only a few small things that needed to be explained. In the effort to do this, big mysteries are encountered. It is a revolution of confusion. Questions are answered but for every question answered, a huge mystery lurks about. The second revolution is the birth of quantum physics.

Chapter 14

The Men of the Second Revolution

This is a revolution of confusion. The first revolution had brought order. We found the world around us consisted of molecules and atoms that made up salt, iron, and the air about us. We could mix things and plan how to do new things. Many of the principles discovered had enabled man to build productive machines. Many thought that physics had answered all the questions of the universe. There were perhaps a few unanswered questions but the assumption was that they could be cleaned up with appropriate attention. However, this was not the case. In efforts to answer these small questions, some major problems popped up.

In this chapter, we are introduced to some key players that attempted to fill in some of the gaps. Progress is made but more confusion emerges. The concept of the ether was beginning to die. An experiment by two men Albert Michelson [1] and Edward Morley [2] finalize its demise. A question remains however, what do light waves wiggle on? Then Max Planck [3] introduces us to a novel concept. Things happen in the universe in small steps called quanta. Albert Einstein [4] applies this concept to light and shows that light manifests quanta-like behavior. Sir Geoffrey Taylor [6] amazes the world revealing that light wave experiments with photons do not act like waves. Earlier Rutherford determined that the electrons whirled around the nucleus but classical physics could not explain how this could happen. Niels Bohr [5] came up with an interpretation combining the orbits with Planck's quanta. However, that only goes so far for it does not explain elements more complicated than a hydrogen atom. These people usher in the second revolution.

Albert Abraham Michelson
(1852-1931), American physicist.

He cooperated with Morley in the famous Michelson/Morley experiment that popularized the idea the ether did not exist.

Michelson was born in Strzelno, Provinz Posen in the Kingdom of Prussia (now Poland) into a Jewish family. He moved to the United States with his parents in 1855, when he was two years old. He grew up in the rough mining towns of Murphy's Camp, California and Virginia City, Nevada, where his father was a merchant. He spent his high school years in San Francisco and lived with his aunt, Henriette Levy.

Michelson married Edna Stanton of Lake Forest, Illinois in 1899. They had one son and three daughters

He was appointed by President Grant to the U.S. Naval Academy. After graduation as Ensign in 1873 and a two-year cruise in the West Indies, he became an instructor in physics and chemistry at the Academy.

He served in several capacities in the military. However, he continued study in Berlin and France. Most of his career was dedicated to teaching physics in universities. Later in life he worked at the Mount Wilson Observatory in Pasadena.

In 1881 he invented his interferometer for the purpose of discovering the effect of the Earth's motion through the then popular notion of the ether. It was a rather large affair. The following shows how it appeared.

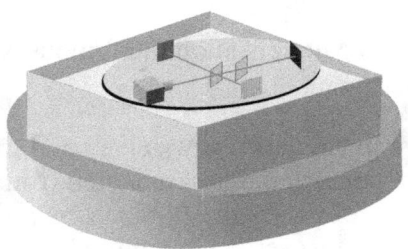

It was built on a large block of granite. There was a large bath of mercury liquid in which floated a round disk. The components were placed on this disk. The purpose of the granite was to cut down on vibration going into the test. The bath of mercury served to further shield the experiment from external vibration and enabled the experiment to turn smoothly as the instrument was set at various angles to the earth's supposed motion through the ether.

In 1882, he estimated the speed of light as 186,320 miles per second. This was the most accurate value then available and remained so for another ten years when he made an even more accurate measurement.

At the request of the International Committee of Weights and Measures, Michelson measured the standard meter in terms of wavelength of cadmium light. He invented the echelon spectroscope and during his wartime service in the Navy, he performed research work on devices for naval use. He developed a rangefinder which was adapted as part of U.S. Navy equipment.

He was President of the American Physical Society (1900), the American Association for the Advancement of Science (1910-1911), and the National Academy of Sciences (1923-1927). He was also a Fellow of the Royal Astronomical Society, the Royal Society of London and the Optical Society.

Edward Williams Morley
(January 29, 1838 - February 24, 1923), American scientist.

He cooperated with Michelson on the famous Michelson/Morley experiment.

Morley was born in Newark, New Jersey and grew up in West Hartford, Connecticut. He graduated from Williams College in 1860. From 1869 to 1906 he was professor of chemistry at Western Reserve College (which was federated into today's Case Western Reserve University).

Morley was the president of the American Association for the Advancement of Science in 1895, and he was the president of the American Chemical Society in 1899. Morley was awarded the Davy Medal, named for the great British chemist Sir Humphrey Davy, by the Royal Society of London, England in 1907. He also won the Elliott Cresson Medal, awarded by the Franklin Institute of Pennsylvania in 1912 for important contributions to the science of chemistry. He had received the Willard Gibbs Medal of the Chicago Section of the American Chemical Society in 1899.

Max Planck
(April 23, 1858 - October 4, 1947), German physicist.

He is considered to be the founder of quantum theory.

Planck was born in Kiel, Holstein, to Johann Julius Wilhelm Planck and his second wife, Emma Patzig. He was the sixth child in the family, though two of his siblings were from his father's first marriage. In 1867 the family moved to Munich. There, Planck enrolled in the Maximilians gymnasium school, where he came under the tutelage of Hermann Müller, a mathematician who took an interest in the youth, and taught him astronomy and mechanics as well as mathematics. It was from Müller that Planck first learned the principle of conservation of energy. Planck graduated early at age 17.

Planck came from a traditional, intellectual family. His paternal great-grandfather and grandfather were both theology professors in Göttingen; his father was a law professor in Kiel and Munich; and his paternal uncle was a judge.

In March 1887 Planck married Marie Merck (1861-1909), sister of a school friend, and moved with her into a sublet apartment in Kiel. They had four children.

After the apartment in Berlin, the Planck family lived in a villa in Berlin. After several happy years, in July 1909 Marie Planck died, possibly from tuberculosis. In March 1911 Planck married his second wife, Marga von Hoesslin. In December his third son Hermann was born.

Upon entry to the University of Munich in 1874 professor Philipp von Jolly advised Planck against going into physics, saying, *In this field, almost everything is already discovered, and all that remains is to fill a few holes.* Planck replied that he did not wish to discover new things, only to understand the known fundamentals of the field.

Max taught extensively, particularly at Berlin University. While president of the German Physical Society (1905 to 1909), he presented a six-semester course of lectures on theoretical physics. It was said his lectures were, *dry and somewhat impersonal.*

In 1894, Planck turned his attention to the problem of black body radiation. He had been commissioned by electric companies to create maximum light from light bulbs with minimum energy.

One of the unexplained factors of black body radiation was

that heated metal displayed various colors at various temp-
eratures.

Color = Temperature

Planck resolved this problem. However, his first attempt failed. Planck then revised his first approach relying on Boltzmann's statistical interpretation of the second law of thermodynamics. As Planck was deeply suspicious of the philosophical and physical implications of such an interpretation
of Boltzmann's approach, his recourse to them was, as he
later put it, *an act of despair ... I was ready to sacrifice any of
my previous convictions about physics.*

This resulted in the concept that electromagnetic energy could
be emitted only in quantized form. Thus, he established the
use of Planck's constant, At first Planck considered that
quantization was only *a purely formal assumption ... actually
I did not think much about it...* Nowadays this assumption,
incompatible with classical physics, is regarded as the birth of
quantum physics and the greatest intellectual accomplishment
of Planck's career.

Subsequently, Planck tried to grasp the meaning of energy
quanta, but to no avail. *My unavailing attempts to somehow
reintegrate the action quantum into classical theory extended
over several years and caused me much trouble.* Several
years later, other physicists like Rayleigh, Jeans, and Lorentz

struggled to explain Planck's constant with classical physics. Planck responded to this by saying, *I am unable to understand Jeans' stubbornness --- he is an example of a theoretician as should never be existing . . . So much the worse for the facts, if they are wrong.*

Max Born wrote about Planck: *He was by nature and by the tradition of his family conservative, averse to revolutionary novelties and skeptical towards speculations. But his belief in the imperative power of logical thinking based on facts was so strong that he did not hesitate to express a claim contradicting to all tradition, because he had convinced himself that no other resort was possible.*

Life after the beginning of World War I presented many pains to Max. During the First World War, the French took Planck's second son, Erwin, prisoner in 1914. His oldest son Karl was killed in action at Verdun. Grete died in 1917 while giving birth to her first child. Her sister married Grete's widower. She died two years later also giving birth to her first child. Both granddaughters survived and were named after their mothers.

When the Nazis seized power in 1933, Planck was 74. He witnessed many Jewish friends and colleagues expelled from their positions and humiliated, and hundreds of scientists emigrated from Germany. He attempted to keep a number of them in Germany and was successful to a limited degree. As the political climate in Germany gradually became more hostile, Planck along with Sommerfeld and Heisenberg was attacked for continuing to teach the theories of Einstein. The Nazi office for science investigated Planck's ancestry, but could only find that he was 1/16 Jewish.

World War II brought similar pain. In February 1944, his home in Berlin was completely destroyed by an air raid. It annihilated all his scientific records and correspondence His youngest son, Erwin, was implicated in the attempt made on Hitler's life in the July 20 plot. Consequently, Erwin died at the hands of the Gestapo in 1945. Although it is said that Erwin could have been spared had Planck joined the Nazi Party, Planck took a stand and refused to join, and as a consequence Erwin was executed. Erwin's death destroyed Planck's will to live. By the end of the war, Planck, his second wife and his son by her, moved to Göttingen where he died on October 4, 1947

Planck was a devoted and persistent adherent of Christianity from early life to death, but he was very tolerant towards alternative views and religions.

Albert Einstein
(14 March 1879 - 18 April 1955), Theoretical physicist, philosopher and author.

Popularized special relativity and explained (among other important things) the photoelectric effect, critical in the development of quantum physics.

Albert Einstein was born in Ulm, in the Kingdom of Wurttemberg in the German Empire on 14 March 1879. His father was Hermann Einstein, a salesman and engineer. His mother was Pauline Einstein. In 1880, the family moved to Munich, where his father and his uncle founded Elektrotechnische Fabrik J. Einstein & Cie, a company that manufactured electrical equipment based on direct current. The Einstein's were non-observant Jews. Their son attended a Catholic elementary school from the age of five until ten. Although Einstein had early speech difficulties, he was a top student in elementary school.

In 1894, his father's company failed. Direct current use gave way to alternating current devices. In search of business, the Einstein family moved to Italy, first to Milan and then, a few months later, to Pavia. When the family moved to Pavia, Einstein stayed in Munich to finish his studies at the Luitpold Gymnasium. His father intended for him to pursue electrical engineering, but Einstein clashed with authorities and resented the school's regimen and teaching method. In the spring of 1895, he withdrew to join his family in Pavia, convincing the school to let him go by using a doctor's note. The Einstein's sent Albert to Aarau, in northern Switzerland to finish secondary school. At age 17, he graduated. With his father's approval, he renounced his citizenship in the German Kingdom of Württemberg to avoid military service, and in 1896 he enrolled in the four year mathematics and physics teaching diploma program at the Polytechnic in Zurich. Einstein's future wife, Mileva Marie, also enrolled at the Polytechnic that same year, the only woman among the six students in the mathematics and physics section of the teaching diploma course. Over the next few years, Einstein and Marie's friendship developed into romance, and they read books together on extra-curricular physics in which Einstein was taking an increasing interest. In 1900, Einstein was awarded the Zurich Polytechnic teaching diploma, but Marie failed the examination

In early 1902, Einstein and Mileva Marie had a daughter they named Lieserl born in Novi Sad where Marie's parents lived. Her full name is not known, and her fate is uncertain after 1903. Einstein and Marie married in January 1903. In May 1904, the couple's first son, Hans Albert Einstein, was born in Bern, Switzerland. Their second son, Eduard, was born in Zurich in July 1910.

After graduating from Polytechnic, Einstein secured a job in Bern, at the Swiss Patent Office. He evaluated patent applications for electromagnetic devices.

In 1905, while he was working in the patent office, the leading German language physics journal Annalen der Physik published four of Einstein's papers. The four papers eventually were recognized as revolutionary, and 1905 became known as Einstein's *Miracle Year.*

Einstein's theory of Brownian motion was the first paper in the field of statistical physics. The theory of Brownian motion was the least revolutionary of Einstein's *Miracle Year.* papers, but it is the most frequently cited, and had an important role in securing the acceptance of the atomic theory by physicists.

One of the other papers was on the theory of special relativity.

The paper on mass-energy equivalence resulted in what has been called the 20th century's best-known equation.

$$E = mc^2$$

Einstein's 1905 work on relativity remained controversial for many years, but was accepted by leading physicists, starting with Max Planck.

In the paper on the photoelectric effect, Einstein postulated that light itself consists of localized particles (quanta). All physicists, including Max Planck and Niels Bohr, rejected Einstein's light quanta. This idea only became universally accepted in 1919, with Robert Millikan's detailed experiments on the photoelectric effect, and with the measurement of Compton scattering.

In 1908, Einstein introduced the photon concept and inspired the notion of wave-particle duality in quantum mechanics.

In 1914, Einstein moved to Berlin, while his wife remained in Zurich with their sons.

By late 1915, he had published his general theory of relativity in the form in which it is used today. One of the major concepts of the general theory of relativity was that the gravity field of stars warped the path of photons flying near the star. In 1911, he had issued a paper suggesting that astronomers at-

tempt to detect this by observing a red shift of light as it passed the sun.

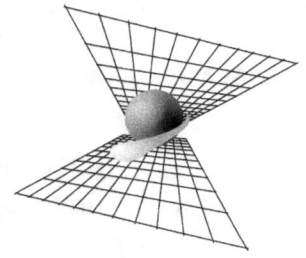

In May 1919, a team of astronomers claimed to have confirmed Einstein's prediction of gravitational deflection of star-
light by the Sun while photographing a solar eclipse with dual
expeditions in Sobral, northern Brazil, and Prķncipe, a
west African island. Max Born praised general relativity as
the *greatest feat of human thinking about nature*; fellow laur-
eate Paul Dirac was quoted saying it was *probably the greatest scientific discovery ever made.*

Marie and Einstein divorced on 14 February 1919, having
lived apart for five years.

Einstein married Elsa Löwenthal on 2 June 1919, after
having had a relationship with her since 1912. She was his
first cousin maternally and his second cousin paternally. In
1933, they immigrated permanently to the United States. In
1935, Elsa Einstein was diagnosed with heart and kidney
problems and died in December 1936.

Einstein became an American citizen in 1940. In his later
years he took up a position at the Institute for Advanced
Study at Princeton, New Jersey, There, he tried
unsuccessfully to develop a unified field theory and to refute
the accepted interpretation of quantum physics. Albert Einstein much preferred the determinism of classical physics
over the probabilistic new quantum physics.

On April 17, 1955, Albert Einstein experienced internal bleeding caused by the rupture of an abdominal aortic aneurysm, which had previously been reinforced surgically by Dr. Rudolph Nissen in 1948. He died in Princeton Hospital early the next morning at the age of 76, having continued to work almost until then.

Einstein's remains were cremated and his ashes were scattered around the grounds of the Institute for Advanced Study. During the autopsy, the pathologist of Princeton Hospital, Thomas Stoltz Harvey removed Einstein's brain for preservation, without the permission of his family, in hope that the neuroscience of the future would be able to discover what made Einstein so intelligent.

In the summer of 1939, a few months before the beginning of World War II in Europe, Einstein sent a letter to President Franklin D. Roosevelt, warning him of the possibility that Nazi Germany might be developing an atomic bomb. Later a reporter named Gosling said, *Albert Einstein interceded through the Belgian queen mother, eventually getting a personal envoy into the Oval Office. Roosevelt initially fobbed him off. He listened more closely at a second meeting over breakfast the next day, then made up his mind within minutes. 'This needs action,' he told his military aide. It was the birth of the Manhattan Project.*

Einstein was a socialist and wrote an article on that subject in 1949 in the Monthly Review. Einstein described a chaotic capitalist society, a source of evil to be overcome, as the *predatory phase of human development.*

His belief in God was questioned often by the media. In a 1954 letter, he wrote, *I do not believe in a personal God and I have never denied this but have expressed it clearly.* In a letter to philosopher Erik Gutkind, Einstein remarked, *The word God is for me nothing more than the expression and product of human weakness, the Bible a collection of honorable, but still purely primitive, legends which are nevertheless pretty childish.* Repeated attempts by the press to present Albert Einstein as a religious man provoked the following statement: *It was, of course, a lie what you read about my religious convictions, a lie which is being systematically repeated. I do not believe in a personal God and I have never denied this but have expressed it clearly. If something is in me which can be called religious then it is the unbounded admiration for the structure of the world so far as our science can reveal it.* ---Albert Einstein

Niels Henrik David Bohr
(7 October 1885 - 18 November 1962), Danish physicist.

He made fundamental contributions to understanding atomic structure and quantum mechanics,

Bohr was born in Copenhagen, Denmark, in 1885. His father, Christian Bohr, a devout Lutheran, was professor of physiology at the University of Copenhagen, while his mother, Ellen Adler Bohr, came from a wealthy Jewish family prominent in Danish banking and parliamentary circles.

Bohr married Margrethe Nųrlund in 1912. They had six sons. Their oldest died in a tragic boating accident and another died from childhood meningitis. The others went on to lead successful lives, including Aage Bohr, who became a very successful physicist and, like his father, won a Nobel Prize in physics, in 1975.

In 1903 Bohr enrolled as an undergraduate at Copenhagen University, initially studying philosophy and mathematics. In 1905, he changed his major to physics and received his doctorate in 1911. As a post-doctoral student, Bohr first conducted experiments under J. J. Thomson at Trinity College, Cambridge. In 1912 he joined Ernest Rutherford at Manchester University and he adapted Rutherford's nuclear structure to Max Planck's quantum theory and so obtained a theory of atomic structure which, with later improvements, mainly as a result of Heisenberg's concepts, remains valid to this day.

On the basis of Rutherford's theories, Bohr published his model of atomic structure in 1913, introducing the theory of electrons traveling in orbits around the atom's nucleus, the chemical properties of the element being largely determined by the number of electrons in the outer orbits. Bohr introduced the idea that an electron could drop from a higher-energy orbit to a lower one, emitting a photon. This became a basis for quantum theory.

In 1916, Niels Bohr became a professor at the University of Copenhagen. With the assistance of the Danish government and the Carlsberg Foundation, he succeeded in founding the Institute of Theoretical Physics in 1921, of which he became its director.

Niels Bohr worked at the top-secret Los Alamos laboratory in New Mexico, U.S., on the Manhattan Project, where he was known by the assumed name of Nicholas Baker for security reasons. His role in the project was important, as he was a knowledgeable consultant on the project.

Bohr believed that atomic secrets should be shared by the international scientific community. He met with President Franklin D. Roosevelt and Winston Churchill to convince them that the Manhattan Project should be shared with the Russians. In the hope of speeding up its results, Roosevelt suggested Bohr return to the United Kingdom to try to win British approval. Winston Churchill disagreed with the idea.

Niels Bohr died in Copenhagen in 1962 of heart failure.

Sir Geoffrey Ingram Taylor
(7 March 1886 - 27 June 1975) Physicist, mathematician, expert on fluid dynamics and wave theory.

He demonstrated that a single photon in Young's experiment would still be diffracted. This was an experiment that challenged the understanding of quantum physics.

Born in St. John's Wood, London, his father was an artist, and his mother, Margaret Boole, came from a family of mathematicians. Taylor followed in the latter's footsteps reading mathematics at Trinity College, Cambridge. As a child, he had become fascinated by science after attending the Royal Institution Christmas Lectures .

Although his first paper dealt with Young's experiment, Taylor's major focus was in air and oceanic turbulence. In 1910, he was elected to a Fellowship at Trinity College, and the following year he was appointed to a meteorology post, becoming Reader in Dynamical Meteorology. His work on turbulence in the atmosphere won him the Adams Prize in 1915.

During World War II Taylor applied his expertise to military problems such as the propagation of blast waves, studying both waves in air and underwater explosions. These skills were put to the service of scientists at Los Alamos when Taylor was sent to the United States as part of the British delegation to the Manhattan project between 1944 and 1945.

He suffered a stroke in 1972 which effectively put an end to his work; he died in Cambridge in 1975.

Conclusion

It is interesting to note that several of the significant experiments conducted by these men produced bigger questions then they were initially supposed to answer. The Michealson-Morley experiment was to measure the speed we moved through the ether. The result was that no ether could be found to measure. Planck attempted to find some correlation between the heat of a bar of metal and the color of light it emitted. He got an answer and was perplexed with it. It did not make sense and thought that what he produced was a mathematical representation that had no real meaning. Interestingly, the confusion was the birth of quantum physics. Likewise, the Taylor experiment was to demonstrate a single photon in Young's experiment would not produce the inter-

ference pattern. However, it did. Then, quantum physics became even weirder.

These examples demonstrate that science is a human thing.
This is contrary to the popular opinion of many that science is
a religion that is based on some fixed belief. Science is subject to change that can be brought about by anyone that has
the willingness to challenge what is said and can demonstrate
the truth of something.

Chapter 15

The Second Revolution

The first scientific revolution occurred during the renaissance when people like Lavoiser began to make careful measurements of the world around them. During that revolution, the Periodic Table of the Elements became a practical tool. With it. the natural elements are accounted for. Then electromagnetism is explained. Likewise, molecules and atoms take their place in our understanding, and the atom has been opened for everyone to see. With access to the very small accomplished, quantum weirdness emerges. In this chapter, we examine the Michelson Morley experiment, Planck's constant, photoelectric effect, and some ideas from Niels Bohr. These open a door to a second revolution.

Michelson and Morley

At the beginning of the twentieth century, the concept that everything consists of air, fire, earth and water as elements has almost died. The only part that lives is the ether that many cultures had added to those four. Physics has needed it to explain how light waves moved freely about us. For over 2000 years, the ether was assumed to be some kind of fine material that occupied the entire universe. However, it had not been seen, tasted, or measured. A man by the name of A. A. Michelson decided to measure its properties. He had already run one set of experiments to do so. He teamed up with Morley to run an exhaustive experiment to measure earth's speed through it.

To grasp what Michelson and Morley wanted to do, consider a boat moving though water. The ether was considered to be somewhat like water. While a boat moves through water it generates waves. The waves generated by a moving boat, move away from the boat. As the speed of the waves is dependent on the motion of the water and not on the motion of the boat, the speed of the boat through the water can be deduced by observing the waves. The waves ahead of the boat will move with a lower speed relative to the boat than the waves away from and behind the boat. This difference of speed can be used to determine the speed of the boat through the water.

Like the water analogy, the scientific community expected light waves to be slower in front of a moving earth and faster behind a moving earth. That is, the earth was assumed to be immersed in the ether as a boat is immersed in water. Thus, Michelson and Morley expected to observe this phenomena when they ran their experiment.

The Michelson Morley Apparatus

The following depicts the important elements of the apparatus for the experiment to measure the speed of earth through the ether. [1]

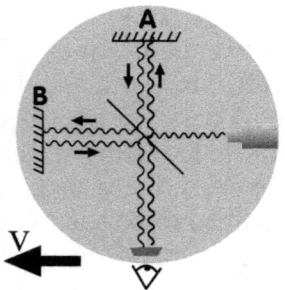

As the apparatus is resting on earth, it is presumed to be moving through the ether at velocity v. Light enters from the right. It hits the half-silvered mirror in the middle and splits in two parts. One part goes up to hit mirror A and reflects back. One part goes to mirror B and reflects back. The light beam going to mirror, A, is like the water wave to the side of a boat. The light beam going to mirror, B, is like the water wave in front of the boat.

Michelson and Morley expected the time for the light to travel from the half-silvered mirror to A and to B to be different. They ran the test many times in many directions and at different times of the year. In each run there was no difference found in the time the beams traveled perpendicular to each other.

A conclusion was that the ether did not exist. [2] This challenged the wave theory of light. How could it be a wave if there was nothing to support a wave?

At this time, this was but one of the weird ideas of this new world many found disturbing. At this time, others emerged.

Max Planck

Max Planck studied electromagnetic emissions coming from radiating bodies. At the time, the community knew that all bodies gave off electromagnetic radiation. The natural bumping and crashing of molecules in matter give rise to these emissions. Likewise, external electromagnetic waves enter bodies and are absorbed constantly. The result is that the body is in equilibrium with its environment. That is, it is giving off as much energy as it is taking in.

Take for example a block of metal. Heat it with a flame or some other heat-producing device. The block heats up and gives off electromagnetic radiation. If it gets hot enough, the electromagnetic radiation gets strong enough to appear as various colors, depending on how hot the metal becomes.

Planck was studying the structure of such bodies to determine how the body radiated electromagnetic energy. His theory was that there were little oscillators in the metal. These oscillators absorbed and radiated energy. Here is a rough example of what this means. The spring and ball on the left in the following picture represents an oscillator. When the ball is pulled down and released, it appears as the spring and ball on the right. The ball bounces up and down.

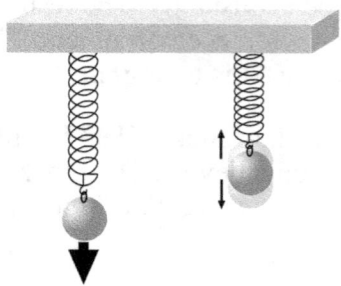

While this is not what goes on inside a piece of metal, it is illustrative. In a sense, the atoms in metal oscillate in a similar way.

At some point, the oscillating ball will emit radiation. That is, the oscillator gives off its energy as depicted in the spring and ball on the left. The energy appears as light.

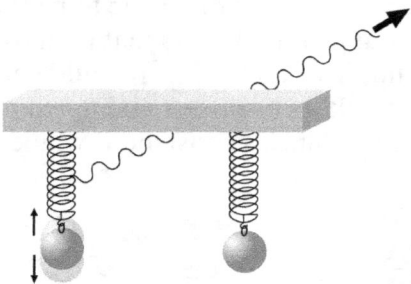

Then the spring and ball stops oscillating, as is shown on the right.

A block of metal contains many of the oscillators that are absorbing energy and radiating it constantly.

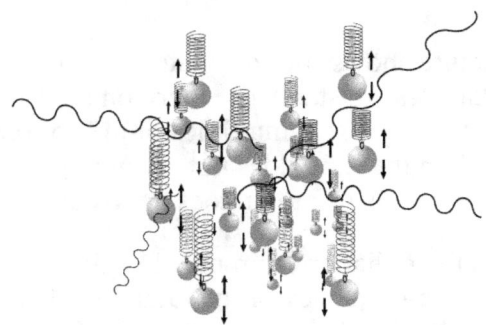

Planck attempted to develop a formula that would predict the energy given off by a number of these oscillators. That is, he attempted to develop a formula that would match the heat actually radiated by a block of hot metal. Planck assumed that the ball could be pulled down at any length and allowed to oscillate at any frequency. This first attempt failed. In exasp-

eration, he attempted other methods. In one method, he based
the energy in each oscillator on some integer number of a discrete value. [3] This method produced an equation that worked. That discrete value eventually came to be Planck's constant. The following picture depicts that if an oscillator oscillated at some multiple of that value, it could participate in the
exchange of energy. If an oscillator attempted to move at
some non-integer of Planck's constant, it would not function.

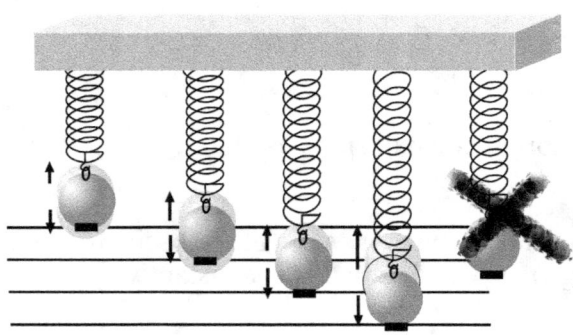

The above picture shows four oscillators bouncing at some
multiple of Planck's constant. The mark on the bottom of each
ball hits lines that represent multiples of the constant. The
spring and ball on the right does not attempt to oscillate at one
of those multiples. Therefore, it does not oscillate.

The energy in the radiation from metal is given by Planck's
constant times the frequency of the radiation. The equation
appears as, $E = hf$, where E is energy, h is Planck's constant,
and f is the frequency of the radiation.

Note that Planck was not interested in the way electromagnetic radiation moved through space. Planck was interested in how the oscillators moved to produce that radiation.
Einstein took up the motion through space later.

The important item to carry away here is that the oscillators in the metal only released energy in multiples of Planck's constant. That came to be known as a quantum. [4]

Now Planck's finding did not get a lot of attention when it was first discovered. However, as time passed, the finding appeared in other places. It seemed to answer questions. However, that success generated even more complicated questions. One of those other places was in Einstein's observations of the photoelectric effect.

Photoelectric Effect

The photoelectric effect occurs when light strikes some metal knocking an electron out of the metal. [5] The following picture depicts this.

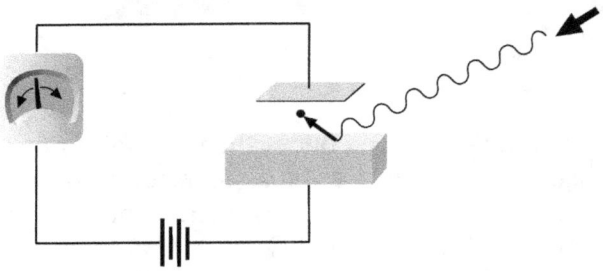

This picture shows an incident beam of light entering the experiment and hitting a block of metal. There is an E field between the block and a plate above the block. When the light hits the metal block, an electron is knocked from the block. The E field pushes the electron to the plate. Then, one electron is allowed to run around the circuit causing the meter to indicate the emission of one electron.

When Planck determined the Planck theory, the photoelectric effect was not a new concept. [6] People were running tests on it in the mid 1800's. Johann Elster (1854-1920) and Hans

Geistel (1855-1923), students in Heidelberg, developed the
first practical photoelectric cells that could be used to measure the intensity of light. This phenomenon was observed by
Heinrich Hertz in 1887. In 1902, Philipp Eduard Anton von
Lenard observed that the energy of individual emitted electrons increased with the frequency of the incident light. By
the time Einstein got around to examining the subject, a great
deal of experimental information was available. He proposed
that the light incident on metals that displayed this effect consisted of some kind of small packets of energy.

Electrons could be knocked out of the material if one of these
packets had enough energy to break an electron free from the
bonds in the material and give the electron some energy to
move away from the material. [7]

These pictures show the impinging packet hitting an electron making it fly away. In this first picture the impinging packet is moving toward an electron in some non-descript atom.

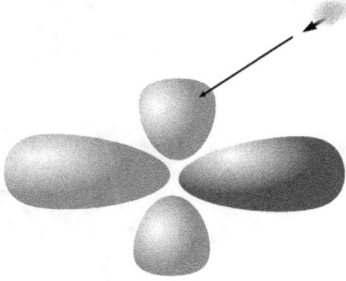

(Note that we are not using the traditional picture of an atom with electrons represented as orbits around the nucleus. We are using the cloud version of electron representation.)

The photon enters the electron field and is absorbed changing the energy level of the electron.

The electron moves away from the atom.

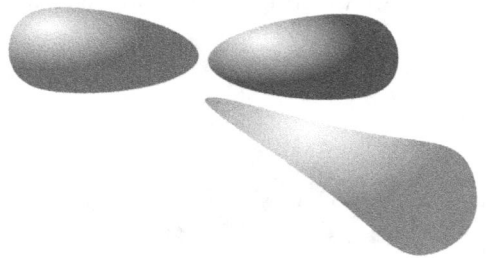

As the electron moves away, the atom readjusts before another electron will replace the one that just left.

There are some noteworthy issues here. First, the incident packet of energy must have enough energy to get an electron to break free of its atom. As the amount of energy in a packet of light is proportional to its frequency, the incident packet must have a high enough frequency for the task. This violated Maxwell's theory of light. That is, it predicted that the intensity of the light should be above some level for such emissions. From the Maxwell theory, the amount of light should be more important than the frequency. The observed phenomena and Einstein's theory appeared to oppose the idea that light is a wave. Instead, this theory proposes that light consists of particles of some kind; each having a specific amount of energy given by Planck's constant. That is, the energy in a packet should be, $E = hf$. As before, E is energy, h is Planck's constant, and f is the frequency of the radiation.

Bear in mind that the concept of the photon was new. Photons had not been named yet and Planck's constant had not received its name either. Some time would pass before the community would honor either of the concepts. In 1915, Robert Andrews Millikan performed experiments that demonstrated Einstein's idea was correct. [8]

The issue to carry away here is that Einstein proposed that the light hitting the metal and causing the photoelectric effect consisted of localized particles made of quanta of energy. That is, light, in this experiment, was not a wave.

Let's move on to another strangeness in this new revolution.

The Rutherford Model has Problems

The Rutherford model, pictured here, presents the idea that the atom is like a solar system. The nucleus is like a sun and the electrons are like planets whirling about a heavy central body.

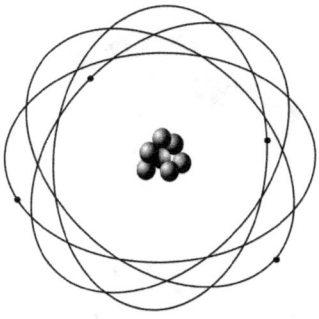

The presumption is that there is an electric force between the nucleus and an electron. The nucleus is positively charged and the electron negatively charged. Therefore, the electrons and nucleus pull toward each other. As the electron is whirling around the nucleus at great speed, there is a centripetal force pulling the electron away from the nucleus. Apparently, this scenario would keep the electron spinning around the nucleus just as the earth does around the sun.

This scenario, however, is in conflict with Maxwell's equations. In that discipline, the electron should be radiating energy as the electron is actually accelerating constantly toward the nucleus. Centripetal force is acceleration. With that in mind, any electron orbiting around a nucleus would immediately collapse into the nucleus because it should be constantly radiating all of its energy. [9] As all the atoms around us appear to be stable, something is wrong with this scenario. Rutherford knew his model was more accurate than the old plum pudding model. However, he did not know how to resolve this problem.

More Strangeness

In 1885, a man named, Johann Balmer, did a great deal of work observing spectra emission lines from hydrogen. He used equipment very similar to Moseley's discussed in an earlier chapter.

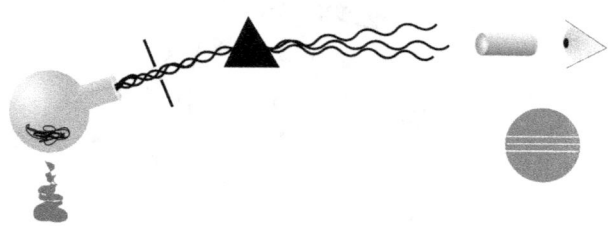

Some material heated in a flask would glow. The glow or light passed through a triangular piece of glass that separated the colors of the light. Those colors would be observed by an human as depicted in the picture above by an eye.

He noticed that the hydrogen spectra appeared as some kind

of regular pattern. The pattern appeared as in the following picture.

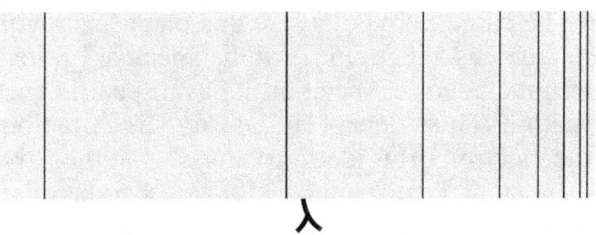

λ

Balmer developed a relationship that correctly modeled the light emission spectra of hydrogen. A man named Ryberg, inspired with Balmer's work went on to define a constant to use in this equation. It is the Ryberg constant. The resulting equation appears as follows.

$$\lambda = \text{wavelength of light}$$
$$R = \text{Ryberg constant}$$

$$\frac{1}{\lambda} = R\left(\frac{1}{2^2} - \frac{1}{n^2}\right) \quad n = 3, 4, 5, \ldots$$

The issue here is that the wavelength of light depends upon the value of, n, which is always an integer.

You may wonder how Ryberg derived the constant. He did it by observing the lines and juggling the math until the equation matched observations. [10] In general, this finding had no meaning until Niels Bohr applied some.

Bohr Model of the Atom

Niels Bohr must have observed the Rutherford model and wondered why the electrons did not fall into the nucleus. The question was, *What kept the electrons in their orbits?* Bohr also must have been aware of the Balmer hydrogen spectra lines. Could he have added the two together? Perhaps he saw that the electrons spun in specific orbits about the nucleus to correspond to the energy of each line of the hydrogen spectra.

Somehow, he came to hypothesize that electrons revolve around the nucleus at certain fixed distances or orbits. He must have concluded that each orbit has a specific energy associated with it. This concept fit the Rdyberg equation well. That is, the electron orbit idea explained what the Rdyberg equation, "meant." Furthermore, Maxwell's equations show that, when an electron changes position, it emits an electromagnetic wave. This would suggest that electrons were switching between orbits and emitting energy based on the orbit it came from and went to.

In 1913, Niels Bohr proposed the Bohr model of the atom. [11]

1. The electrons can only travel in special orbits: at a certain discrete set of distances from the nucleus with specific energies.

2. The electrons do not continuously lose energy as they travel. They can only gain and lose energy by jumping from one allowed orbit to another, absorbing or emitting electromagnetic radiation with a frequency f determined by the energy difference of the levels according to the Planck relation:

$$\triangle E = E_2 - E_1 - hf$$

where h is Planck's constant.

3. The frequency of the radiation (f) emitted at an orbit of period T is as it would be in classical mechanics; it is the reciprocal of the classical orbit period:

$$f = \frac{1}{T}$$

The following depicts an electron in a Bohr atom moving from a higher orbit to a lower orbit.

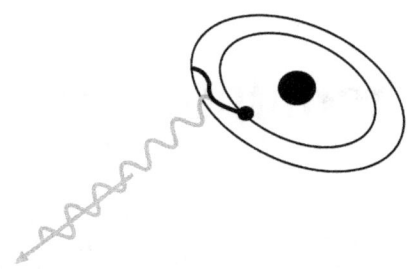

When the electron changes orbit, it emits some electro-magnetic radiation. Note, the word photon is not used here because that concept was not developed at this time.

Henry Moseley used this information in 1913 to establish the number of electrons in elements. The chapter on Atomic Theory discussed his experiments and focused on completing the Periodic Table of the Elements.

Conclusion

Here we have discussed the beginning of the second revolution. In this beginning, some explanations have been offered but more mysteries have been found. The mystery of light grew. It lost its means of getting around, the ether was gone, and Einstein was suggesting it was not a wave but a particle. Planck wondered why the little oscillators inside metal would not vibrate at any frequency and Bohr decided that electrons had fixed orbits around the nucleus inside the atom. This confusion ushered in the second revolution.

Part 5

A Paradigm Shift

When quantum physics burst onto the scene, a universe of questions erupted. One of the burning questions was whether light was a particle or a wave. Then, a man by the name of Louis de Broglie comes up with an idea. He shows the world that particles are waves just as light. Reality suddenly changes before our eyes. The hard table that we put our drinks on apparently is made of waves. That is, matter waves. The goal of this part is to begin with this revelation and trace a thread from there to cover a number of weird things associated with quantum physics.

Chapter 16

The Paradigm Men

Quantum physics has been born and its weirdness is impinging on the physics community. A man from France, Louis, 7th duc de Broglie [1], submits a thesis for his Ph. D. that his professors do not understand. They turn to Einstein for help. With Einstein's help, the man from France gets his Ph. D. and a revolution of confusion blossoms. Matter waves are born. Max Born [2] attempts to clear up a few things by showing the matter wave is some kind of probability field describing the particles they represent. With these concepts and the work of Heisenberg [3] and Schrödinger [4], the paradigm of quantum physics changes.

Louis, 7th duc de Broglie
(15 August 1892 - 19 March 1987), French physicist.

De Broglie introduced matter waves to quantum physics creating a new field in physics called wave mechanics.

Louis de Broglie was the youngest child of five of Victor, 5th duc de Broglie. He was a French nobleman from Dieppe, Seine-Maritime. The family consisted of diplomats and politicians. Louis was home schooled as were the wealthy of the time. His eldest brother Maurice became a scientist studying X-rays and radioactivity. Their father died when Louis was fourteen. Maurice sent Louis to the Lycée Janson de Sailly in Paris. Louis did not marry.

Although he started in humanities, he switched majors and got a degree in physics. With the outbreak of the First World War in 1914, he joined the army in radio communications.

In 1919, after the war, de Broglie continued working at his brother's laboratory. There, with the activities in his brother's laboratory and a study of the work of Max Planck and Albert Einstein on light, he considered whether matter might exhibit dual properties like light. In 1924, de Broglie presented his conclusions in his doctoral thesis at the Sorbonne. It consolidated three shorter papers he had published the previous year. A conclusion of this paper was that instead of the wave and particle characteristics of light and matter being at odds with one another, they were the same behavior observed from different perspectives. The thesis examiners, unsure of the material, passed his thesis to Einstein for evaluation who endorsed his wave-particle duality proposal wholeheartedly. As a result, de Broglie was awarded his doctorate. De Broglie thus created a new field in physics.

De Broglie devoted much of his career to teaching and to developing his theory of wave mechanics. Austrian physicist Erwin Schrödinger also grasped the implications of de Broglie's work and used it to develop his own theory of wave mechanics. Max Born interpreted the waves as probabilities.

In 1933, he accepted a specially created chair of theoretical physics at the Henri Poincaré Institute. He held this position for the next twenty-nine years. He established a center for the study of modern physical theories. That same year, he was elected to the Académie des Sciences, becoming its Life Secretary in 1942. He used his influence there to urge the Académie to consider the harmful effects of nuclear weapons as well as to explore the philosophical implications of his and other modern theories.

De Broglie continued his research publishing over twenty books and numerous research papers. He studied a wide variety of subjects including cybernetics, atomic energy, particle accelerators, wave-guides, X-rays, gamma rays, atomic particles, optics, and a history of the development of contemporary physics. De Broglie died of natural causes on March 19, 1987, at the age of ninety-five.

Max Born
(11 December 1882 - 5 January 1970), physicist and mathematician.

Interpreted de Broglie's matter wave function as a probability density function.

Max was born in Breslau, Poland. He was one of two children born to Gustav Born, an anatomist and embryologist, and Margarethe Kauffmann from a Silesian family of industrialists. She died when Max was just four years old, on 29 August 1886.

Initially educated at the König-Wilhelm-Gymnasium, Born went on to study at the University of Breslau followed by Heidelberg University and the University of Zurich. During study for his Ph.D. and Habilitation at the University of Göttingen, he came into contact with many prominent scientists and mathematicians. In 1908-1909 he studied at Gonville and Caius College, Cambridge.

Born married Hedwig, née Ehrenberg, on 2 August 1913. She was of Jewish descent on her father's side and was a practicing Lutheran. Born converted to the Lutheran faith in 1914. The marriage produced three children, including G. V. R. Born. His daughter Irene was the mother of British born Australian singer and actress Olivia Newton-John.

From 1915 to 1919, except for a period in the German army, Born was a professor of theoretical physics at the University of Berlin, where he formed a life-long friendship with Albert Einstein.

In 1921, Born became ordinarius professor of theoretical physics and Director of the new Institute of Theoretical Physics at Göttingen. While there, he formulated the probability density function interpretation for which he was awarded the Nobel Prize in Physics in 1954, some three decades later.

In a letter to Born in 1926, Einstein made his famous remark regarding quantum mechanics, often paraphrased as *The Old One does not play dice.*

In 1933 Born emigrated from Germany. He took up a position as Stokes Lecturer at the University of Cambridge. From 1936 to 1953 he was Tait Professor of Natural Philosophy at the University of Edinburgh. He became a British subject and a Fellow of the Royal Society of London in 1939.

Max and Hedwig Born retired to Bad Pyrmont (10 km south of Hamelin) in West Germany, in 1954. Born died in Göttingen, Germany. He is buried there in the same cemetery as Walther Nernst, Wilhelm Weber, Max von Laue, Max Planck, and David Hilbert.

Werner Heisenberg
(5 December 1901 - 1 February 1976), German theoretical physicist.

He made significant contributions to quantum mechanics including the uncertainty principle and matrix mechanics of quantum theory.

Heisenberg was born in Würzburg, Germany to Kaspar Earnesta August Heisenberg and Annie Wecklein. Kaspar was a secondary school teacher of classical languages. While a young man, Heisenberg was a member and Scoutleader of the Neupfadfinder, a German Scout association and part of the German Youth Movement. He was a Lutheran.

From 1924 to 1927, Heisenberg was a Privatdozent at Göttingen. From 17 September 1924 to 1 May 1925, under an International Education Board Rockefeller Foundation fellowship, Heisenberg went to do research with Niels Bohr, director of the Institute of Theoretical Physics at the University of Copenhagen. He returned to Göttingen and with Max Born and Pascual Jordan, over a period of about six months, developed the matrix mechanics formulation of quantum mechanics. On May 1, 1926, Heisenberg began his appointment as a university lecturer and assistant to Bohr in Copenhagen. It was in Copenhagen, in 1927, that Heisenberg developed his uncertainty principle while working on the mathematical foundations of quantum mechanics.

After Adolf Hitler came to power in 1933, Heisenberg was attacked in the press as a "White Jew" by elements of the German Physics movement for his insistence on teaching about the roles of Jewish scientists. He had to do some political maneuvering to stop the attacks.

Heisenberg enjoyed classical music and was an accomplished pianist. In January 1937, Heisenberg met Elisabeth Schumacher at a private music recital. Elisabeth was the daughter of a well-known Berlin economics professor. They were married April 29. The fraternal twins, Maria and Wolfgang, were born to them in January, 1938. They had five more children over the next 12 years: Barbara, Christine, Jochen, Martin and Verena. Jochen became a physics professor at the University of New Hampshire

In June 1939, Heisenberg bought a summer home for his family in Urfeld, in southern Germany. He also traveled to the United States in June and July, visiting Samuel Abraham Goudsmit, at the University of Michigan in Ann Arbor. Heisenberg refused an invitation to immigrate to the United States.

In the same year, shortly after the discovery of nuclear fission, the German nuclear energy project, also known as the Uranium Club, was formed. Heisenberg was one of the principal scientists leading research and development in the project.

On February 26, 1942, after the Army withdrew most of its
funding, Heisenberg presented a lecture to Reich officials on
energy acquisition from nuclear fission. The Uranium Club
was transferred to the Reich Research Council in July 1942.
On June 4, 1942, Heisenberg was summoned to report to Albert Speer, Germany's Minister of Armaments. They discussed the prospects of converting the Uranium Club's research toward developing nuclear weapons. During the meeting, Heisenberg told Speer that a bomb could not be built before 1945, and would require significant monetary and labor
resources. Five days later, on June 9, 1942, Adolph Hitler
issued a decree for the reorganization of the Reich Research
Council as a separate legal entity under the Reich Ministry for
Armament and Ammunition. The decree appointed Reich
Marshall Göring as the president.

In February 1943, Heisenberg was appointed to the Chair for
Theoretical Physics at, what is today, the Humboldt University to Berlin. In April, his election to the Prussian Academy of
Sciences was approved. That same month, he moved his family to their retreat in Urfeld as Allied bombing increased in
Berlin. By 1945, Heisenberg had moved most of the staff of
the Kaiser Wilhelm Institute for Physics to the Black Forest.

Operation Alsos was an Allied effort during the war to determine if the Germans had an atomic bomb program and to
capture any German related resources for the United States.
Heisenberg was captured on May 3, 1945. Germany sur-
rendered just two days later. Heisenberg would not see his
family again for eight months. Heisenberg was moved to
England on July 3, 1945 with nine other detainees.

On January 3, 1946, the 10 detainees were transported to Al-
swede, Germany. Heisenberg settled in Göttingen. There,
in July, he was named director of the Kaiser Wilhelm Institute
for Physics, then located in Göttingen. Shortly thereafter,
it was renamed the Max Planck Institute for Physics for political reasons. In 1958, the institute was moved to Munich. At
the same time, Heisenberg became a professor at the Univers-

ity of Munich. Heisenberg resigned his directorship of the institute in December 1970.

Heisenberg died of cancer of the kidneys and gall bladder at his home, on 1 February 1976.

Erwin Schrödinger

(12 August 1887 - 4 January 1961) Austrian theoretical physicist.

He made significant contributions to quantum physics and is famed for the Schrödinger Wave Equation.

In 1887, Schrödinger was born in Vienna, Austria to Rudolf Schrödinger, a botanist, and Georgine Emilia Brenda. His mother was half Austrian and half English. Schrödinger learned English and German almost at the same time due to the fact that both were spoken in the family household. His father was a Catholic and his mother was a Lutheran. In 1898, he attended the Akademisches Gymnasium. Between 1906 and 1910, Schrödinger studied in Vienna. At an early age, Schrödinger was strongly influenced by Schopenhauer. As a result of his extensive reading of Schopenhauer's works, he became deeply interested throughout his life in color theory, philosophy, perception, and eastern religion, especially Hindu Vedanta.

Between 1914 and 1918, he participated in war work as a commissioned officer in the Austrian fortress artillery. On April 6, 1920, Schrödinger married Annemarie Bertel. The same year, he became the assistant to Max Wien, in Jena, and in September 1920, he attained the position associate professor in Stuttgart. In 1921, he became a full professor in Breslau, Poland.

In 1921, he moved to the University of Zurich. In January 1926, Schrödinger published a paper about wave mechanics that became what we know as the Schrödinger Wave

Equation. A third paper in May showed the equivalence of his approach to that of Heisenberg.

In 1927, he succeeded Max Planck at the Friedrich Wilhelm University in Berlin. In 1933, however, Schrödinger decided to leave Germany as he disliked the Nazis' anti-semitism. He became a Fellow of Magdalen College at the University of Oxford. His position at Oxford did not work out as his unconventional life style was not met with acceptance. In 1934, he took up a position at the University of Graz in Austria in 1936.

In 1939, Schrödinger had problems because of his flight from Germany in 1933 and his known opposition to Nazism. He went to visiting positions in Oxford and Ghent Universities.

In 1940, he received a personal invitation from Ireland to help establish an Institute for Advanced Studies in Dublin. Schrödinger stayed in Dublin until retiring in 1955. During this time, he fathered two children by two different Irish women. He had a life-long interest in the Vedanta philosophy of Hinduism, which influenced ideas at the close of his book, *What is Life?* These were speculations about the possibility that individual consciousness is only a manifestation of a unitary consciousness pervading the universe.

During this period, Schrödinger turned from mainstream quantum mechanics' definition of wave-particle duality and promoted the wave idea alone causing much controversy. Schrödinger decided in 1933 that he could not live in a country in which persecution of Jews had become a national policy. Alexander Frederick Lindemann, the head of physics at Oxford University, visited Germany in the spring of 1933 to try to arrange positions in England for some young Jewish scientists from Germany. He spoke to Schrödinger about posts for one of his assistants and was surprised to discover that Schrödinger himself was interested in leaving Germany. Schrödinger asked for a colleague, Arthur March, to be offered a post as his assistant.

The request for March stemmed from Schrödinger's unconventional relationships with women. Although his relations with his wife Anny were good, he had had many lovers with his wife's full knowledge who had her own lover. Schrödinger asked for March to be his assistant because, at that time, he was in love with March's wife Hilde. Many of the scientists who had left Germany spent mid-1933 in the Italian province of Bolzano. Here Hilde became pregnant with Schrödinger's child. On November 4, 1933 Schrödinger, his wife and Hilde March arrived in Oxford.

In early 1934, Schrödinger was invited to lecture at Princeton University and while there he was made an offer of a permanent position. On his return to Oxford he negotiated about salary and pension conditions at Princeton but in the end he did not accept. It is thought that the fact that he wished to live at Princeton with Anny and Hilde both sharing the upbringing of his child was not found acceptable. The fact that Schrödinger openly had two wives, even if one of them was married to another man, was not well received in Oxford either. Nevertheless, his daughter Ruth Georgie Erica was born there on May 30, 1934.

On 4 January 1961, Schrödinger died in Vienna at the age of 73 of tuberculosis. He was buried in Alpbach, Austria.

Conclusion

The names Heisenberg and Schrödinger are almost household names. De Broglie should be part of that show but for some reason he did not make the list. After all, those two men built their reputations on de Broglie's theories. This author considers de Broglie to be the man that brought this second scientific revolution into full bloom with his matter waves.

Chapter 17

A Paradigm Shift

The Frenchman, Louis de Broglie, developed the concept that particles have wave properties just as photons or light waves. He developed a matter wave function. This is similar to electromagnetic wave functions. The matter wave function is the basic tool that enables us to determine what electrons do in an atom. Heisenberg took the matter wave concept and determined how to represent states of particles with matrices. These are manipulated to produce the whereabouts of some particles existence. Then, Schrödinger took de Broglie's matter wave concept and produced the Schrödinger Wave Equation. It was a generalized wave function formula. With it, one can develop many wave functions that solve many problems in chemistry.

The goal here is to point out that de Broglie's concept was the basis for Heisenberg's work and Schrödinger's work. We will not address the matrices or wave functions developed by these men. We only observe the basic wave function to grasp is meaning. The primary point here is that this is all about wave functions.

The Paradigm Shift in a Nutshell

De Broglie observed the electromagnetic wave function Maxwell developed in 1861.

$$y = A \sin\left(2\pi \tfrac{x}{\lambda} - 2\pi \tfrac{t}{T}\right)$$

The community understood that electromagnetic waves were

photons moving through space.

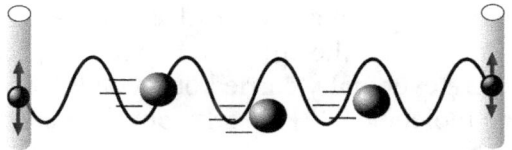

De Broglie must have thought there would be a similar wave function for matter particles. Observe, however, that the electromagnetic wave theory consists of wavelength and frequency. These are the basic variables in wave theory. They are not normally associated with particles. Particles consist of mass, velocity, momentum and energy. However, de Broglie must have also observed that there is a relationship between energy and frequency given by Planck's law,

$$E = hf$$

In addition, he must have observed the relationship between energy and mass as shown in Einstein's equation.

$$E = mc^2$$

That is, there seemed to be a clear correspondence between the wavelength of a photon and the energy of a photon. Recollect that energy consists of velocity and mass while momentum consists of velocity and mass. De Broglie must have concluded that what was true for photon particles was true for other particles. Thus, the wavelength of a particle could be calculated in a similar way.

$$\lambda = \frac{h}{mv}$$

De Broglie's matter function was the same as Maxwell's shown above. The values of wavelength and frequency were derived by these relationships. De Broglie's theory was experimentally confirmed in 1927 when physicists Lester Germer and Clinton Davisson fired electrons at a crystalline

nickel target and the resulting diffraction pattern was found to
match the predicted values. That is, matter particles behaved
like waves in that they generated a diffraction pattern demonstrating a wavelength predicted by de Broglie. [1] Similar
experiments were eventually carried out with full sized atoms
and even some molecules. [2] Clearly, physical particles
behave like light. Matter waves diffract and interfere just as
light waves.

Conclusion

Recollect, from earlier discussions in this book: radiation
demonstrating an interference pattern is a wave. This began
with Huygens and Young during the Renaissance. As de
Broglie's math and subsequent experiments demonstrated,
electrons, atoms, and even small molecules are waves. The
world of physics was turned upside down.

Chapter 18

PSI

When de Broglie released his theory about matter waves, it seemed others wanted to manipulate it in their own way. The concept was incredibly powerful and the author does not believe de Broglie got the credit for this that is deserved. It was a significant paradigm shift.

Max Born decided that the matter wave function was a probability function. [1] In this chapter, we continue to look at the significance of the de Broglie matter wave function. Here, we continue along the line of thought established by Max Born. What this means for you is that you must look at this along with everyone else and treat the matter wave function is as a probability function. However, you would be wise to hold some thought that there might be a more practical point of view. Moreover, there is a possibility that the use of probability may be clouding the issue and the reason why no one truly understands what is happening.

All of this is very confusing. First, it does not make sense. One must realize that these phenomena are the result of observations in experiments. They are not just theoretical thoughts dreamt up by someone with their feet on a desk. The mathematics is capable of producing real life results. The attempt, in this chapter, is to glean the essence of what it means. This attempt is to gain understanding of how the matter wave function is interpreted. Keep in mind, the interpretation here is slanted toward treating it as a function of probability.

Clarify What the Matter Wave Is

The matter wave function appears as the following.

$$y = A \sin \left(2\pi \frac{x}{\lambda} - 2\pi \frac{t}{T} \right)$$

The curve under the equation represents the plot of that equation. That plot is a wave function and appears like the electromagnetic (light) wave function and the water wave function.

Now, consider an experiment where a small particle such as an electron is shot out of a gun.

The matter wave function is to describe the motion of an electron emitted from the electron gun. At first glance, this does not make sense. If one takes this at face value, the equation seems to imply the particle is bouncing up and down as it moves. The equation says nothing about some force causing this to occur.

To make some sense of this we need to consider the Heisenberg uncertainty principle. Consider two experiments. In one,

a bullet is shot from a rifle. In the other, electrons are shot
from an electron gun.

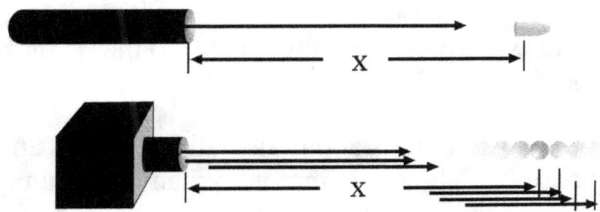

When we shoot a bullet, as depicted in the above picture, we
can know the velocity of the bullet as it leaves the barrel.
Then, in time t, we can calculate the distance x the bullet will
be from the end of the barrel. We cannot do that with very
small particles shot from a particle gun as depicted in the
lower part of the picture. After time t, the particle might be at
position x or it might be at position x plus or minus some
small distance.

What could the variable y represent in the following equation?

$$y = A \sin \left(2\pi \frac{x}{\lambda} - 2\pi \frac{t}{T} \right)$$

Observe that the equation is attempting to tell us something
about the particle at distance x. In a water wave, y was the
height of the wave. After some thought about the small particle experiment, the y was thought to have something to do
with where the particle could be. It need not be at x. In fact,
the Heisenberg uncertainty principle indicates it will not be at
x. Notice that the shading in the picture suggests that the particle would be near the position x. This represents the idea
that the y implied something about how close the particle is to

the x position. That is, y became a value to indicate how
accurate the equation is. There is a great deal of mathematical
hand waving done over all this. The most conclusive reason
to have it this way is that it works. There are mathematical
explanations of why this is so but getting into that is far beyond this book.

In an effort to consol doubts about the validity of this approach, here are two phrases common to many quantum physics books: *No one understands quantum physics,* and, *It
works.* The goal here is not to debate the issue but explain it
in some way that you can get a working idea of what is going
on.

The y in the above equation is replaced with the Greek symbol Psi. You may wonder why the Greek symbol came to be
in quantum physics. That symbol is commonly used in probability and statistics. If you look up PSI on the internet, you
will find it is an acronym for Probability, Statistics, and
Information.

With all this, we end up with the equation appearing as
follows.

$$\psi = A \sin \left(2\pi \frac{x}{\chi} - 2\pi \frac{t}{T} \right)$$

We will attempt to explain subtle parts of this equation. Right
now, let's go over the parts of this equation quickly. We know
that the above equation is a wave equation. The A is the maximum height of the wave. Then we take the sine of the term in
the parenthesis. This will produce some fraction that multi-
plies A. This supplies the various heights of the wave. Remember the wave represents something mathematical, not
something that actually exists in the real world. The first two
factors of interest in the parenthesis is x and lambda. X is

some position along the wave or some distance from the end of the gun. Lambda is the wavelength of the wave. The wavelength is represented as some fraction of 2 pi. Thus, the fraction x over lambda is a fraction that when multiplied by two pi gives us some fraction of the wavelength. The next items of interest are T and t. T is the period of the wave. That means it is the time the wave takes to go through one wavelength. Then, t is the time the particle is to travel from the end of the gun to where we think it should be. Finally, Psi, the pitchfork looking symbol, is the probability amplitude of the particle being where the equation says it should be.

To gain more understanding, we need to pull this equation apart, isolate various aspects and get some idea of what each part means. The above function suggests the wave represents a particle coming out of the gun as follows.

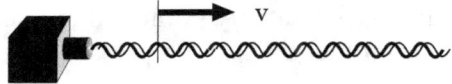

Notice that the function depends upon x and t. You can also notice that the wave is moving to the right. That is due to the t independent variable. We can examine the wave a bit better if the wave would stop moving to the right. We can stop the wave's motion to the right by selecting a constant value for t. In essence, we stop its motion so we can look at it. A convenient value for t would be zero. Then the equation appears.

$$\psi = A \sin\left(2\pi\frac{x}{\lambda}\right)$$

Then, the wave seems to oscillate up and down as it moves across the page. With that, we can look at it closely.

Notice that there is no limit on x. From this, we can assume it

can have any value. The following picture depicts the wave
going from minus infinity to plus infinity.

$-\infty$ $-\infty > X < +\infty$ $+\infty$

While this is unreasonable, the math suggests that this is true.
However, the math appears to be true when the distances are
very small as in some distance in the immediate vicinity of an
atom. The experiment here, shooting electrons out of a gun, is
a mental exercise in the attempt to clarify these ideas. It is not
often done.

The next issue is to understand, with clarity, is what the wiggly curve means. The curve has come to represent (thanks to
Max Born) the probability amplitude of the existence of the
particle as it moves from the left to the right. The probability
of existence of the particle at any point is then the square of
the probability amplitude.

$$\text{Probability} = \psi^2$$

The most notable consequence of doing this is that the negative parts of the matter wave become positive. When we
square Psi, the curve appears as follows.

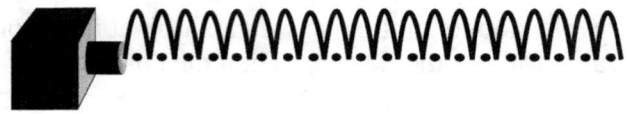

This curve represents the probability of where the particle
could be as it moves to the right. The dots under the curve
represent where a particle would be during this trip. At first
glance, it does not make sense. In this scenario, there is only
one particle. Why does it seem it can be at all places? This is
not the point of the curve. The curve represents where the

particle can be as a function of x. Let us look at part of the curve closer.

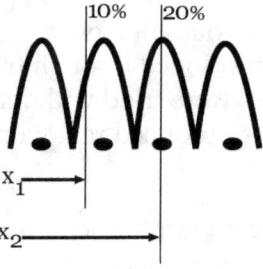

Here the wave function implies that at position x sub 1, the probability of the particle being there is 10%, *when the value of x in the wave function is x sub 1*. The probability of the particle at x sub 2 is 20%, *that is when the value of x in the wave function is x sub 2.*

Bear in mind, the curve does not represent the particle moving to the right. It is a representation of the position probability as the particle moves to the right.

Let us attempt to clarify this a bit more.

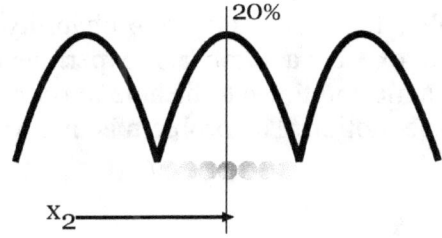

When the particle is at position x sub 2: the probability of the particle being there is 20%. However, as the picture attempts to depict, the particle may be farther than that or less than that.

Again, this does not make sense in a common sense way.
This is observed in the world of the very small. This is quantum physics. If you try to follow a logical path of thought a-
bout why this was done, you will have a great deal of difficulty. This path was developed through trial and error. At
some point, the concepts were tried with real world experiments. The experiments produced real world answers.
Therefore, *It works!*

We need to simplify the picture some more. Begin with the
function without time dependence.

$$\psi = A \sin\left(2\pi\frac{x}{\lambda}\right)$$

Again, the plot of this equation is the curve.

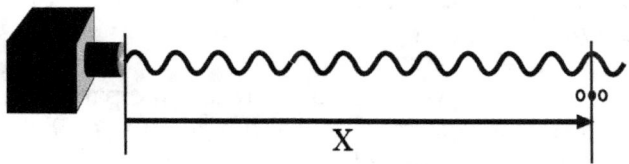

There is a problem here. To study the probability particle motions, we want to execute a number of experiments. We may
want to do this a million times to be able to observe an overall
pattern of particle motion. The problem is that we cannot ob-

serve the precise position x because the curve does not start off at a consistent point. The following picture depicts this.

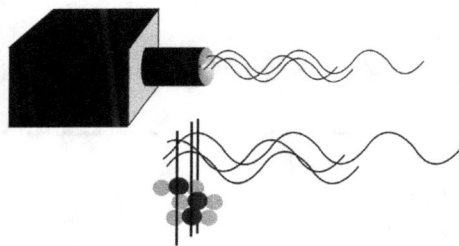

In this picture, the waves all begin at different positions. Thus, from shot to shot, the x will be different. If the x is the same spot on each shot, we can better observe the result. To solve this problem, we use a special particle gun that shoots every particle out with the curve starting at the same point within one wavelength. It would appear as follows.

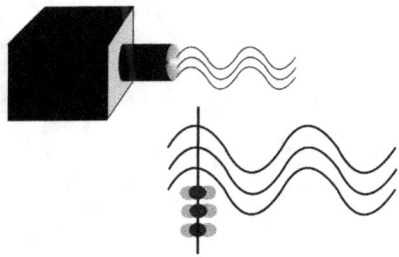

Now, we can shoot a number of particles and can compare the effect of each shot with each other shot.

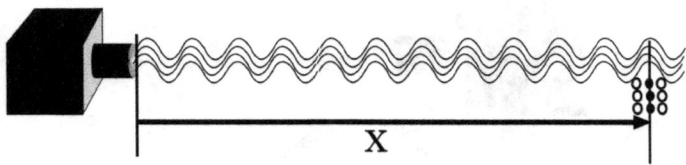

X

Observing the Flow of Many Particles

With this new device in place, we can more accurately observe what is happening as particles emerge from the gun. The following represents the matter wave functions just presented.

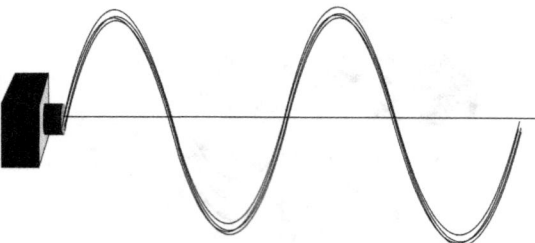

The following represents the flow of very small particles emitted from the gun.

Note that where the probability amplitude has a large value, the beam is darker indicating more particles are there. The light areas represent a smaller number of particles present. Again, keep in mind that this is the result of shooting many, many particles. With many such events, the probability of appearance increases so much that the combined probabilistic occurrences seem to become solid. That is, if a million electrons shoot out of the gun each second, when you put your finger in the middle of the black area, you will find an electron.

Conclusion

What should we walk away with? Although confusing, we have seen that the matter wave is to represent some kind of probability of where the particle is. At first, this does not make sense. However, if you consider that in an atom the electron makes many millions of passes at the same spot, the probability suddenly seems to be 100% something will be there if the wave probability is high. The following chapter clarifies this some more.

Chapter 19

Boundary Conditions

Recollect from the chapter on electromagnetism that determining the precise electric function of an electric field depended upon the boundary conditions of the field. Here are
some examples.

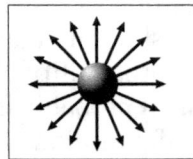 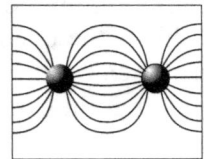 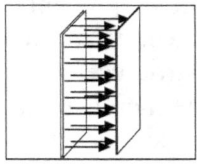

In the one on the left, the boundary condition is the surface of
a single sphere. The resulting E field simply radiates out from
this sphere. The middle image boundary conditions are the
surfaces of two spheres. Appropriately, the E field is complex. In the right image, the boundary conditions are formed
from two flat plates. In this example, the E field is simply a
number of straight vectors.

A proposition in this chapter is that the shape of a matter
wave function can be determined by boundary conditions
similar to the way boundary conditions determine the shape of
E fields. Heisenberg and Schrödinger were instrumental in
formulating ways to develop matter wave functions that fit
boundary conditions formed by the nuclei of complex molecules and atoms. Werner Heisenberg developed a matrix
method to describe particles as wave functions and developed
the Heisenberg Uncertainty Principle. Schrödinger introduced the Schrödinger Wave Equation, a generalized

wave function. Note that Heisenberg and Schrödinger produced things requiring very heavy mathematics to understand. This book will attempt to avoid the heavy math and attempt to explain what the math was trying to say. Thus, not much will be said about the math involved.

Matter Wave Function Boundaries

Our task is to apply boundaries to the matter wave function as with E fields. Up to now, we have let the particle fly free through space. Electrons, normally, do not have this freedom. Most are limited in motion around an atom. Our task is to view this in a simple way.

One Dimension

The first is the simplest. The particle will be allowed to bounce back and forth in a single dimension in a box with walls of infinite hardness. That is, the particle will bounce off the walls without energy loss so the particle will bounce back and forth constantly. The variable x is greater than zero and is less than the wavelength lambda. These are the boundaries for this experiment.

$$\psi_x = A \sin\left(2\pi \frac{x}{\lambda}\right) \quad x \geqslant 0 \quad x \leqslant \lambda$$

The following shows a plot of the wave function over the path the particle is to take. Note that the solid straight line is the actual path the particle will follow. The curve represents the probability the particle will be at a particular point on the line.

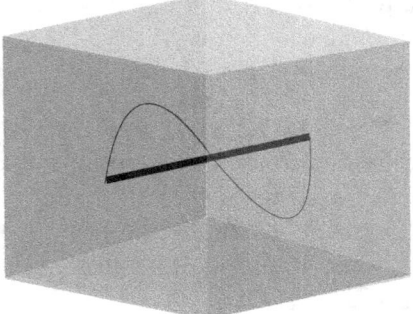

The possible places the particle can be appear as follows.

The dark areas in the path indicate where the particle is likely to be and the lighter areas show where the particle is less likely to be.

Two Dimensions

The next experiment is in two dimensions. We need two functions: one for the x direction as was just done and one in the y direction.

$$\psi_x = A \sin\left(2\pi \frac{x}{\lambda}\right) \quad x \geqslant 0 \quad x \leqslant \lambda$$

$$\psi_y = A \sin\left(2\pi \frac{y}{\lambda}\right) \quad y \geqslant 0 \quad y \leqslant \lambda$$

As before, the following picture illustrates the path of the particle with the wave function superimposed on top to show the relationship.

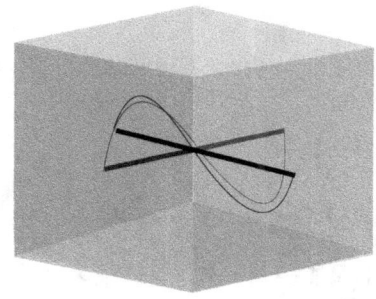

The possible positions of the particle are represented in darker shades in the following picture.

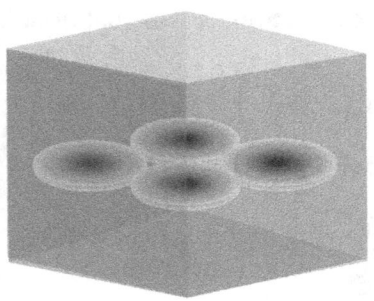

Three Dimensions

The next experiment is in three dimensions so we need three functions: one for the x, y, and z directions.

$$\psi_x = A \sin\left(2\pi\frac{x}{\lambda}\right) \quad x \geqslant 0 \quad x \leqslant \lambda$$

$$\psi_y = A \sin\left(2\pi\frac{y}{\lambda}\right) \quad y \geqslant 0 \quad y \leqslant \lambda$$

$$\psi_z = A \sin\left(2\pi\frac{z}{\lambda}\right) \quad z \geqslant 0 \quad z \leqslant \lambda$$

The following picture illustrates the path of the particle with the wave function superimposed on top.

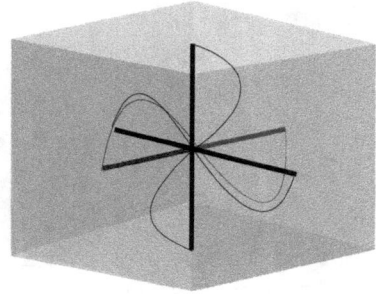

The possible places the particle can be are as follows.

More Complex Matter Waves

The examples offered have served to show the principles involved. To enable these examples to be simple they were selected to have a constant energy and use boundaries defined by the wavelength of the wave functions. Constant energy was achieved by having the particles bounce off infinitely solid walls so they changed direction but not energy. Here is an experiment in which kinetic and potential energy are not constant. We use a negatively charged particle suspended above a positively charged plate. The charge will move up and down as if on a spring.

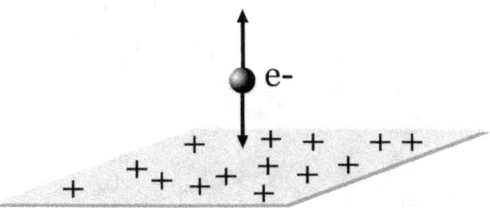

Gravity pulls the charged particle down. The positive charges push the particle up. The result is that the particle will bounce up and down. We cannot begin to analyze this mathematically for the math is far beyond this book. However, you will real-ize that motion of the tiny particle is governed by quantum physics principles. This means that there is a probability wave function describing where the particle is as it oscillates up and down. This is depicted in the following picture.

There are some other features of this experiment that are of note. One is that the distance the particle moves up and down was determined by the amount of energy put into this system when it was initialized.

Another feature of this experiment is that the energy of the particle is constantly changing. You may be aware that there are two types of energy. One is potential (not moving) and the other is kinetic (moving). You experience potential en-ergy when you hold a brick above a tabletop. Note, the brick is not moving. The force of gravity is pulling it down. There is a definite amount of potential energy in the brick relative to the top of the table. If you let go of the brick, it will fall to the tabletop. During that fall, the potential energy when you held the brick changes into kinetic energy just before the brick hits the table. Just before the brick hits the table, it is moving with some velocity. That is kinetic energy.

The particle under study is no different. At the bottom of the up and down motion, the velocity of the particle is zero. The positive charge pushes the particle up. Half way through it's up and down motion, the particle is moving at its maximum velocity and the potential energy is zero but its kinetic energy is at its maximum. Gravity is constantly pulling down on the particle so it will eventually stop moving upward. At this highest point, kinetic energy is zero and potential energy is again maximum.

This introduces yet another feature. The velocity of the particle is constantly changing. This is significantly different from previous experiments.

You will also notice that a wavelength of the probability curve is not defined. The distance up and down has not been chosen but was randomly selected by the energy put into the system. In fact, there will be several possible wavelengths.

Thus, the appearance of the probability cloud could appear as follows.

Atomic Probability Waves

You can now imagine the complexity of electrons moving around atoms. Consider an electron near a positively charged nucleus. In the following picture, the electron is the smaller dot.

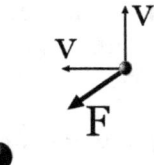

There is a force F, pulling the electron toward the nucleus depicted as the larger dot or ball in this picture. The electron is moving with some velocity in all three directions. The matter waves for this situation are very complex. A probability cloud could appear in very complex ways.

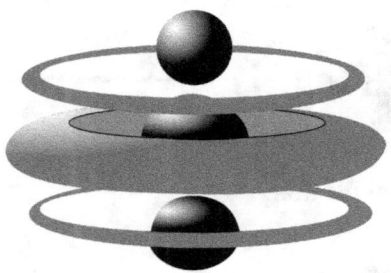

Note, that this example represents a single electron. Consider the complexity of many electrons in an atom.

Now, Look Even Closer at a Particle

As mentioned, experiments have shown that they behave in mysterious ways. This has been outlined above. Here, we want to attempt to look inside a small particle such as the electron and see what it is. We really cannot do that but we will attempt to see what quantum physics today thinks it is.

Let us begin again with de Broglie's matter wave function.

$$\Psi = A \sin \left(2\pi \frac{x}{\chi} - 2\pi \frac{t}{T} \right)$$

This essentially says that:

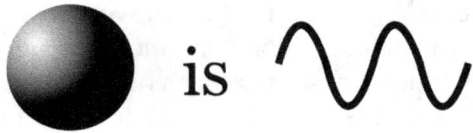

How can this be? Consider the emission of an electron from a gun again. This picture of the particle flying across the room

is to be replaced with this.

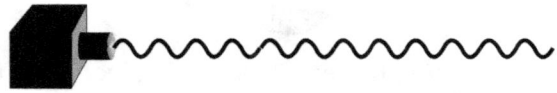

If, in fact, the de Broglie matter wave does look like that, what must we do to make it look like an electron shooting across the room? Here is a solution.

Begin with the perspective that we do not know what an electron is. However, whatever it is, it behaves like a wave. If the wave were represented by the matter wave function offered, it would appear to go from minus infinity to plus infinity. This was suggested earlier. This does not make sense but here we intend to stick with the idea that an electron really is a wave.

While studying wave phenomena, one discovers the superposition of waves. The interesting thing is that waves can add and create an apparently new wave. Some guitars utilize this phenomenon. Such guitars have two strings very close together. The strings are tuned so their vibrating frequency is slightly different. When one is plucked, you hear one tone. When the other is plucked, you hear another tone. When both are plucked, you hear those two tones but you can also hear a much lower tone that is caused by the addition of the two tones together. It is very magical. Here is a picture showing how this appears with waves. The lower two waves represent the two strings that are tuned to be similar.

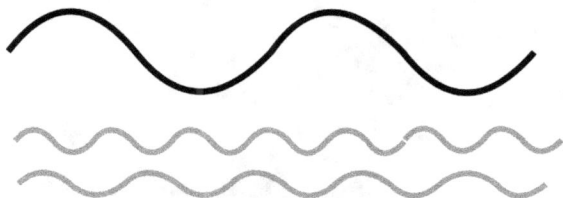

The upper wave is an addition of the lower two. It has a much lower frequency. Now apply this concept many times to many waves that are close together in frequency. They could superimpose to create a tight group of waves moving through space. This is the idea behind the structure of an electron being shot from a gun. The picture here shows the electron coming out of the gun is actually a number of waves adding together to produce a wave group of a particle.

The waves cancel out except at that one position. Then, it appears (mathematically) as a particle flying through space.

We can repeat the experiments showing how the particle bounces from wall to wall in a cube.

Now, the electron cloud around the nucleus of an atom would be the many de Broglie matter waves that would contribute to the existence of the electron.

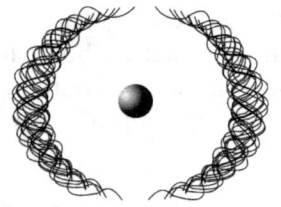

A few details about this structure bear repeating. There is no explanation what these waves mean. We cannot see them. Essentially, they only exist in some mathematical space. The traditional thought is that when we measure the presence of an electron, something magical happens. We must review this.

Measuring Small Particles

In general, the act of measurement requires we hit the object measured with a photon. The photon bounces off the object and enters our eye or some other instrument that detects the reflected photon. The following picture depicts this.

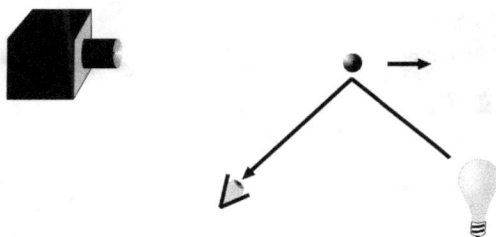

However, the theory here is that the electron is some kind of unknown quantum wave space we cannot see much less understand. So what is going on with this picture?

Mathematically the wave space can be observed at some particular point in time through a mechanism called wave collapse. Consider the wave flying along and the light bulb shines some light on the particle.

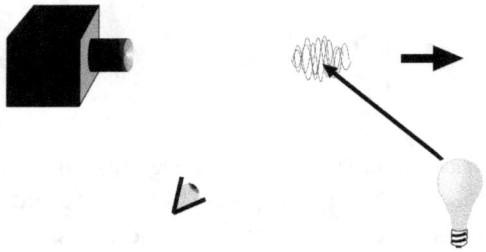

When a photon from the light bulb hits the particle for us to observe, the wave space at that instant collapses into a particle we can measure.

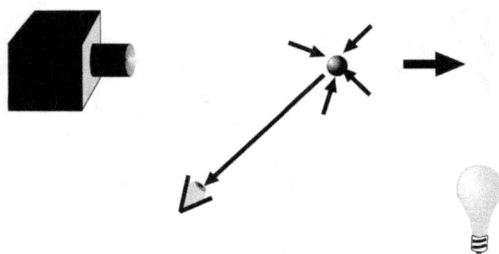

At that moment, it collapses. The electron reflects the photon to our eye or instrument to measure the existence of the electron. After that, the electron expands back to its wave space and continues moving through space.

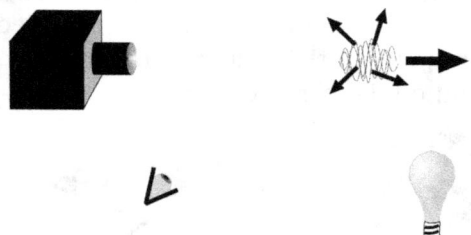

This concept of a particle existing as a group of waves moving through space is well accepted. There is a great deal of deep mathematics surrounding it. This includes Heisenberg's matrix representations of the properties of matter and the Schrödinger equation. The Schrödinger equation is a device that produces a mathematical description of the wave space or probability cloud discussed in this chapter.

Right now, we need to move on to other aspects of quantum physics.

Why is Quantum Physics Important?

Understanding the value of quantum physics is important. As mentioned elsewhere in this book, eyelets and hooks at one time were thought to be the bonds that held molecules together.

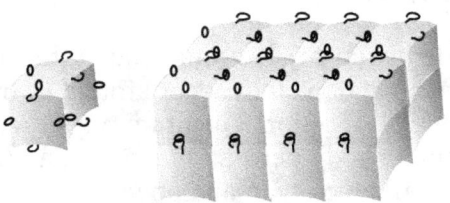

Iron, smooth and hard

Now, through quantum physics, we know something about how electrons exist around a nucleus. From the forgoing, we have seen that probability clouds or wave space envelopes represent electrons.

The clouds in the following picture show how molecules stick together. The top figure shows a separate hydrogen atom and chlorine atom. The chlorine has space for an extra electron. The hydrogen can share its single electron.

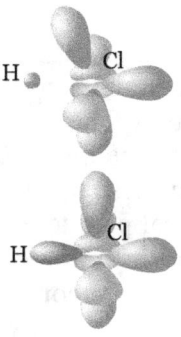

The electron cloud of hydrogen atom reaches into the chlorine atom to complete the outer electron level of the chlorine atom binding them together. Thus, the single electron from the hydrogen atom becomes part of both atoms keeping the hydrogen chlorine atom together.

Instead of seeing hooks and eyes holding atoms together, we see the cloud of an electron reaching over to another atom and keeping the two together.

In essence, the behavior of electrons around the nucleus describes the chemical actions that occur around us every day. Baking is a chemical reaction. Quantum physics explains that process. The human body performs many chemical reactions to process food, utilize oxygen and power the human brain. Quantum physics reigns over all these areas. Quantum physics principles led to the development of transistors. LED's are a result of quantum physics. Even standard light bulbs depend on quantum physics principles. In fact, that is what started it all in the first place. As mentioned in another chapter, a light bulb manufacturer hired Max Planck to determine how to reduce current through light bulbs to reduce energy requirements and produce more light. In the process, he developed Planck's theory and Planck's constant. That marks the beginning of quantum physics.

Conclusion
Quantum physics is difficult to understand because the principles involved are not clear and depend on sophisticated mathematics for explanation. The first step in understanding it all is to understand that few people really understand it. Feynman claims that no one understands it. This means the desire to follow some solid logical path to enlightenment will not be strong. This book is proposing another path. Here we attempt to understand what the science people did and how they got their conclusions. Hopefully, this provides a workable understanding of what happened and how it all fits together.

The key points to take away from this chapter are that subatomic particles appear as waves that are described by wave functions and the particle can be located when such functions are observed or measured. Complex atoms consist of many of the wave functions that seem to cling together. Their study reveals much about the behavior of such molecules and the study is of great use in chemistry.

Chapter 20

More Quantum Weirdness

The demonstration that particles such as electrons, protons and even small molecules produce a diffraction pattern just as light, shook up the physics establishment and produced significant growth in quantum physics. Quantum physics is full of surprises beyond that. The goal in this chapter is to go over some of the additional phenomena that show two things. First, these items display the very strange nature of quantum physics. Second, these items, in their weirdness, underline the fact that today we do not understand the workings of quantum physics.

Here is a list of the items.

> More Slit Experiments
> Heisenberg Uncertainty
> Wave Function Collapse
> Quantum Entanglement
> Faster than Light Communication
> Bell's Theorem
> Quantum Computers
> Virtual Particles

We will take up each of these and discuss them.

More Slit Experiments

The diffraction experiments showing particles were capable of displaying wave properties caused confusion. Further slit experiments produced more excitement when they demonstrated that the particles, light or otherwise, could produce the wave phenomena even when the particles passed through the slits one at a time.

Single Photon Experiment

In 1909, Geoffrey Ingram Taylor demonstrated that interference patterns of light were generated even when the light energy introduced consisted of only one photon. [1] The following depicts that if one photon at a time is allowed to go through the two slits in the double slit experiment, the wave interference pattern still appears.

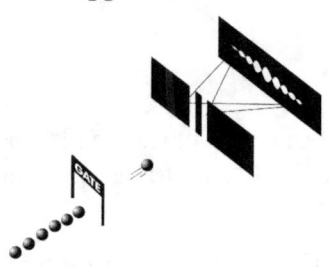

The question arises, which slit does the photon go through. If one slit is blocked allowing the photons to go through one slot, the pattern disappears. To date, there is no accepted explanation for this.

Electron Double Slit Experiment
In 1961, Clauss Jönsson performed Young's double-slit
experiment for the first time with a particle other than photons
by using electrons and with similar results. [2]

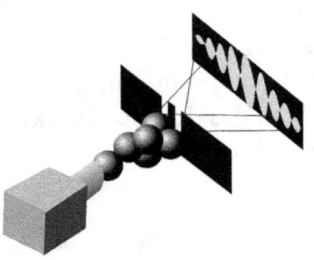

This confirmed that massive particles also behaved according
to the wave-particle duality that is a fundamental principal of
quantum field theory.

Single Electron Double Slit Experiment
In 1974, Pier Giorgio Merli performed the double slit experiment by shooting single electrons at a two slit barrier. [3]

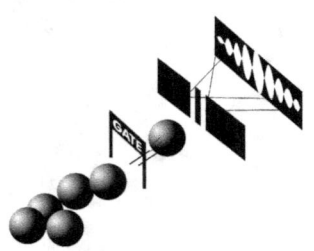

Again, the experiment demonstrated, as with single photons,
that a flow of single particles would still produce the interference pattern. And again, covering one slit would cause the
pattern to disappear. The mystery builds. Which slit does a
particle go through?

Heisenberg Uncertainty

Heisenberg's uncertainty principle states that there are complementary measurements of the motion of a very small particle that are mutually unpredictable. The most commonly discussed are position (x) and momentum (p). We consider the motion of an electron. If an electron is emitted from an electron gun, we can at some t time later, measure its position and its momentum. The Heisenberg uncertainty principle implies that we cannot measure both accurately. [4] One or the other will be greater than some small value, which is equivalent to the value of Planck's constant. Note that momentum is the mass of the particle times its velocity.

Let's explain this a little bit before looking at this closely. There is some uncertainty in each measurement. The uncertainty is delta x and delta p. Delta is a symbol commonly used in physics to denote difference or change. Heisenberg's uncertainty principle states that if we multiply these uncertainties, the product will always be larger than or similar to the value of Planck's constant. Note that this relationship has been refined since Heisenberg proposed it, but we will use the original version as it satisfies our immediate purposes. Here are the factors of this principle.

p = momentum (mv)
x = distance
$\triangle$ Pronounced delta
$\triangle x$ Uncertainty of the x measurment
$\triangle p$ Unerctainty of the p measurment
h = Planck's constant

Heisenberg Uncertainty Principle
$$\triangle x \triangle p \gtrsim h$$

This implies that if the uncertainty if the measurement of x is small, then the value we measure for momentum p is larger than h. Likewise, if we measure p with accuracy, x is larger or similar to h.

Let's look at this closer with some pictures to better understand what is going on. Consider something familiar outside
of the quantum universe. The following depicts a rifle
shooting a bullet four different times.

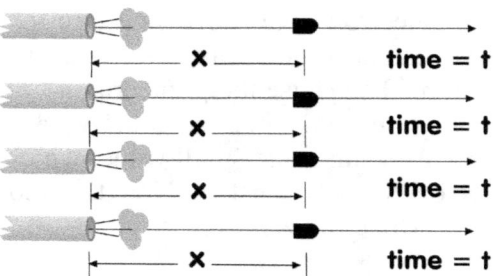

Note that the bullet travels x distance during time t with each
shot. That is, if each bullet weights the same, is shaped the
same, and is shot with the same amount of force; each bullet
will travel the same distance in the same amount of time. This
is what we are accustomed to.

This is not true for small particles such as electrons. The following depicts this situation. We assume each electron (the
dots in the picture) is expelled from the electron gun on the
left with the same force. At the end of time t, each electron

will have traveled a different distance. This is quantum phenomena.

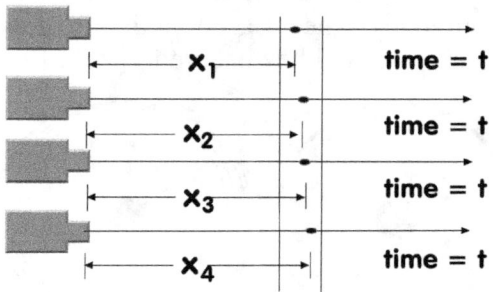

Note that, although the distances are different, they are relatively close to each other.

Right now, we are not offering an explanation why this is true. At one time, there was a thought that the process of observation using light changed that, which was observed. The thought was that, during the process of observation, photons used to do the observing would change the position of the electron just a bit. While this is true it is not the reason the distance is different each time we measure the position of the electron. It is a fact that the electron will be at a random distance from the electron gun when we measure it. And that distance will be within some small difference of distance. This is a manifestation of the weirdness of quantum physics.

There is another factor that must be kept in mind. Random action and probability play an important part in quantum physics. Observations in the world of hard ball physics are not random. In the above experiment shooting a bullet, each try was the same. A similar experiment with electrons used four attempts that did not produce the same result. Each of the four was different from the other. Realize that such an experiment is not normally performed. Measuring the motion of a single electron and measuring where it is later, is not practical. Thus, the experiment shown with the electron gun is a mental experiment. This phenomenon is important inside an atom. The electron is whirling around the atom very fast. The electron repeatedly passes the same point many times where it may be observed many times. Based on the theories presented to this point, each pass will be observed to be different.

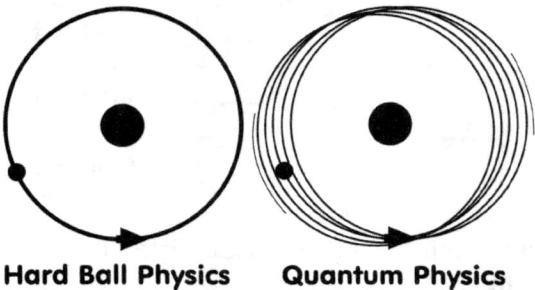

Hard Ball Physics   Quantum Physics

Thus, we must repeat some action we are measuring several times to observe what is going on.

Calculating Momentum and Distance

For this closer look, let's do an experiment to show that when position is not known with accuracy, the momentum is known with good accuracy. The following depicts a mental experiment. We assume that we have sensors that can tell us when an electron is at a specific position and the time it is at that position. The assumption is that the electron will travel some distance x during some time t plus or minus some delta t. Note the assumption here is not about some outside agent causing the electron to move erratically. The assumption here is that the electron actually moves erratically. In a sense, this is the heart of quantum physics.

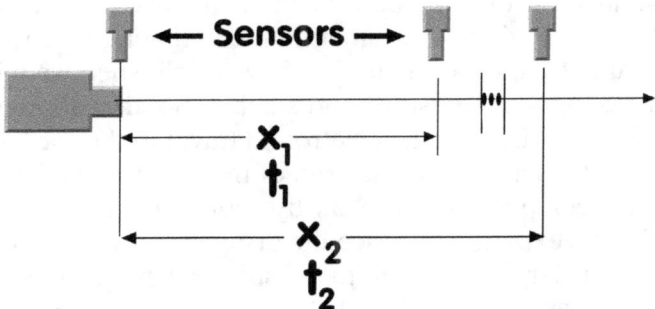

Electrons are emitted from the gun on the left. Sensors are placed at the end of the gun barrel, and at distance x sub 1 and distance x sub 2. When an electron leaves the barrel of the gun, a timer is started. When the electron passes the second sensor, time t sub 1 is noted. When the electron passes the third sensor, time t sub 2 is noted. The small dots representing electrons might fly to somewhere between time one and time two. Note the electrons as depicted, travel different distances but remain close together.

Our task is to calculate the momentum of the electrons between x sub 1 and x sub 2. The following represents the necessary calculations.

$$v = \frac{distance}{time}$$

$$v = \frac{x_2 - x_1}{(t_2 \pm \triangle t) - (t_1 \pm \triangle t)}$$

$$v = \frac{x_2 - x_1}{t_2 - t_1}$$

$$p = mv$$

First, velocity is some distance divided by some time. In this experiment, distance is x sub 2 minus x sub 1. You can see the representation of time is a bit complicated. Normally, we would simply use t sub 2 minus t sub 1. However, as we have seen from the previous experiment, time will not be the same for each run of the experiment. We have established we will take a measurement at x sub 1 and x sub 2. So, that does not change. But the time for the electron to travel to those exact positions will change with each run of the experiment. Thus the time at each position will vary by some time that we call delta t. In this experiment, since the distance between x sub 1 and x sub 2 is large compared to the space where the electrons always are at an average of t sub 1 and t sub 2, we can consider delta t very small and ignore it. This means the time on the bottom of the equation for velocity is simply t sub 2 minus t sub 1. Finally, the lower equation gives us the moment p. Note that due to the variation of delta t, the value of p will never be precise. However, it is vastly more accurate than x.

If we run this experiment over and over; we will find that the resultant momentum is very accurate compared to the inaccuracy of the position. The inaccuracy of the position is revealed by x sub 2 - x sub 1.

Now, let's increase the accuracy of x and see if the accuracy of momentum decreases. The following depicts this.

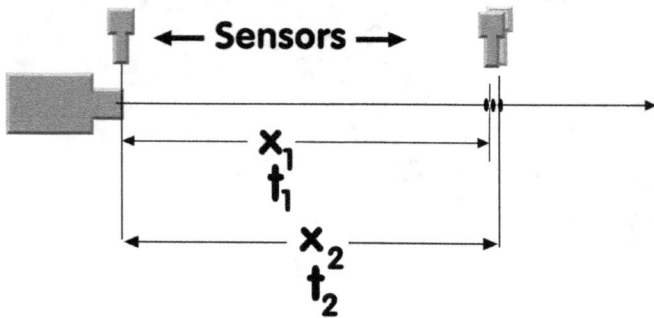

The two sensors that were measuring the time at x sub 1 and x sub 2 have been moved. The distance x sub 2 minus x sub 1 is now much smaller. Again, the experiment is executed several times and we get values for t sub 1 and t sub 2. Let's do our calculations again and see what happens.

$$\Delta v = \frac{x_2 - x_1}{(t_2 \pm \Delta t) - (t_1 \pm \Delta t)}$$

$$\Delta p = m\, \Delta v$$

Now the difference between t sub 1 and t sub 2 is not as great as it was before. The delta t is comparable in size to t sub 2 - t sub 1. So now, we must leave delta t in the equation. Then for each execution, the velocity calculation will produce a significantly different answer. Thus, the results for momentum p will be significantly different.

The difference x sub 2 - x sub 1 will be considerably smaller than before. The difference of positions of the sensors can be very small so the measurements for x are very accurate. But the error of momentum gets larger as x gets more accurate.

Discussion
The point of this mental experiment is that small particles, such as electrons, move some x distance within varying amounts of time. Because of that we can either measure the position of the electron accurately or the momentum accurately but not both at the same time. This demonstrates that for atomic particles one can have either of the following two.

Know position with accuracy and
not know momentum accurately.

Not know position accurately and
know momentum with accuracy.

The experiment shows that if we allow a long distance to measure velocity, the precise position of the electron is uncertain but the momentum measured is very accurate. However, if we use a short distance to measure veloicity, the position is known with accuracy but the momentum is uncertain.

This represents Heisenberg uncertainty principle.

Wave Function Collapse

In quantum physics a particle moving through space is described as a wave function. At the time of this writing, the physics community has not settled on what the wave function is. Some believe that it is representative of some real world artifact that we do not yet understand. Others believe that it is some kind of historical information gathered while running an experiment. Both are represented with this wave. However, most believe that when the wave is observed or measured, the wave collapses into something that gives an answer that the experimenter is seeking. [5] The following depicts this.

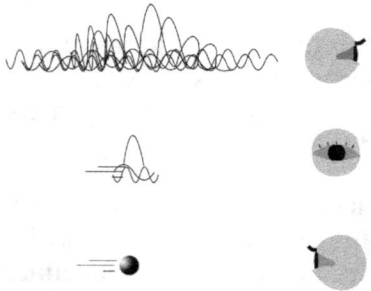

The top depicts a wave function not measured. The eye is not looking at the wave function. As we move down in this picture, the eye turns to observe the wave function. As the eye turns, something changes in the universe. Part of the existence of the wave function is changing. When the eye turns totally to view the wave function, it has collapsed into a particle that can be measured.

The point here is that the way the wave function collapses determines the position or momentum of the particle when it collapses. Collapses never happen the same way twice. One idea is that the wave function represents probabilities of where the particle is at any given time. When the eye focuses on the wave, the observation is of a particular wave and that one folds into itself and becomes the object of investigation.

A factor of this that is worthwhile noting is the rate at which the collapse occurs. Present theory has it that the collapse is instantaneous.

Quantum Entanglement

Quantum entanglement is observed presently when some complex sub atomic particles decay. This decay may occur in a high energy particle accelerator in which particles are slammed into each other at high speeds or when natural radioactive elements decay. When certain particles decay under these circumstances they will emit multiple particles of the same kind but different states. For example, two particles may be of the same kind but two different spin states; one up and one down.

Spin is an attribute that many atomic and sub-atomic particles share. Those that have a plus or minus electric charge (electrons have a minus charge) generate a magnetic field when they spin. This produces a magnetic vector that runs through the particle. The spin may be up or it may be down.

The idea is that the particle that split consisted of an unob-
served wave function.

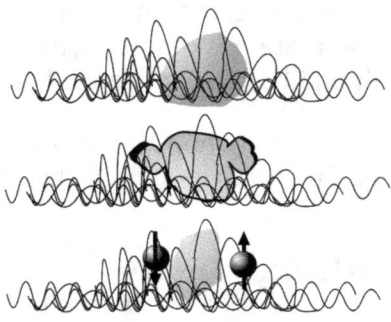

After the split, the parts of the split were still part of some
kind of wave function. Then we have a situation in which two
particles are independent of each other but are still part of the
same wave function. [6] The two particles are said to be
entangled.

Faster than Light Communication

This is a consequence of quantum entanglement. In the just
presented concept of entanglement, the two particles can
move far from each other while in the wave function state.

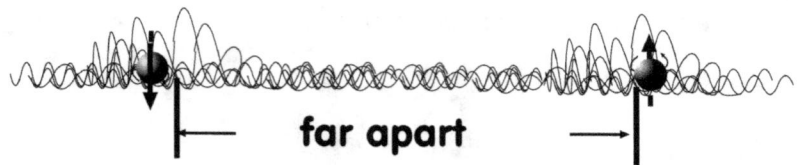

far apart

When far apart, one can measure some property of one of the

entangled particles. According to this theory, the wave function will collapse at the time of that measurement.

far apart

One of the particles is measured. Because the wave function collapses, the other particle will become real as well as its wave function has collapsed. Conservation laws require that one be spin up, the other be spin down when they exist together. The assumption is that they are up and down in the wave function state. Thus, the spin of the particle that was not measured will be discovered as it must be different from the one measured. The two particles can be far apart so the speed of communication of this spin state is apparently communicated instantaneously or at least faster than the speed of light. [7]

Note that this description is disputed by many. The most popular concept is presented here.

Bell's Theorem
John Bell's theorem is a statement that the above situation is true. It proposes experiments that can be run to demonstrate it. [8] At the time this book is written, this author cannot find any experiments that conclusively agree with the above.

Quantum Computers
This is a concept based on the idea that a particle in wave function form or in a form before being observed, consists of an infinite number of possibilities. This concept suggests that those states or possibilities can represent data in the form of an infinite number of 1's and 0's. [9] Claims have been pre-

sented that experiments have been performed that show this is possible. As the data within modern computer systems consist of 1's and 0's, this suggests that enormous amounts of data could be stored in the wave function of, say, a single electron. This would suggest that enormous amounts of data could be stored in something the size of the tip of a needle. There is also the suggestion that the possibilities somehow interact with each other suggesting that the data could be added and compared. This suggests that data processing could occur within a wave function that would be very fast. This is very speculative but people are pursuing it.

Virtual Particles

The concept of virtual particles seems to be a way to gather a diverse set of phenomena under one umbrella. Several unrelated phenomena seem to have been placed into the category of virtual particles. Photons are categorized into virtual and real groups. In addition, the phenomenon where particles materialize out of nowhere is one of these. [10] Let us look at some examples.

Virtual Photons

Nuclear scientists have noticed that photons responsible for electric forces are short lived. They appear to reach out to generate the electric force between particles but rapidly disappear. Recollect that two negative electrons will repel each other. The belief is that photons travel between the electrons causing them to move apart. The scientists can detect the travel of the photons but do not see where they come from or go. Hence, they are referred to as virtual photons.

The same has been observed in antenna transmitting electromagnetic waves. When the antennas are far apart, the scientists are aware that one antenna transmits photons that the other picks up to complete the transmission. However, when the antennas are near to each other, the transmission still occurs but the photons causing it cannot be detected. These are also called virtual photons.

Virtual Materialization of Particles

When cloud chambers were introduced, particles would occasionally appear out of nowhere. A cloud chamber is a glass case that contains air saturated with water vapor. Then the vapor is cooled. This creates an atmosphere that is super-saturated. That is, the air in the chamber contains more water vapor than it holds normally. The plan is that any disturbance in the chamber will cause water droplets to form around the disturbance. When a high energy particle flies through this atmosphere, the particle will leave a trail of water drops suspended in the air. At that point, a picture is taken just before the entire chamber becomes filled with droplets or a cloud due to the disturbance. There are large magnets on top and bottom of the chamber to cause charged particles to move in circles. The picture below gives an idea what the chamber looks like. It depicts two virtual particles appearing out of nowhere.

The appearance of these particles was observed in cloud chambers during tests of high-energy particles coming from outer space. This was also observed when such chambers were set up for some sub-atomic high energy particle tests and the above event suddenly occurred. The two particles appear, react a bit with the environment, and then disappear. The particles were measured and appeared to be a bit of mat-

ter and anti-matter. They appeared then annihilated each
other.

Theoretically, this is a phenomenon that occurs everywhere in
the universe. Hawkins used it in a theory that is about black
holes slowly disintegrating. Here is how it works. The following picture depicts a black hole.

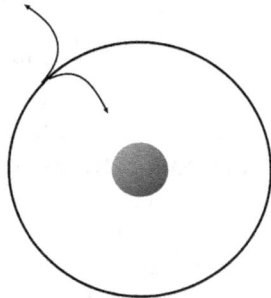

The circle about the black hole is called the event horizon.
Particles outside the line of the circle can move away from the
black hole. Particles that happen to fall inside the line are
sucked into the black hole. When virtual particles appear on
the line one particle can fly to the outside of the circle while
the other particle falls into the hole. If the one falling into the
black hole is anti-matter and the other normal matter, energy
is conserved. The anti-matter particle interacts with matter in
the black hole to annihilate each other. This causes the black
hole to diminish in size. Over a long period of time this will
have a lasting effect. Hawkins believes that all black holes
experience this effect and are decaying.

Conclusion

One of the principle arguments for the acceptance of the
theories of quantum physics is that it works. Well it does and
has produced miraculous products for our society. They range
from atom bombs to LED junk jewelry. We clearly have mastered the beast and have chained it to our personal uses. However, these little things remain unexplained. Consider what

happened when Planck wondered why a block of metal changed color as it was heated. What would happen if we suddenly had insight about what was going on with all of these little things?

Chapter 21

What is Quantum Physics?

The promise of this book is that you will understand quantum physics if you read it with a mediocre amount of effort.

If you have followed the theories of quantum physics presented over the years or listened to those that expound on universal meaning, you will have heard Feynman's quote, *Those that claim to understand what quantum mechanics is, do not understand quantum mechanics.* In this chapter, an explanation is offered why this is true. Second, this chapter makes a statement that you now have the power to understand what quantum physics is because of what this book has presented.

First, we consider a reason we cannot know.

We Are Taught We Cannot Know

When this author was in school working toward a degree in physics and beginning a class on theoretical mechanics, the professor announced the following to the students on the first day of that class. He said that up to this point in the classes on physics: little arrows, wheels, ropes, and pictures represented events happening in the real world. He went on to say these mental artifacts would no longer be useful in a continued study of physics. His point was that the only way to study physics was mathematically.

This mirrors the way Einstein special relativity is taught. It is explained with a mathematical formula. Whenever this subject appears in a physics text, the comment is added that special relativity is not intuitive. The appendix in this book titled, "Special Relativity" is presented in this manner. The reader is told that there is no way to understand special relativity other than with the mathematics of it.

This seems to be a precedent in higher level physics. Using wheels, pulleys and arrows is fine for science day at school and when classical physics about visible subjects are presented. However, when discussing the building blocks of the universe, it seems the rule is that normal man is not allowed to view it unless they acquire a Ph. D. in physics and master an arcane language.

This author believes that Niels Bohr was a strong supporter of this precedent if not an instigator of this precedent. Somehow, Niels Bohr attained a power position in physics. He successfully described the radiation coming from the hydrogen atom utilizing information from Dirac and Rydberg. He strongly believed that pictures and drawings of universal structures were useless and demanded that any explanation of the real world had to be represented with math only. Moreover, he seemed to suggest the math must be incomprehensible to all except those that had proven themselves in the field.

A few incidents revealed his domination and control of the situation.

When de Broglie produced the idea of the matter wave, Niels Bohr grabbed it and essentially made it his own.

The author has come to this conclusion after studying the material in this book and reflecting on his classes in physics during the 1970's. Part of this reflection is in regards to a textbook used in a modern physics class. The class centered on modern aspects of quantum physics. The de Broglie concept appeared early in the textbook for that class. [1]

The book then rolled forward with in-depth discussions of quantum physics. De Broglie's name appeared briefly when referring to matter waves. At the time this author first read

that book, the author wondered why de Broglie's name
appeared at all in these in-depth discussions. After some
study of the subject and understanding grew after years passed, the author came to believe that the real development in
quantum physics was due to de Broglie's concept. Yet, de
Broglie had no position in the physics community that reflected that achievement. A chapter in the book is devoted to a
Niels Bohr development. The development depends on de
Broglie's concept. A statement was made in the book that the
theory was no longer valid but was of educational value. It
seemed to overshadow the de Broglie concept. As the discussion of de Broglie matter waves continued in the book,
they seemed to be a construct of Niels Bohr. In addition, the
development depended heavily on the matter wave being
interpreted as a probability function. This was a contribution
by Max Born. His name nor contribution did not appear anywhere in the book. This author was led to believe the matter
wave concept and its interpretation as a probability function
were due to Niels Bohr. Only careful reading revealed that
Niels Bohr did not develop the matter wave function and did
not develop the concept of the wave function being a probability function.

Furthermore, several instrumental in the development of
quantum physics did not agree with Niels Bohr. De Broglie
did not agree with the idea the matter wave function was a
probability function. [2] Likewise, Schrödinger, who used
the de Broglie concept to develop the Schrödinger Wave
Equation, came to believe the basis of quantum physics was
not necessarily probability. [3] Einstein agreed with de Broglie and Schrödinger. His statement, *God does not play
dice,* demonstrates this. [4] De Broglie and Schrödinger
got intense pressure from the physics community to stuff their
ideas. Einstein attempted to answer questions of this sort for
the remainder of his life. His thoughts were essentially
ignored.

Niels' attitude was revealed in a meeting where Richard
Feynman presented his work on Quantum Electrodynamics.

When Feynman drew his pictures of photon and electron interactions on a blackboard, Niels Bohr leaped out of his seat screaming indignation. In fact, the entire audience rebuked Feynman for using pictures to explain anything. Dirac called Feynman an idiot and said Feynman knew nothing about quantum physics. [5]

This author believes this is the reason that everyone believes quantum physics cannot be understood. An audience is expected to observe real world events and hammer them into a structure defined by Niels Bohr and his disciples. The issue here is not whether Bohr is right or wrong. The point is that the community is not allowed to investigate reality unless that regime is followed. Clearly, quantum physics is not completely understood. Unless the community has the ability to take up other avenues, progress seems doubtful. This leads to the comment, *Those that claim to understand what quantum mechanics is, do not understand quantum mechanics.* Right now, you have enough information to put together an explanation of the subject that is as good as anyone else has.

Keeping a Promise
This is the answer promised at the beginning of this book. The de Broglie matter wave function describes the behavior of very small particles. When these particles are electrons residing near a nucleus forming an atom, the matter wave function describes the behavior of the electron. Most importantly, it describes interactions with other electrons in other atoms.

To gain a better perspective of what this means, we should consider how we got to this point.

The Wave Theory
The wave theory of light was established over a period of 2000 years. In the 1800's the wave theory prevailed. The following people supported it.

Aristotle
Huygens
Descartes
Young
Fresnel
Foucault

A Particle Theory
Many others viewed light as some kind of a particle behavior.

Vaisheshika Hindu School
Titus Lucretius Carus
Dignāga & Dharmakirti Indian Buddhists
Pierre Gassendi
Democritius
Newton
In 1905, Einstein got a Nobel Prize for demonstrating light is
a particle.

The de Broglie Concept
Louis de Broglie observed the following and derived a matter
wave function.

First, Planck shows there is a relationship between the wavelength of light and some definite amount of energy called a
quantum. From this, de Broglie must have gotten the idea that
energy in waves could be related to mass. Einstein shows,
with

$$E = mc^2$$

that there is a relationship between energy and mass. From
this, using Planck's concept, de Broglie converts the mass of
a particle into a value for energy. The value for energy is converted to some value for wavelength and produces a wave
function for particles called matter waves. Essentially, this
means de Broglie found a way to determine the wavelength of
an electron moving at some specific speed through space.

Electrons are inserted into diffraction and interference experiments with the size of the slits similar to the de Broglie wavelength. The moving particles produce the traditional diffraction and interference patterns observed in corresponding light experiments.

This is the heart of quantum physics. An electron has a wavelength as it moves through space.

No More Hard Balls

Quantum physics is not represented by the following picture.

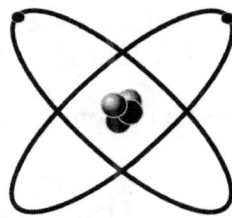

A better picture is as follows.

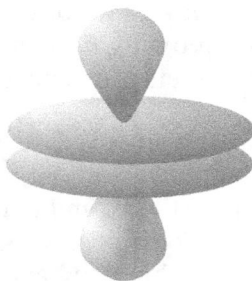

A little hard ball spinning around a hardball nucleus no longer represents the electron. The electron is now a matter wave function that oscillates around another matter wave function that is the nucleus.

So, What Is Quantum Physics?

The primary answer is that light and matter appear to be particles and waves. In general, observing both from a distance, they appear to be particles. Light, in experiments such as the photoelectric effect, appear to consist of particles. If you look at an electron zipping through space from a distance, it appears as a solid particle.

When looking at both up close, they appear to be waves. When a light wave is constrained to move through a small slit, it manifests wave-like characteristics. Look at an electron up close and it seems to oscillate like a wave.

This wave-like motion has a profound effect when we study the interaction of atoms in molecules. This is a small space and we are forced to look closely at electrons as they spin about the nucleus in a tight orbit-like motion. The wave passes a single point so many times a second; the electron appears to be solid at that point. This affects the behavior of the electrons in atoms that associate with other electrons in other atoms. This affects the behavior we observe in the macro world. LED's make their glowing light and chemical reactions depend upon the electron's nanoscopic interactions.

At the level of science today, this is as close as we can get to an understanding of what it is about. At this point, the best we can do is say that we subject light and matter particles to a given set of tests. Those tests are designed to label the subject a wave or a particle. From what you have seen, those tests are based on the results found by observing water waves. Attempting to compare light waves and matter waves to water waves is a very dangerous thought process. Note, while light and matter can appear as particles and waves, water waves do not have corresponding particle phenomena. Yet we visualize particles of matter and photons to be like water via these tests. Our own assumptions block our own understanding.

The question remains, close up, what do a photon and an

electron consist of?

Conclusion

We can understand quantum physics up to a point. Now, you have an understanding as well as anyone else what it is about.

The ensuing material in this book presents an attempt to break away from Niels Bohr's idea that the matter wave function is a probability wave. Instead, a path is taken that Einstein, de Broglie, and Schrödinger: wanted to follow. On this path, we find something we can visualize in our head.

Part 6

We Are Beings of Light

Do you see the wax and wane of knowledge? The attempt to understand the world around us is like a candle flame. At one time, everything is bundled and explained. At one time, all was fire, earth, water and air. The candle flame of knowledge pursuit waned. These four elements seemed to explain it all. Then, Lavoisier began categorizing something called elements. That led to chaos in the understanding of the world around us. The flame of effort to understand waxed. However, the developments in the first scientific revolution presented us with Newton's Laws of Motion and Maxwell's Equations. The pursuit of knowledge waned. A professor told Max Planck that all in physics had been done. Physics was at a dead end because we understood everything. Max Planck went on to make the flame of knowledge flare up with confusion. He triggered the emergence of quantum physics. Many questions have been posed and have been answered. The answers are leading to questions that are more complicated. The best we can do is claim we have the mathematics that does not make sense but works. The flame is waxing.

In this part of the book, we take a close look at light. Our goal is to determine with some precision, the behavior of photons. A list of the properties of light is presented, an interesting theory is proposed, and that theory is tested. Then, we use Einstein's Theory of special relativity to determine what else is in the universe.

The philosophy in this book now changes. Max Born interpreted the de Broglie matter wave function as a probability wave. Here, we side with de Broglie, Einstein, and Schrödinger. They believed the wave is physical, not a probability function. We continue with that assumption.

Chapter 22

What is Light?

Look how far we have come. We once believed that everything was made of fire, water, air and earth. That disappeared
when Antoine Lavoisier began listing the elements around us.
The mighty ether that lasted for over 2000 years falls to dust.
The organization of the Periodic Table of Elements enables
us to see inside the atom. A young Max Planck is told that
there few new developments to be had in physics. Yet, he
creates a scientific revolution called quantum physics. De
Broglie discovers matter waves and expands our understanding of light and matter. At the same time, quantum
weirdness brings more confusion.

In this chapter, we focus on the properties of light. The goal is
to list the known properties of light along with some behaviors of light that we do not understand. Then, an idea of the
structure of the photon is presented. If that structure explains
the properties of light and its strange behaviors, we might
conclude that the idea is correct. To this end, we begin with a
review of light properties and its behavior that are not now
understood.

A Review of the Properties of Light

The goal here is to simply list the properties and not-yet
understood observations. One could generate this list by going over any physics text that discusses light and quantum
physics. The goal is to isolate each item and present each in a
concise way. Here is a list without explanation. Then we will
explain each briefly.

- It is some kind of packet moving through space. [1]
- Its wave equation has the following form. [2]

$$y = A \sin\left(2\pi \frac{x}{\lambda} - 2\pi \frac{t}{T}\right)$$

- It can be polarized [3]
- It manifests diffraction [4]
- It manifests interference [5]

Here is a review each of these so they are fresh in our mind.

A Photon Is Highly Localized

The photon is a highly localized phenomenon. It is not supported by something like water that supports water waves. A photon is something that occupies some well-defined volume of space. The following picture represents some area of space that contains a photon. The arrow indicates it is moving through space in a specific direction at the speed of light.

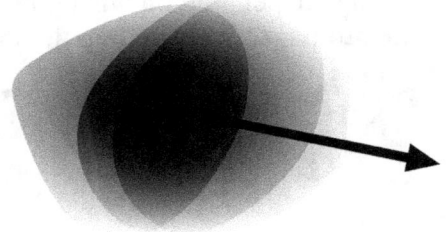

A Photon is Represented by a Wave Function

Light has a wave function equation of the following form.

$$y = A \sin\left(2\pi \frac{x}{\lambda} - 2\pi \frac{t}{T}\right)$$

Let's recollect what this means. If there is an oscillating cur-

rent in a wire, as depicted in the wire on the left in the following picture, the space around the wire is changed. [6]

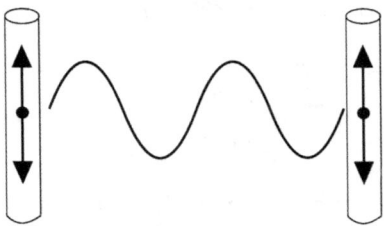

The change appears as an oscillating E field and is represented by the above equation. The effect of that field on another wire (here depicted on the right) is that it experiences a
changing electric current as shown. [7]

Recollect the effect of an E field. The E field consists of E
vectors. These are illustrated in the following picture as arrows. Note they change depending on where they appear on
the curve of the electromagnetic wave function. [8]

The wave function and E fields are a mathematical description of space around the wire. We observe the presence of the
E field by its effect on electrons. The following picture shows
a ball to represent an electron. The E field causes the electron

to move as is depicted by the arrow attached to the ball in the picture. [9]

Maxwell concluded that the electromagnetic wave was indeed a flow of light. That is, the flow between the wires consisted of photons. [10]

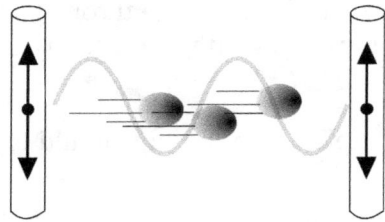

This appears as something very difficult to explain.

Light can be Polarized

The following depicts that light is polarized. [11] The picture shows a variety of light waves entering a slit in a barrier. The light coming out is linearly polarized. This implies the E field only appears in the x-y or vertical plane.

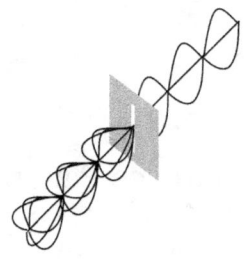

Likewise, there should be some explanation for the existence of circularly polarized light.

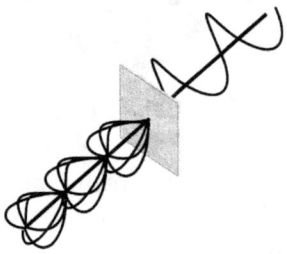

This picture shows that light can go through some kind of crystal filter and moves on with a constantly rotating E field. [12]

Once again, any explanation of light should explain this phenomenon.

Light Displays Diffraction Patterns
If the new concept for light has value, it will explain why it forms a diffraction pattern as it moves through a single narrow slit. [13]

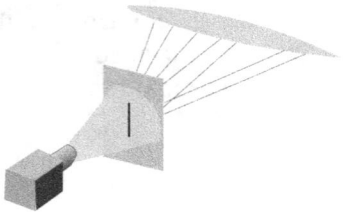

Light Displays Interference Patterns
If our new explanation is to truly represent light, then it must explain the interference pattern when light is thrown against two thin slits in a barrier. The following shows a source of

photons hitting a barrier with two slits. The photons going through the slits will make an interference pattern. [14]

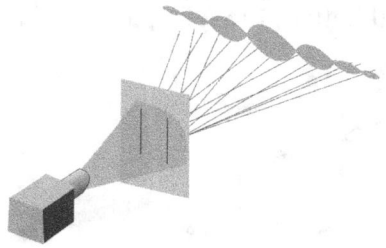

Likewise, if we have a good explanation for photons, our theory will also explain why the pattern disappears when we block off one of the slits.

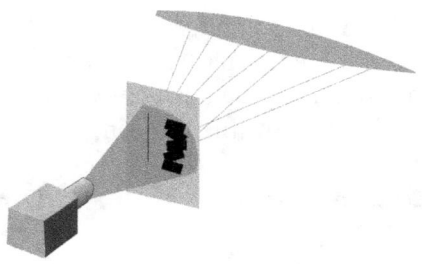

The above picture has the right slit covered with black tape. The result is that the pattern disappears. [15]

Introducing the Helical Photon Concept

If one goes over and over this list, a concept for the photon emerges. Let us review this concept, then take up each property of light and see how it measures up.

In a Nutshell

The main issue in this chapter is that a photon is a point that moves in a helical pattern though space.

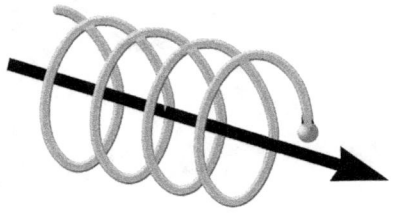

Here are some details of this concept.

The following is kind of an "under the hood" look at a helix. This picture defines the helix parts. The left image is of a helix photon moving out of the surface of this media. The right image is a side view of this motion.

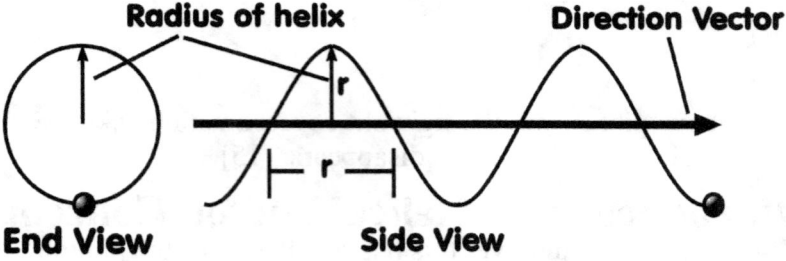

The direction vector is the center of the helix motion. The r is the radius of the helix pattern. Note that half the cycle length of the helix pattern is equal to the radius.

The projection on the x-y plane of a helical function has the following characteristics.

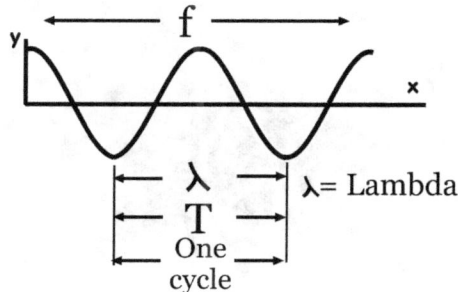

Lambda is the wavelength. T is the period or time of one cycle of oscillation. The small f represents frequency or number of cycles per second.

The photon advances at the speed of light along the direction vector. Here is what this looks like from a 3/4 view.

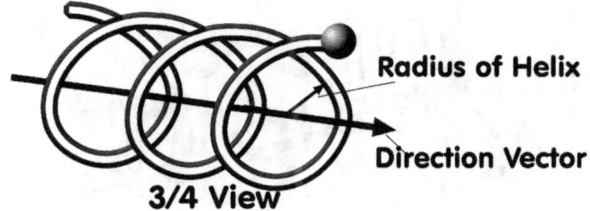

Does this Concept Fit the Properties?

The remainder of this chapter seeks to show that the helix structure fits the properties of light presented above.

Is the Helical Photon Highly Localized?

The first question is, *Does this device appear as a localized packet of something that moves through space?* The answer is clearly yes. A simple point moving through space satisfies the requirement the photon is localized. If we put a thought box around the point and the "wave motion" it generates as it moves, it certainly matches the concept of a packet moving through space.

Is there a Helical Wave Function?

The graphic representation of a water wave function and a graphic representation of a helix pattern projected onto the x-y plane appear the same.

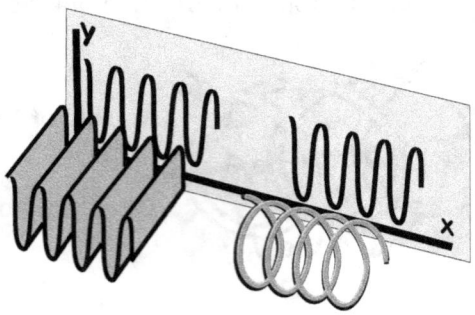

The following equation is the wave function for water waves, Maxwell's equation for electromagnetic radiation (light) and the path of a photon moving in a helical path in the x-y plane.

$$y = A \sin\left(2\pi \frac{x}{\lambda} - 2\pi \frac{t}{T}\right)$$

For a water wave, y is the height of the wave. For a light wave, y is the intensity of the E field at x. For a helix, y is the height of the path projected on the x-y plane. The following is a plot of this function when time t=0;

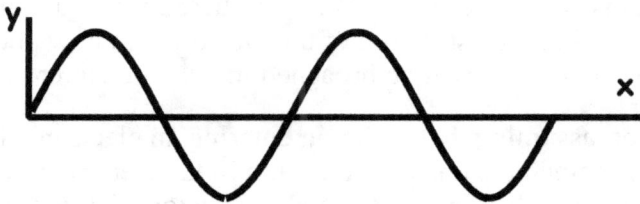

The following is a plot of this function when time t=T/4.

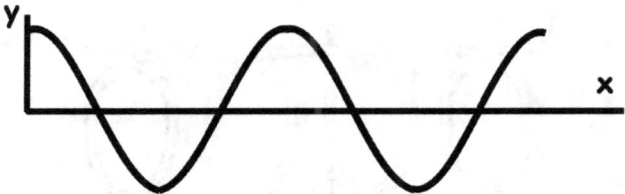

Thus, as projected onto the x-y plane, the same equation works for water, light, matter, and helix point motion.

Is there an Electric Vector?

We must show that this new explanation has something like the E vector which causes electrons to move when the photon encounters one. The photon in this scenario interacts with electrons in a way that resembles the traditional E field.

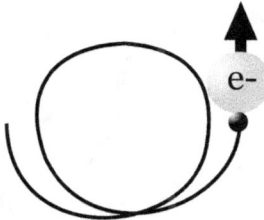

In this scenario, the photon encounters an electron. The collision results in motion of the electron. That is, the vector of motion of the photon appears to have a similar effect on electrons as E fields. In this respect, they are considered the same. This cannot be covered in detail here for we lack information about the structure of the electron and how photons relate to them. This will be broached in a later chapter.

However, assuming that a photon entering an electron will cause an immediate change of motion in the electron in the same direction the photon is traveling, vectors of this change would appear as follows.

Note that the incident angle of attack is different from how photons are traditionally thought to approach electrons. The traditional approach is thought to be along the direction line of the helix. Here the approach is along the line following the helical path. Observe this in the previous picture. From a side perspective or as a projection against the x-y plane, the change of velocity or E field vectors from helix photon motion appear as in the following picture.

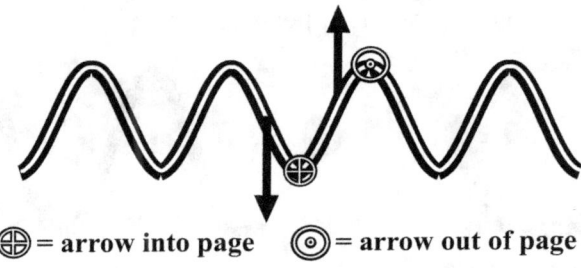

⊕ = arrow into page ◎ = arrow out of page

Compare this to a traditional picture of E vectors in the traditional wave function.

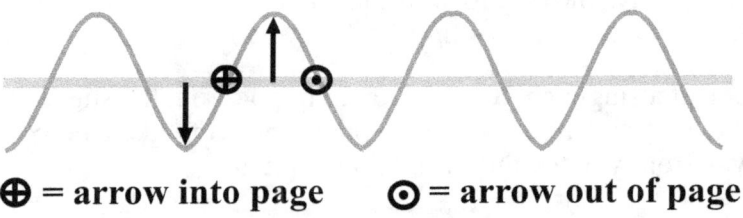

⊕ = arrow into page   ⊙ = arrow out of page

They appear much the same. The difference is that the traditional E vectors emanate from the axis of the curve whereas the helix vector counterparts emerge from the helix itself away from the axis. Is the traditional form of the E vectors some representation of a real phenomena or an artist's way of representing the vectors?

Does It Manifest Polarization?

Let's begin by looking at one photon moving through space.
The picture below offers two perspectives of this.

   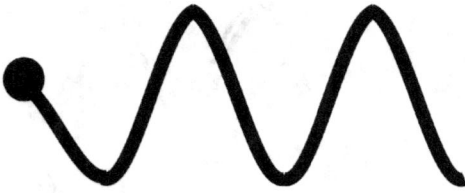

Moving at you   Moving to the left or side view

In the above picture, the left image is of a photon moving
directly at you. The image on the right is a view of the motion
of a photon from the side. The key point here is that the photon is always moving forward but also moving horizontally
and vertically as it goes through its cycle.

Look at a single photon attempting to go through a slit. The
following picture is to depict a photon moving like a helix
away from you toward the slit in the picture.

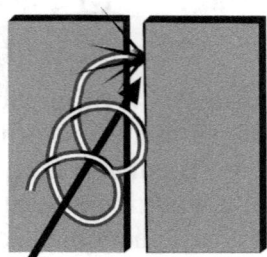

In the above example, the part of the helix that is traveling
horizontally does not go through the slit. Here, the photon
enters the slit but collides with the inside of the barrier stopping its forward motion. The vertical part of photon motion
behaves differently.

Again, the following picture is to represent a photon moving away from you.

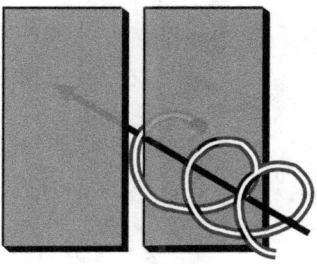

In this example the vertical part of the helix pattern slips right through the slit and keeps traveling. This phenomenon is responsible for polarization.

There is of course more than one photon attempting to slip through the slit. Consider a number whirling about a common axis. The left image in the following picture shows several photons moving directly at you. The right image shows a side view of this.

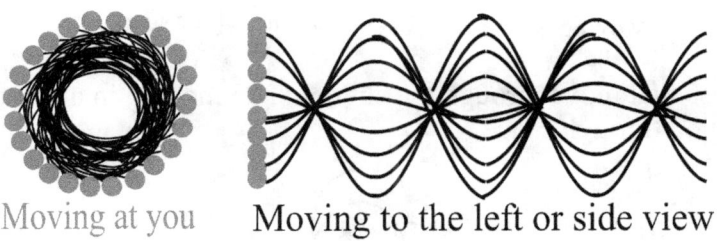

Moving at you   Moving to the left or side view

Imagine this whirling mess heading toward a slit in a barrier. In this experiment, the photons are aimed so some subset of photons are moving vertically when the photons get to the opening in the barrier.

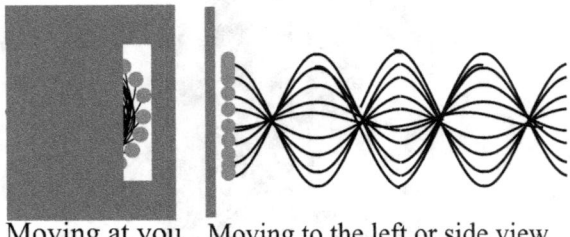

Moving at you   Moving to the left or side view

Those that are moving vertically when they reach the barrier slip through the slit. The others are reflected or absorbed by the barrier.

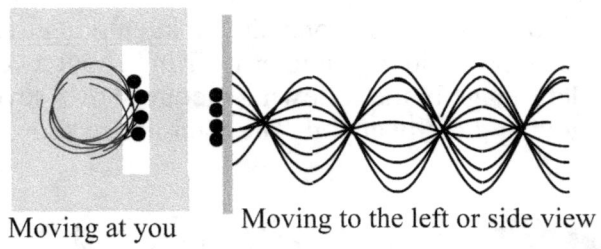

Moving at you   Moving to the left or side view

Those that move through the slit continue moving in a spiral fashion through space.

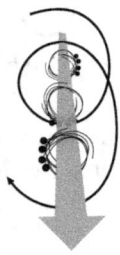

This is an example of circular polarization.

In this case, the photons approaching the slit were all rotating in the same direction. There is no reason for this to be a common case as photon helical patterns move in both directions. Let's look at this from that perspective.

The following picture represents two clusters of photons. The one on the left is rotating counter clockwise. The one on the right is rotating clockwise.

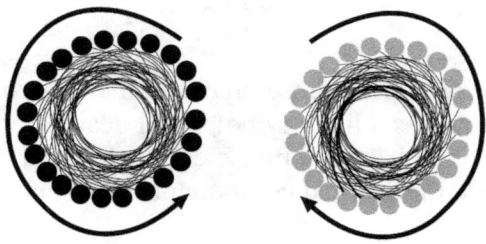

The following picture depicts photons moving vertically from both rotations slipping through a slit in a barrier.

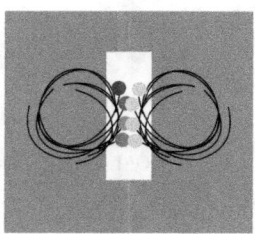

As these photons continue to move through space, the re-

versed motion of both will produce E fields that alternately
add to each other and oppose each other.

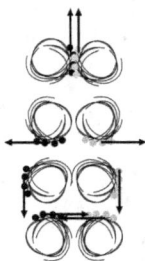

The two E fields superimpose on each other to produce an E
field that appears as a linearly polarized beam of light. Note
that the horizontal vectors oppose each other and cancel each
other out while the vertical components add to each other.

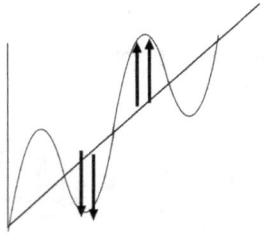

This is the polarization that most think about when discussing
polarized phenomena.

Does the Helical Photon Exhibit Diffraction?
This is probably the biggest challenge of any light theory. After all, when an electron was shown to display diffraction as
predicted by de Broglie's matter wave function, the whole of
the physics community was turned upside down. Suddenly,

particles were also waves. Look closely at light traveling through a slit. The following depicts a wide slit.

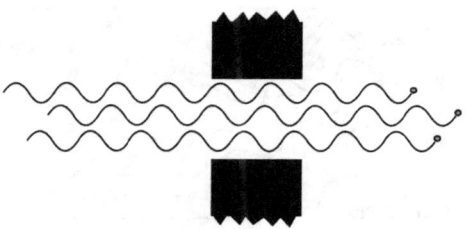

The wavy lines are helical photons moving through the slit. If the slit is wide, the photons will pass through without interruption. The spot they cast on a wall will resemble the size of the slit.

However, if the slit is narrow, the helix photons will encounter many collisions with the edges of the slit as depicted in the following picture.

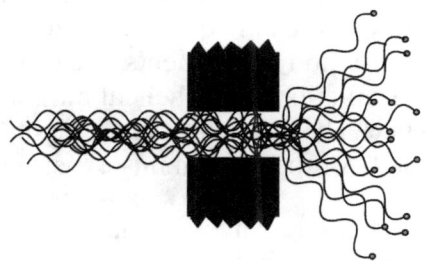

The result would be many photons bouncing through the slit in a wide variety of angles.

There is another factor to realize about this experiment. Because the slit is narrow, the photons moving through it will be polarized. This means the photons that make it through the slit will also be in phase with each other. The following picture attempts to indicate how this will shape the motion from the slit to a barrier. In the following picture, the gray bar to the right of middle represents the barrier the photons ultimately hit. Only half the experiment is drawn for clarity.

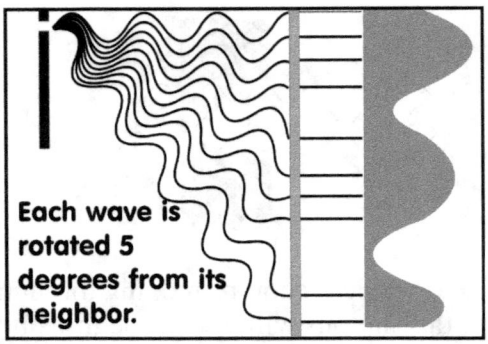

Each wave is rotated 5 degrees from its neighbor.

The waves in this picture have been drawn to come out of a slit in phase and each is rotated five degrees from its neighbor. Note, that due to the helical motion of each, the waves terminate at various positions on the vertical gray line. A horizontal line marks the point where a photon hits the gray line. Note the positions are not equal from each other. This suggests that in diffraction experiments, the diffraction pattern is due to the oscillation of the helical motion. It is not due to wave interference. The gray wave at the right of the picture suggests the intensity of the light in this experiment.

Here is a picture with more waves included. The waves emanate from a single slit. Each path from the slit includes two wavy lines implying two photons rotating in opposite directions. Each pair is five degrees from the other.

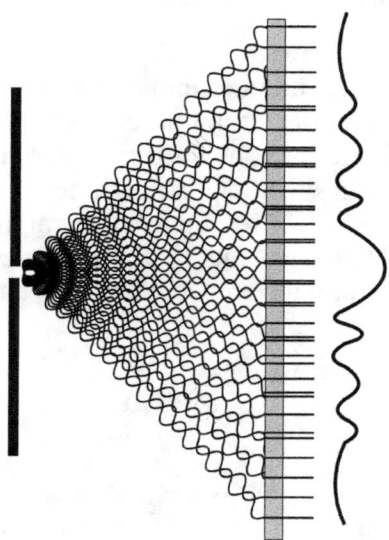

The primary issue to observe here is the spot where the waves hit the gray bar. A horizontal line has been drawn at each point. Again, note the variation where the lines fall. The curve on the right attempts to show the intensity of the line positions.

This device would also explain why experiments done with single photon shots produce the same effect. As noted in earlier chapters, experiments were conducted in which single photons were allowed to pass through a single slit. After a single photon was allowed to pass alone, another was allowed some time later. Then, there was a mystery. If the photons could not interfere with each other, how could the diffraction pattern appear? In the theory presented here, the photons do not interfere with each other. The photons move randomly through the slit and the helical pattern of motion produces the appropriate fluctuations.

The conclusion here is that a photon moving as a helix will display the characteristic pattern of light called diffraction.

Does It Exhibit Interference?

Water waves generate an interference pattern when hitting two openings in a barrier. Light produces the same phenomena when hitting two openings in a barrier. As we have seen, the mathematics for both is the same. Let's take a quick glance at this. The following represents straight water waves striking a barrier with two slits in it. The waves move through the slits and form an interference pattern on another barrier.

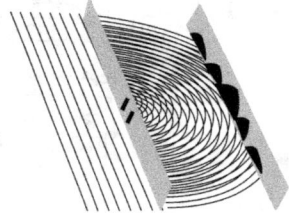

Light behaves the same way. This is demonstrated in the following picture where light shines on a barrier that has two slits in it. The light moves through the slits and forms a similar pattern on an opposite barrier.

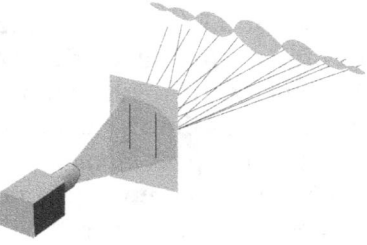

The theory in this book proposes that the patterns are not generated by interference but by the helical motion of the photons striking the barrier at various angles. This was just demon-

strated in the single slit experiment. Here, we superimpose
the helix wave patterns from two single slit experiments to
demonstrate the double slit phenomena. The goal here is to
show that the superposition of the resultant energy curves
shown on the right of the picture appear as in normal double
slit experiments.

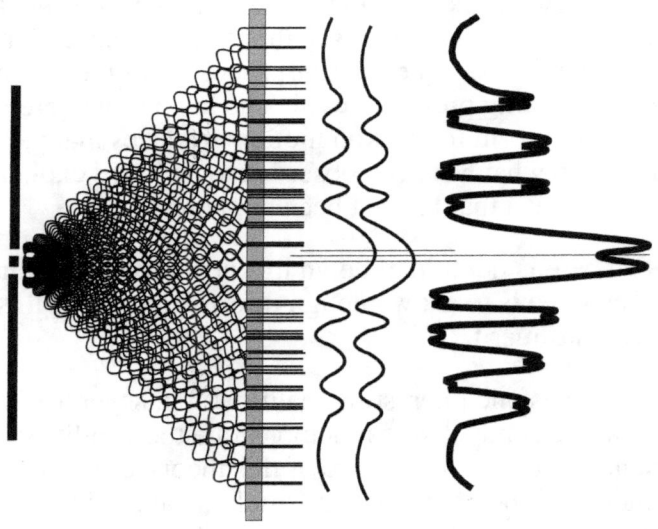

Because the two slits are separated, the energy distribution
curves are out of phase with each other. The combination of
the two creates the more complex energy distribution curve.
Note that the resultant curve is much stronger than a single
slit energy distribution as we are adding two curves together.

As demonstrated in many experiments, if one slit is blocked
off, the large energy interference pattern disappears and is replaced by a diffraction pattern.

The helical photon motion also explains why single photon
emissions traveling through the slits still produce the final
double slit energy distribution. The photons are not inter-

fering but finding their way randomly to the barrier. The primary explanation for this is that the polarization effect causes the photons to leave the slits with the same phase and spread out equally to hit the barrier. The varying locations are due to the helical motion.

Conclusion

There are a few problems using a helical pattern for the motion of a photon. However, this author believes the differences may be due to some unclear properties of the measurement devices being presently used. The main issue here is that using a helical form for photon motion produces many, many similarities to what has been observed to date and explains some phenomena that has not been explained to date.

The hope here is not to nail down this new theory conclusively. The hope is to show enough information that will show the theory has merit.

This theory has one interesting feature. First, several experiments can be conducted that demonstrate the validity of what is presented. Second, the nature of this theory enables computer models to be built that can simulate the motion of the photons moving in a helical pattern. Such an activity would yield valuable information and could be tweaked to simulate reality.

In a sense, an experiment has already been done to demonstrate the validity of what is presented here. In polarization, barriers have already been erected at various angles to demonstrate that the photon moves in this helical pattern.

Consider some of the problems with this theory. First, the idea of a helix is strange. What would cause the photon to move in such a way? Intuitively, we think it should simply move in a straight line. Again, this represents the way the human mind works. We compare apples to oranges and get fruit. This may bother some people. It may bother many people. However, many in science are becoming accustomed to strangeness as quantum physics, no matter how you slice it, is strange. Later in this book, in Part 10, reasons for the helical motion will be presented.

The point is that, while there are a few problems here, the similarities between the helix theory and observed phenomena are so striking, further investigation is warranted.

Finally, let's make one point clear. This does not explain what light is. The above explains how light manifests itself. A theory of what light is unfolds as this book unfolds.

Chapter 23

All is Light

In this chapter, we learn that special relativity has been un-
wound. The consequence of this is quite spectacular. This author has always thought that if one understood special relativity, one would understand the universe. What other phenomena could alter time, distance and mass? It is one of the
most proven theories in physics. Here, we look closely at one
of the postulates Einstein used to formulate special relativity.
The postulate is that the speed of light is observed to be constant whether you are moving or not. This does not make
sense. Nevertheless, there is an explanation for it.

The material in this chapter is quite tedious. Jumping to the
end to get the conclusion of it all would be acceptable. Then,
if you have the desire to see how this conclusion was drawn,
you can come back and read the details.

We face several barriers in understanding the material here.
The first is the widespread belief in the physics community
that there is no explanation for special relativity. In every
textbook in every classroom in the world that discusses special relativity, you will find a similar statement. That statement is that one cannot see any logical reason special relativity is the way it is. The statement goes on to point out that,
the mathematics explains the phenomena and there is no logical system or model to visualize the phenomena. This barrier
stands from the lowest to the highest levels of academia; we
are told, we cannot understand it.

Then, there is another barrier. It is about what we are accustomed to observing. We have spent out lives viewing the

world around us from a certain perspective. To exist in this complex life, we have established many mental foundations of life we must assume are absolute. As humans, we compare what we observe with what we understand. We do this to deal with our reality. Special relativity simply, does not compare with anything in our normal life or understanding.

This chapter attempts to deal with both of these mental barriers.

Right now, let's state what the purpose of this chapter is. The purpose of this chapter is to show we live in a simple space that contains only and many photons all of which move at the same speed. That speed is the speed of light. An observer immersed in that space will see all photons move at a constant speed irrespective of his motion in that space. The reason is that measurement is a process of comparing the speed of one thing to the speed of another thing. As the only thing in the universe is the photons zipping around, any measurement will yield the fact that photons move with a constant speed called the speed of light. Because photons cluster into particles that appear different, we believe we are comparing apples to oranges. However, because both are made of photons we are actually comparing acorns to acorns. The speed of one acorn will always appear to be the same speed of another acorn.

The ultimate purpose of this is to point out that the universe consists only of photons.

Our intention here is not to study Einstein's special relativity but rather to examine one of the postulates that support it. That postulate is that the speed of light is observed to be constant irrespective of the speed of the observer. [1]

Note that some liberty has been taken with the definition of this postulate. Normally, this is presented with the phrase, *The speed of light is constant.* This has been replaced with the phrase, *The speed of light is observed to be constant.* This is critical and drives to the heart of the issue. This author hopes that this becomes apparent as this discussion proceeds.

As we move forward in this chapter, some very complex material is covered. The author regrets the depth to which one must think about all of this. The author has spent several years denying what is here and then coming around to believe it. The issues are not simple. Every effort has been invested in attempts to make the concepts presented clear.

In a Nutshell

Understanding that the speed of light appears to be constant is a problem. Let us look at it as if it were an assignment for a high school physics experiment. We will list givens, provide a statement of the problem, and offer a solution.

Givens:

1. We assume we live in a Euclidian space.
Euclidian space consists of three dimensions: height, width, and depth. Moreover, we assume that time marches consistently forward. [2]

2. We do not live in a Euclidian space.
We live in a non-Euclidian space. If we are not moving or if we are moving, we see light move at the same speed. [3]

3. We believe this does not make sense.
We cannot understand why light appears to move at the same speed in our non-Euclidian space. However, we are capable of showing a mathematical relationship between a Euclidian space and a non-Euclidian space that explains it mathematically. The equation is called the Lorentz Transformation. [4]

Statement of Problem:

The givens do not answer the question, *Why is the speed of light constant?.*

Solution:

The solution is to locate a physical transformation that mirrors the mathematical Lorentz Transformation. The theory here is that there is such a physical transformation. The parts of this solution are as follows.

1. We live in a Euclidian space that contains point particles that all move at the same speed. (Note, the term point particle was selected carefully. You may think of these as dimensionless particles that move. This is dealt with extensively in a later chapter.)

2. The appearance that the space is non-Euclidian is because an observer will always measure the speed of all point particles as the same.

3. This last is because measurement is an act of comparison. When the speed of one of these point particles is measured, the only comparison is to another similar point particle. Thus, the observation is that every point particle in the space moves at the same speed as every other point particle.

4. From this, we see that the speed of light is constant from the point of view of all frames of reference moving at constant speed relative to each other.

The Spectacular Result

The introduction to this chapter mentioned that the consequence of this material would be spectacular. The consequence is that our universe consists of only one kind of particle (or point particle) of which there seems to be an infinite number. This suggests that the universe consists of nothing but point particles. These are photons.

Discussion

The text "In a Nutshell," is intended for those that understand special relativity. If you are new to this, it will probably not make sense. You would be wise to read the appendix titled,

"Special Relativity," of this book. Right now, we will go over "In a Nutshell" for those that are new to these concepts.

Given 1
A Euclidian space is essentially that which we observe around us every day. Mathematicians will define Euclidian space with more rigor than is done here. We view our space as three dimensions: height, length and width. Likewise, we can have a value for each of these dimensions to locate a specific point in the Euclidian space. To do this one must define a zero point for height, length and width. Then the values of the dimensions will identify a precise point away from that zero point. Another feature of this space is that time in it marches consistently forward. This is a fourth dimension we are accustomed to dealing with. We can identify this with a value and further identify a point in space and time. There is no magic here. This is common knowledge to twelve year olds.

There are a few other things that one should understand to make this discussion complete for our purposes. We need to know how we observe moving things in different circumstances. The following depicts one of interest.

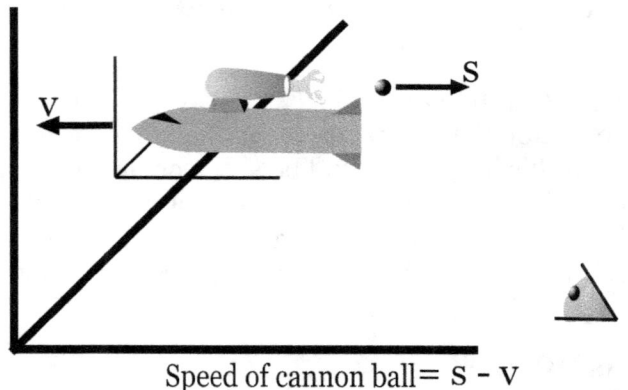

Speed of cannon ball= S - V

It is of a space ship moving through space. There is a cannon strapped to the top of the ship and it was just fired. The ship is moving at velocity v relative to an observer stuck on the page. That is, the observer is not moving. If we assume we live in a Euclidian space, we will assume that the speed of the cannon ball relative to the moving ship is s. However, we are astute enough to know that the speed of the cannon ball relative to our rest point of view is s-v. [5] That is, if the cannon ball is fired at us at a speed of s relative to the space ship, we should subtract the speed of the space ship moving away from us. This would be a characteristic of a Euclidian space.

Given 2
Given 2 attempts to say we apparently live in a non-Euclidian space. Describing this non-Euclidian space is difficult. In general, it appears very similar to the Euclidian space we observe every day. There is a difference and it is most easily observed when studying things that move very fast relative to a rest position. The picture of the space ship is handy for this. We can visualize a space ship moving away from us at a very high speed. This time however, there is a flashlight strapped on top of the ship. It is on and shooting photons behind it. We will study just one.

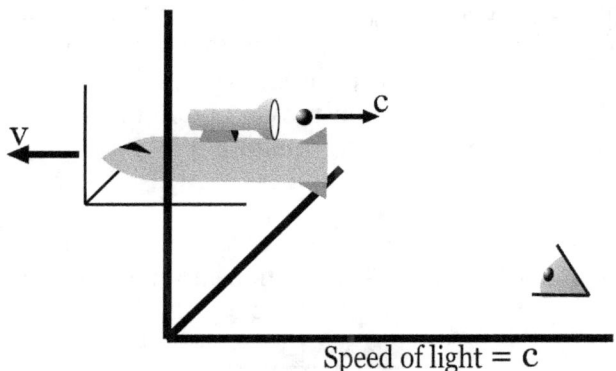

Speed of light = c

If we assume we live in a Euclidian space, we would say the photon is moving with speed c-v. That is what we are accustomed to in a Euclidian space. However, the observer is familiar with special relativity. Thus, the observer, at rest, knows that the photon is moving away from the ship, relative to the ship, at speed c. The observer also observes that the speed of the photon, relative to his non-moving point of view is c, the speed of light. The speed of light is the same in both reference frames. [6] This is not the same as the experiment with the cannon ball.

We must stop now and reflect on this. Your first thought is probably that this does not make sense. Your next thought would probably be that this to complex for you to understand. The response to these thoughts is yes and no. The first point here is that there is no more to it than that. Someone sitting on a rocket ship moving away from Earth very fast will observe a photon of light the same as someone standing on Earth. Both will see it move at the same speed, the speed of light. It does not make sense but that is all there is to it. For years, the most sophisticated physicists in the world have just accepted this. From every aspect one can think of, this has been proven to be true.

Given 3
In this given, the task is to point out that the most sophisticated physicists in the world explain this confusion mathematically. In the Euclidian space, the values for the x dimension can be moved from a non-moving point of view to a moving point of view with the equation. [7]

$$X_{moving} = x - vt$$

In the non-Euclidian space, we observe the values for the x dimension can be moved from the non-moving point of view to the moving point of view with the equation,

$$X_{moving} = \frac{x - vt}{\sqrt{1 - \frac{v^2}{c^2}}}$$

This is called the Lorentz Transform. [8] It mathematically

transforms dimensions in our non-Euclidian space from a non-moving reference frame (the page) to a moving reference frame (the moving space ship). Studying this math now is not important. The point is that this was the only solution until a physical transform was discovered.

Statement of Problem
In the givens, there is no logical explanation of why the speed of light is constant. This statement is to recognize that the Lorentz Transform explains why the speed of light is constant mathematically but does not provide a model or something a logical human mind can grasp. [9] However, the givens describe the territory of the problem.

Solution
A euclidian space that is filled with point particles moving in random directions but moving at the same speed, will appear as the non-euclidian space we are studying. We need to discuss this in detail and from various perspectives.

More about the Solution
The solution is that our space called universe consists of many of some kind of common thing. To understand the solution we must understand some tools used in the analysis of the problem. Then we can apply these tools to perspectives of the problem and, perhaps, get a grasp on all of this.

The tools referred to here are those items that are customarily used to measure the world around us. This includes time, distance, and speed. The following touches on the important aspects of each of these.

The Practice of Measurement
To understand the concept presented requires understanding of how we measure time and distance. The people of science do not understand what time and distance are. They are required to understand how they are used and how to analyze the world around us and produce practical answers. Those that only lecture and pontificate on the universe around us do

not need to produce results. They desire only to generate ap-
plause from their audience. This comment is offered with the
hope you will understand the difference between those that
must produce real world results and those that only wish to
sound erudite. The needs of these two groups are different.
The men of science locate some standard thing from which to
compare to either time or something of length. The point is
that a scientist measures something by comparing a known
length to some unknown length to determine the length of the
unknown. [10] In days gone by, a stick would be held against
the tip of the king's nose and a mark made on the stick at the
tip of the king's fingers. That distance would be called a yard.
[11] That stick then would become a standard to measure
lengths in all the land. Copies of the stick would be made and
distributed throughout the kingdom for this purpose. Today
we use the wavelength of light of a particular color. So many
wavelengths of light represent some specific length. [12]

The same is true for time. Perhaps the most common standard
of time is the speed of the sun through the sky. This has been
compared to the number of drops of water that drip from some
vessel. [13] Today, the standard of time is the oscillation of
some subatomic particle. [14]

The primary point here is that measurement of time and distance is accomplished by comparing one thing to some other
thing. We choose one of these as standard.

The Practice of Measuring Speed
Speed is defined as the distance something travels during
some duration of time. Distance and time must be measured to

determine speed. The following picture depicts this. A ball is moving from x to y.

The observer is watching this happen. He is holding a stop-watch in his hand. When the ball started at x, the observer pressed the stem on the stopwatch and started the timer. When the ball reaches y, the observer will press the stem on the stopwatch and stop the time. This will give him the time. He knows the distance from x to y is d. Thus, the speed of the ball is $v = d/t$.

The Practice of Observation
The following picture depicts the most common device used in our society to observe.

●

That spot represents the place where photons are allowed to enter our eyes. This poignantly makes the statement that we only observe when we sense photons that occupy the same point in space and time as we do. To observe things away from this point of view, the observer must perceive a photon that interacts with the real world and enters this local point of

view. Then the observer must use logic to decipher what
transpired. Recollect that when Einstein discussed all of this
he was quite emphatic that all was from one point in space
observing another point in space. All of this will fall apart if
not taken seriously. That is, you must understand that observations are made by interpreting the photons STRIKING
THE POINT OF OBSERVATION not the point that reflected or generated the photons.

Consider this, the sun or a light bulb shines light on the objects around us. Light reflects off these objects and strikes our
eyes. We make mental calculations of where the light must
have traveled and deduce the shape and distance of that which
we observe. We do this automatically and are not aware it is
going on. The following picture shows photons coming from
everyday objects around us, striking our eyes.

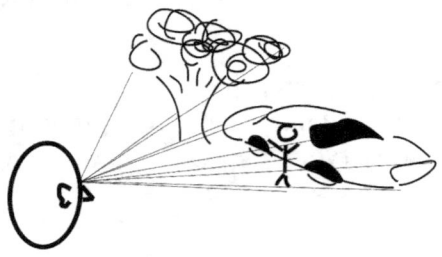

Once we have gathered this data, we mentally build a picture
that we view in our head. Once that image is in our head, we
behave as an omniscient observer looking at the subject as if
we were in another dimension. In the logic we are studying
here, we should not do this automatically for we are studying
the process itself.

This point strikes at the heart of confusion about the principles presented here. We as humans are very accustomed at
viewing images we build in our head. When we do this, we
have an omniscient view of the subject at hand. Unfortunately, the image in our head is not identical to the real
structure in the outside world we are imaging. We are not

consciously aware of this. We can be if we put our attention on it. This mechanism comes into play when viewing the pictures in this chapter describing what is actually going on with our process of observation. That is, we take the information from the page and build an image of it in our mind. If we are not careful, we interpret this image in our head as an omniscient observer and ignore the details on the page.

The correct way to view the experiments presented is as if you are sitting in the seat of the experiment. Somehow, one must learn to observe the subject from two points of view. The two points of view are what you know and what the point of view knows. Magicians, musicians and computer programmers understand this process very well. Magicians must do something unknown to the audience to produce some magic effect. What the audience observes is drastically different from what the magician did. The good magician can grasp these two points of view accurately. The audience can only view one accurately. Musicians find themselves in the same position. They must study a variety of sound frequencies and learn how to manipulate an instrument in a variety of patterns. They logically construct these patterns. The listener is not aware of the technology involved but hears some sound that is pleasing, inspiring or someway emotional. The logic producing the sound is mechanical but the thing produced is emotional. Computer programming is similar. Programming consists of three logical actions: compare, branch and step. The users of computers, however, see none of that. Depending on the program, users see something that can talk to them or even appear to think. In each of these disciplines there appears to be two perspectives that seem dissimilar. Yet, one exists because of the other. These three examples are presented in an attempt to demonstrate something can be observed from two different perspectives.

Perspectives

Now that we have gotten some understanding of the tools that we must use in this analysis, the associated perspectives of the task are examined. A problem with this examination is that the student will constantly flip back to the idea our universe consists of many kinds of material. If there is any thought that something else exists, that will give rise to the thought that there is another way to measure light particles. In addition, the student will constantly flip back to the feeling he or she has an omniscient view of the universe. As mentioned, this is due to the unconscious action of viewing the pictures in our head instead of consciously focusing on the data that is input from the world outside of our head. Perspectives are about focusing on the fact that everything we regard in the real universe is made of these common photon particles. Furthermore, we cannot see these common things at a distance. Therefore, a few perspectives are about how we must manipulate common things to bring information to the little dot that is our point of observation.

Photon Clocks

To repeat, any construct in the universe will consist of the particle points also called photons. Therefore, a clock used to time any event must be made of photons. An elementary clock is presented here. While it is not like any clock or stopwatch you are familiar with, any clock or stopwatch will have, within its basis of operation, the device you are about to see. Therefore, to make some progress in this explanation, here is a clock made of photons to be used in these elementary discussions.

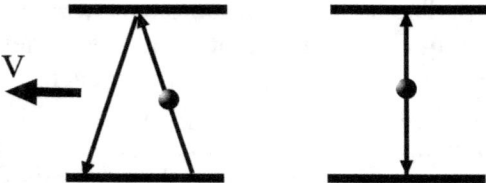

The above is a picture of two identical clocks. The left one is moving to the left. The right one is not moving. The clocks consist of two plates that act as mirrors. A photon is placed between the plates that bounce up and down off the mirrors. The dot in the middle of the plates depicts this photon. As the right clock is not moving, the photon is bouncing straight up and down. As the clock on the left is moving to the left, the photon must bounce at an angle so it moves with the clock as the clock moves.

Now, we cannot describe the structure of the mirrors. For now, we must accept that they are made of some magic material that causes a photon to bounce in the opposite direction when hit. The important issue is that a photon bouncing up and down is used as a clock.

Time is measured with this clock by counting the number of times the photon bounces up and down. This, of course, requires that the distance between the mirrors is some fixed distance relative to our motion. That is, the distance is determined when the clock is at rest relative to our frame of reference.

We Cannot See Photons

Now that we have a clock that fits the universe we are studying, let's use it. The following picture shows a photon flying though space, an observer, and our new clock. This picture is an attempt to duplicate the process of measuring the speed of a ball as was done just a moment ago.

X ————— d ————→ Y

However, we cannot do that with what is shown in this picture. The observer cannot see the photon in the picture for there is nothing between the moving photon and his eye to

give the observer information about where the photon is. Photons are only observed when one is occupying the observer's
position. This repeats several earlier statements and needs to
be stressed heavily. To measure the time a photon flies would
require some kind of apparatus that appears as follows.

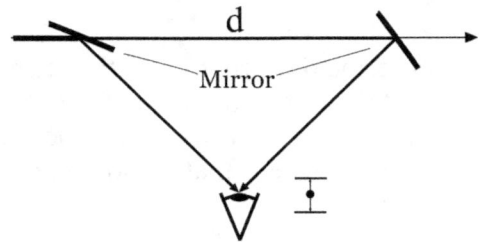

Our desire is to measure the time a photon moves from the
mirror on the left to the mirror on the right. The mirrors are
half silvered so part of the beam is reflected and part is passed
on. When the photon (or photons in this case) hit the left mirror, part of the beam is sent to the observer. When the observer sees this, he can start counting ticks in the photonic clock
to his right. When the photons hit the right mirror, some other
photons are reflected to the observer. When he gets these, he
can stop counting ticks in the clock. He will thus have a time
of travel and already knows the distance from mirror to
mirror. He can calculate the speed of light from this.

Understandably, this experiment would be difficult to perform. The time is very fast and getting it accurate would be
difficult. In addition, getting the photons to fly side by side
might present a problem. Notwithstanding these problems,
the point of this argument is that in a normal Euclidian space,
the speed of the particle points would always be measured as
the same and would appear as the speed of light. That is, the
measurement produced by this experiment would be the same
if sitting still on earth or flashing through space half the speed
of light away from earth. Please reread this paragraph. To
clarify, this experiment done in a normal Euclidian space, at
various speeds, would result in an identical measurement.

This is the strange thing. Because of the structure of this device, the speed of the photons measured will always be the same. The velocity of the photons traveling to the observer will eliminate the effects of the device moving at any speed in any direction. An astute student might say, *Well, we can mathematically calculate the effect of the photons to the observer and get a correct value for the speed of the photon under observation.* The problem is that process requires an omniscient view of the entire experiment. A judgment would be made about the photons being used for observation while you are studying the exact thing that has already been judged. That is kind of like using a word to define itself. Cats chase rats because cats chase rats.

Removing the extra travel of the photons from the mirrors to the observer would be desirable.

However, the main point here is that we cannot see a photon from a distance. To do so we need to construct some mechanism to bring the information to the observer. When that is done, the measuring system will nullify the velocity of the apparatus. Any such apparatus will always measure the speed of the particle point the same as any other particle point.

How Do We Measure Photons?
We can structure the above experiment to measure speed directly. Consider the following apparatus. This picture shows

a photon gun that shoots a photon to the left. A mirror is on
the left that reflects the photon back to the observer.

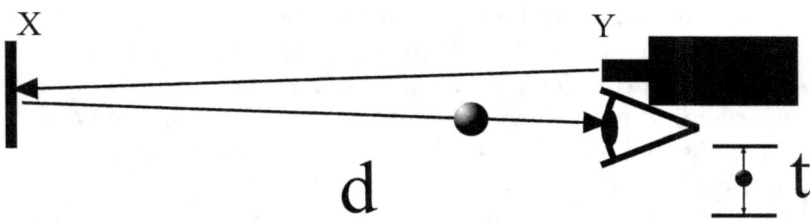

The observer knows the distance d from x to y. He can start
the clock when the photon leaves the gun and stop it when the
photon hits his eye. From this, the observer can calculate the
speed of motion of the photon.

Here are some problems with this scenario. The observer cannot observe the photon as it moves away and toward him. In
the theory presented, this experiment would occur in a Euclidian space. Let us assume the apparatus is moving with
some velocity to the left. The following images depict the motion of the photon as it makes this back and forth trip. The

three images are to show what happens to the photon as it travels.

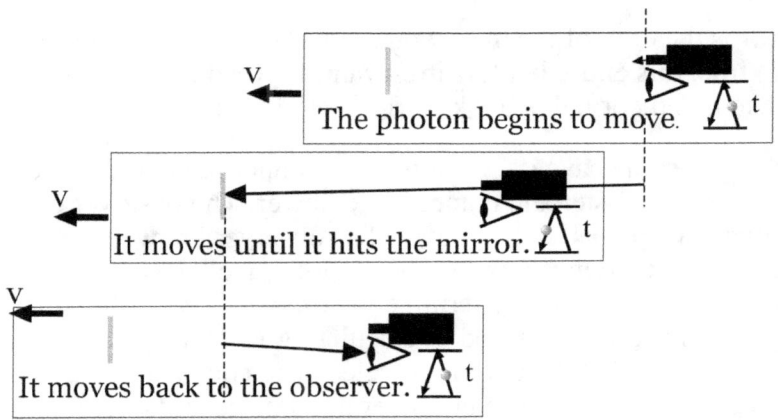

The photon begins to move.

It moves until it hits the mirror.

It moves back to the observer.

The top image shows the photon just emerging from the gun. The middle image shows the condition of the experiment when the photon hits the far mirror. Note that this path is very long as the apparatus is moving also. The lower image shows the photon going back to the observer. Note the path is very short. Then, look at the clock. The photon in it is moving at an angle caused by the motion of the apparatus. This has the effect of slowing time down as the apparatus moves. Remember, the measurement of time is the number of times the photon in the clock bounces up and down.

Bear in mind that this experiment is done in a Euclidian space. As expected in this space, the velocity of the photon and the apparatus is additive. Notice that the motion of the photon in both directions cancels out the effect of the motion of the moving apparatus. Furthermore, note the clock. As the motion of the apparatus affects the path of the photon in the clock, the time measuring device is also changed. As a result, this apparatus will measure all point particles as moving at the same speed. This will occur at any speed of the apparatus.

The observer will conclude that he is living in a universe that is non-Euclidian where the speed of light is constant.

This is the crux of the theory presented here. If this theory is not true, this entire book is irrelevant. If you do not understand this theory, this book is irrelevant for you.

Here are some caveats. The observer cannot see that the photon goes out faster and comes back slower. The observer can only measure the time of travel. The observer knows from previous experience how far the photon traveled. His "standard" clock is used to determine the time of flight. Then he applies the known method of calculating velocity to determine how fast the photon traveled. The conclusion is that it traveled just as fast as every other point measured at various speeds of the apparatus. In other words, the speed of light is constant.

The Reason
The reason there can be no other things other than photons is that any other thing would provide some other device to measure the speed of photons. Those other things would or could have some dimension to them that could be compared to the motion of photons. In addition, photons must not have any dimension for that would provide another device to measure a fixed distance and, thus, measure the speed of photons. If there were objects other than photons in the universe, the appearance of a non-Euclidian space would disappear. The theory is, since we observe this non-Euclidian space, it is populated with only photons that are dimensionless.

The ideas in this entire book unfold from this concept.

Observing the Observation
Here is another way to explain the phenomena. To grasp this, the reader must come to understand that this works when one realizes that we are not only observing a common particle, we are using that common particle to observe all other particles.

To get started, let's build a universe from nothing instead of using some established universe. Mentally, we can build a small experimental universe. Our universe consists of a basic rectangular coordinate system. This universe has points in it that define locations in this three space. The points can be identified as x, y, z, and t. The t is to represent time.

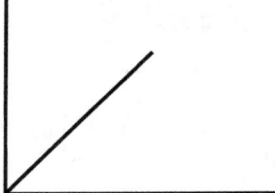

Put an observer in that system.

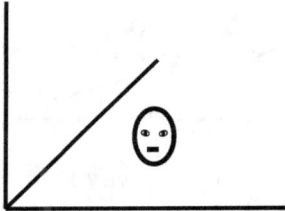

Fill the system with point particles all of which move with a constant speed relative to the universe's rectangular system. The observer can only observe a moving point particle when

one coincides with the observer's point of view. The observer
does not know where he is in this universe.

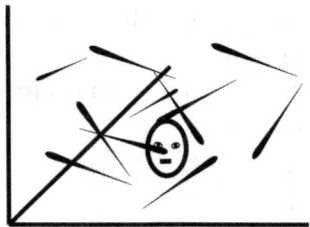

The observer cannot observe a point particle some distance
away from him as in the following depiction of this universe.

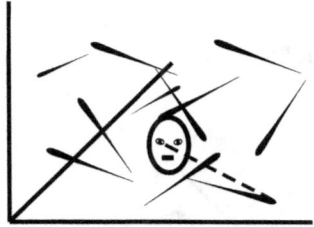

To get information about things away from his point of view,
the observer must evaluate those point particles that enter his
point of view and calculate what happened to the point particle some distance away. This is done in a couple of ways.
First, he can allow a point particle in his point of view, go
out, interact with other point particles and return to his local
point of view. Alternatively, he can allow the multitude of
points to interact with remote objects and randomly enter his
point of view so they can be processed.

With respect to the speed of the point particless, the observer
can only determine the speed of one point particle by
comparing it to another similar point particle. Consider the
observer in the following picture. The top observer is not
moving. He is attempting to measure the speed of the point

moving directly at him. Another point is passing just below him. He considers that moving point his standard. When compared, they appear to be moving at the same speed.

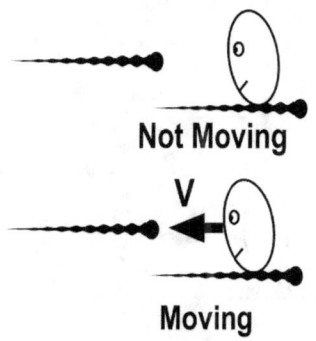

Not Moving

Moving

Now consider the lower observer. He is moving. Again, a point is moving toward him. Again, there is a moving point just below him. While we can understand that, due to his forward motion, the point seems to be moving faster toward him than in the previous example, the lower observer compares the speed to his standard, which is just below him. His conclusion will be that both points are moving the same speed. The conclusion would be that the speed of the points is constant.

Consider another aspect of this. In the picture below, the left observer is not moving. He is observing a point coming straight down on him at speed w. The other point is a standard to measure the other. The result is that the one observed will always be the same speed as the standard used.

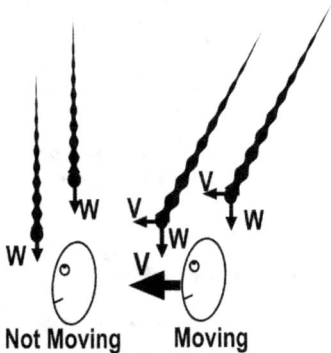

Not Moving   Moving

Now, consider the observer on the right. He is moving with velocity v to the left. The points coming down on him are not only moving down with velocity w, but the points are also moving to the left at velocity v. The observer will see both points coming down on him directly from above. His impression of the point motion will be the same as the observer on the left. He will also see that both are moving at the same speed. The conclusion will be that the speed of the points is constant.

One can perform a number of similar mental experiments as these. Each will demonstrate that if any point used as a standard to measure the speed of any other point, the conclusion must be the same. All points move at the same speed regardless of the motion of the observer.

There will be those that say that the speed is not constant. They will be quick to point out that the speeds v and w are different. These people are taking the point of view of an omniscient observer. They have not learned that someone "In the system," or "In the seat," cannot observe the omniscient point of view. Switching to the omniscient view destroys the point being made here.

All of this is to show that an observer in a sea of points moving at the same speed will perceive that all the points are moving at the same speed, regardless of the motion of the observer through the sea. If these points are considered photons or light, the speed of light is constant.

The point made here is that in reality, light moves at different speeds depending on the motion of the source and destination. Because we cannot observe the details of this motion but only observe the end points of the motion, we will always observe that the velocity of the motion is the same.

Conclusion

The conclusion is that our universe is filled with dimensionless photons and nothing else. The fact that photons are dimensionless and there is nothing else in the universe to compare against manifests Einstein's second postulate of special relativity: the speed of light is observed to be constant.

While this conclusion is very unusual, it seems to unfold to describe all natural phenomena in our universe. The following chapters develop this.

It would be nice if this new theory, that all consists of photons, could be tested in the real world in some way to verify that it is true. In a way it has been. Consider the result of high-energy particle experiments in which atomic and subatomic particles are slammed into each other. There is always an effort to discover what makes up such particles. However, isn't the result clear? The wide array of particles appear suddenly with a variety of properties and lifetimes. However, they do not just disappear. The energy they had is always measured. That energy is manifested as electromagnetic radiation. That is, photons. The result of all these collisions is a shower of photos. Isn't that a demonstration that all is made of photons?

Dirac presented another example of this. He observed that

when a high-energy photon passes by a very heavy particle, there is a possibility a virtual electron will appear that will disappear sometime later in a burst of energy. [15] This is again, a shower of photons. This is another suggestion that all matter consists of photons.

Part 7

Subquantum Physics 101

What have we learned from the previous part of this book? We have been presented with two paradigm shifts. The first is that light does not move in straight lines. If you believe this or at least are willing to consider the concept, you can see how quantum physics would be difficult to understand. If you believed that photons moved in a straight line and looked for it along some such straight line, you would go batty for the photon would seem to jump from one place to another without existing at any point in between. In a sense, this matches the reality of what is observed today.

The other paradigm shift is that photons are the only things that make up the physical universe. That is an even bigger paradigm shift. This seems wholly illogical. However, if you follow this line of thought and could see that cars, amoeba, mountains, and seas follow from this, would you become a believer?

The author believes that photons move in the helical pattern. He also believes that the physical universe consists of photons. However, if this is true, how do these items manifest the universe around us? In an effort to demonstrate how this could be, this book offers guesstimates how the universe would unfold from these two paradigm shifts. The goal is to provide a possible path so some naysayer cannot say, *This is ridiculous. The moon is not made of cheese or photons. There is no possibility this could lead to what we see around us.* Hopefully, the material presented here will cause one to ponder the ideas and perhaps look closer at the possibilities. In this vain, these possibilities will be carried forward in this book to complete a full cycle of description of the universe. You may call this science fiction if you wish.

With that said, the purpose of this part of the book is to bridge from the classical concepts of a universe of hardball things to this universe made of light. This is an introduction to sub-quantum physics. Subquantum physics is the study of how photons interact to manifest particles. During this study, we discover the background sea of photons. This feature of the universe supports the existence of particles. With this new concept of what particles might be and how they persist, we go back to our understanding from the old hardball physics concepts and clear up some mysteries. If this new concept "explains" old quantum weirdness, we might have more re-spect for this new concept. The purpose of this is to shed the old ideas and bridge to this new concept. Once the old clothes have been shed, we examine the heart of the universe, sub-space and the Super Matrix.

Chapter 24

Introduction to Particles

In general, quantum physics refers to behavior of very small, subatomic particles. This book purports that these particles have a structure that consists of groups of photons. Then, where quantum physics deals with particle dynamics: sub-quantum physics deals with the interaction of the photons. This chapter serves to introduce how the photons work together to enable particles to exist. We start with a bit of history.

A Bit of History

Historically, matter consisted of atoms that were indivisible. [1] As science advanced, atoms were found to consist of even smaller parts. [2] Due to intelligent thought and some interesting experiments, people like Rutherford and others discovered that electrons were some small part of an atom and the nucleus, with protons and neutrons, were another part. [3] There was a possibility that these were the smallest particles that made up the universe. However, nuclear physicists began experiments in which various elements were accelerated at high speeds and caused to collide with one another. The result was that the atom was cracked open and revealed that there are many other things besides electrons, protons and neutrons. The experiments continued and the number of things shaking out of the atom began to be counted in the hundreds. Organization ensued and something called the Standard Model emerged. [4]

The Standard Model is an attempt to organize the proliferation of subatomic particles somewhat as the Periodic Table of the Elements organized molecules. The Standard Model is also a theory to explain what matter is and what holds everything together. Many believe that this theory is not complete but, on the other hand, many feel it is useful in understanding the structure of matter. [5] In the Standard Model, there are 12 fundamental particles. [6] They include 6 leptons and 6 quarks. The scenario is a bit more complicated than this. However, that is beyond the material in this text. Essentially, the theory proposes that the universe consists of combinations of these particles.

Just to be a little more complete, here is a word about leptons and quarks. Leptons are subatomic particles that are similar to electrons. An electron is a lepton. Leptons do not interact with strong forces. Strong forces hold the nucleus together. Quarks are subatomic particles that, in general, form protons and neutrons. They experience the strong forces inside a nucleus. [7]

Hardball Physics
In general, the concept of hardball physics persists today. The point of view is that the universe consists of small hardball thingies. As you have seen, present day quantum physics replaces the hardballs with wave phenomena. However, note that the wave phenomena is manipulated with something called measurement or observation. This is to collapse the wave

into one of these hard balls during observation. [8] These small hardballs are surrounded by fields.

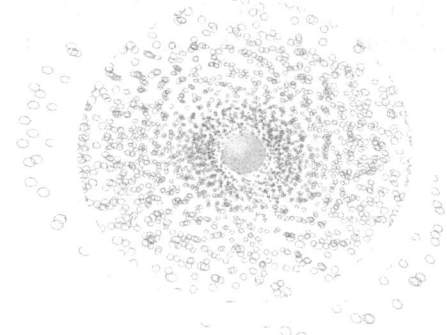

The small hardballs exert force on each other via these fields. However, when this author studies these fields closely, the fields appear to consist of even tinier hardball thingies.

In the presentation that follows, there is a lot of similarity to the hardball concept except there is no hardball. Here, what appears as a field IS the particle. In addition, the field consists of photons. However, let us not morph from the past to this idea. Let us start from where we left off in the previous chapter.

The previous chapter proposed the concept that the universe consists only of photons. Then, what is the stuff we see around us? Based on simple logic; photons. In this chapter, this idea is advanced. A few ideas about particles are introduced such as existence, persistence and motion.

In a Nutshell

A particle is a cloud of photons bouncing off each other. The following picture attempts to illustrate this. The squiggly lines represent photons moving in and out of a particle.

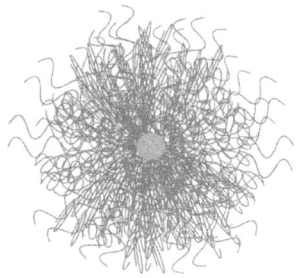

An electron, for example, may consist of a million, million, million photons. The photons in such a cloud are constantly colliding and bouncing off each other to maintain the form of an undulating cloud. The motion of photons in such a cloud appears to move in some kind of cohesive pattern. The photons are not held together with any kind of container. They have randomly acquired a common motion together. The motion is chaotic and hangs together like a cloud floating in the air.

A Universe of Dots

In this theory, electrons and protons are the only particles that exist permanently. Neutrons can only exist when in an atom. All other particles are unstable and eventually decay into smaller particles or dissipate into free photons that fly randomly in space. This would include neutrinos that, although they can persist for a long time, eventually decay.

Accepting the idea that real matter consists of clouds of photons is very difficult. We, in the macro world, see a single photon as something that barely exists. We have difficulty in accepting the idea that the table in front of us is constructed of something so frail. Consider however, that everything is made of these fluff balls. When we hit our hand on the table, we are not moving something hard from outside the universe. We are moving photon cloud against photon cloud. In some way, the clouds generate electric fields that react with each other.

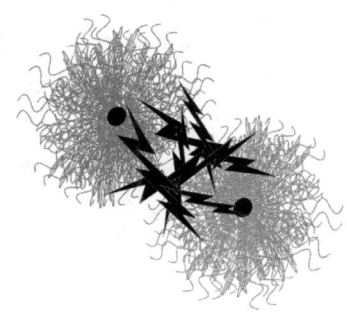

This can generate as much force as two 50 lb steel balls bouncing off each other.

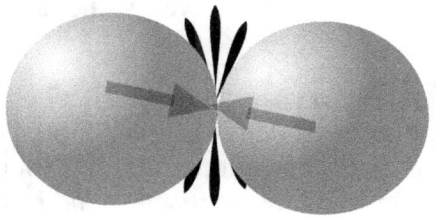

As we watch these clouds interact, we perceive they are hard or soft. This of course depends how the clouds react to each other. The difference between the littleness of photons and solidity of say a tree is somewhat like looking at a picture of the tree. A picture is made of small tiny dots. If we get a magnifying glass, we can see the dots.

As we look at the dots, we cannot see what the picture is. If we pull back, the dots shrink and become invisible but the picture they form comes into focus. The photons, then, are like the dots. To us they are invisible. However, when we look at the picture they produce, we see the universe around us.

Photon Collisions

The existence of particles depends upon photon collisions. Traditional theory purports that photons cannot collide. Understand that this book purports they can. Logically, photons cannot collide because they are points with essentially zero size. This will be dealt with later. In free space, collision would be very unlikely. Within a particle, the density of photons is so high, many collisions occur all of the time. In fact, the existence of a particle depends on many photon collisions.

The little star bursts in the middle of the following picture indicate where the collisions many be occurring.

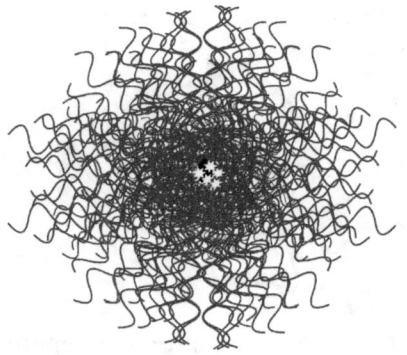

A later chapter presents more about the structure of photons and explains why photons can collide. In addition, photons only interact when colliding.

There is No Action at a Distance
There is no interaction between photons when moving through any space. Even when they come close, there is no interaction.

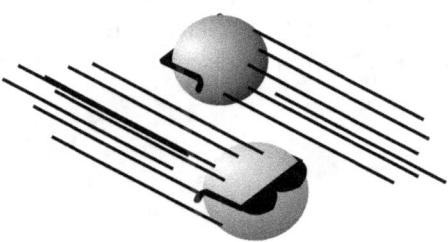

This is a very important point. No particle or photon exerts any kind of force, control, presence on any other particle or photon in the universe. Particles consist of many photons.

Collisions Rule

Particles only interact when constituent photons of one particle contact constituent photons of another particle.

Thus, the entire structure of the universe depends upon some
set of rules that photons obey when they collide. These are referred to as the Photon Rules of Interaction.

Particle Persistence

A first question could easily be, *How do these photons hang
together?* That is, how do groups of them exist and persist.
There are several possible aspects to this. Our next task is to
examine a couple possibilities.

Photon Collisions Cause Bounce Back

A cloud of photons consist of a large number of photons that
are bouncing off each other back and forth. The following

picture shows two photons that hit each other in the middle of such a cloud.

Then, as a photon travels away from the particle, it collides with a stray photon and bounces back to the particle where it can again bounce out of the photon.

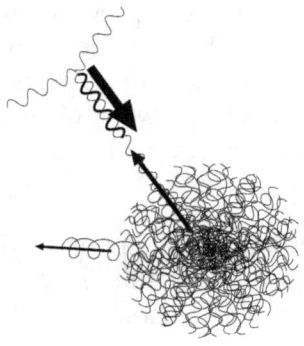

The concept is that once a photon collides inside a particle, the photon moves out of the particle and randomly collides with another photon that causes the outward moving photon to bounce back to the middle of the particle. The repetitive collisions chaotically result in the cloud of photons apparently hanging together.

Photons Return Automatically

Another possible device that could cause a photon that bounced out of a particle to return is an automatic return system. When a photon is kicked out of a particle, it might be programmed to stop and return to the particle that emitted it.

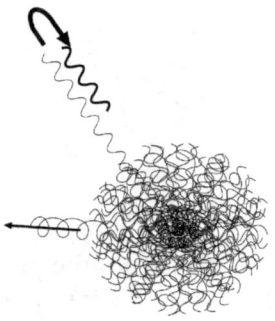

Later in this book, the concept of programmable photons is presented. In this concept, photons obey some algorithms that define their behavior while moving through free space. One set of rules govern collisions between photons. Another set of rules govern how a photon moves through space when it does not collide.

In the automatic return scenario, a photon will exit from a photon during its normal oscillation, in and out of the photon cloud. The internal data in the photon directs it to reverse direction after traveling some pre-defined distance. This enables photons to move away from the center of a particle then return a moment later. As many photons are doing this, the photons seem to hang together as a cloud of photons.

Photons Collide in a very Small Space

The photons associated with a particle tend to collide at a small point common to many photon paths. Such a scenario appears as in the following picture.

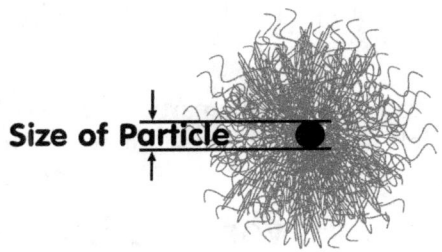

The photons here collide in a small space at the center of the cloud. There, the density would be very high supporting a high degree of collisions. The bounce back motions of the photons bring them back to this center. When we measure the size of the particle, we tend to look at where the particle is most dense. That would be the space of high collision.

The Magic Number

This theory depends upon random chance action. This theory purports there is some specific number or magic number of photons in a particle that enables a particle to exist. This number of photons produces a very large number of photon combinations that continue to bounce off each other. Should the number in the particle fall below this magic number, the particle would quickly absorb random photons around the particle to keep the magic number of photons constant. If the particle grows with photons over this magic number, the particle will vent or release photons to drop the photon count to the desired level.

Antimatter

You are aware there are two kinds of photon motion. A helix can be left-handed or right-handed. That is, a helix can rotate to the left or to the right. In a right hand helix, the thumb points in the direction of motion of the photon. The half-curled fingers point in the direction of the circular motion of the spiral. The left-handed motion is opposite. From this perspective, some photons are left handed and some are right handed. In the theory presented here, one particle can consist of all left-handed photons. Then, its antiparticle would consist of right-handed photons.

Part of this scenario is that some particles consist of some combination of left and right-handed photons. These particles have no charge.

Spin, Charge, Flavor, and So On

The varieties of particles that exist depend upon how many photons make up a given particle. The random motions of the photons manifest the particle properties that are observed in each combination. The photons in a particle are forever moving. They move in such a way that produces a particular spin or magnetic moment.

The Electron and Proton

Electrons and protons are combinations of some magic number of photons. The number for protons is considerably higher than that of electrons. The Photon Rules of Interaction provide for two levels of photon interaction. When the density of photons is very high, collisions of photons may be much higher or collisions of three photons at one time may occur.

All other particles have an unstable magic number of photons. The number of combinations that enable persistence depends upon the number of photons. Thus, particles of various sizes will experience different decay rates or half-lives.

Particle Motion

Next, we consider particles in motion. We stick to the theory that there is no interaction between photons as they fly through space. They only interact during actual collisions. The figures below show what happens when each collision on the right is balanced by collisions on the left. Likewise, all motion after a collision that moves to the left is balanced by motion after a collision that moves to the right. The aggregate motion of the entire particle is then zero. The particle is not moving relative to our reference frame.

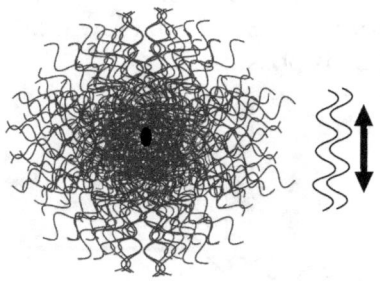

The squiggle image in the right of the picture suggests the particles motion is up and down without contributing motion to the entire particle.

The following picture suggests what might happen when each collision of photons in the particle consistently angles to the left.

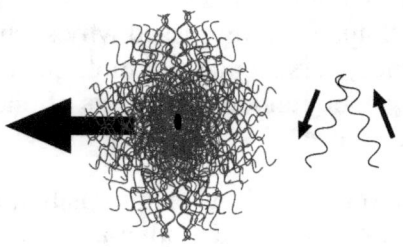

The two waves in the right of the picture depict a collision that causes the particle to move to the left.

A Position in Space

During presentations of thought experiments earlier in this book, a position in space was not defined. This was necessary because in the floating universe proposed, there was no reference point to, *Hang one's hat on.* Now that we have introduced particles, this is possible.

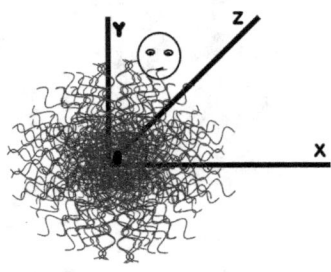

An observer can find a particle that is at rest relative to his point of observation and call the reference frame of that particle the observer's position in space.

Conclusion

In this chapter, the idea has been presented that particles in the real world consist of clusters of photons whirling about each other. This chapter has discussed why such particles are capable of existence, persistence and motion. A major issue is that it all depends upon random collisions of photons. There is no action at a distance.

The primary idea to grasp in this chapter is that particles consist of photons bouncing off each other around a common location in space. The study of how photons cluster and stick together as a group is the subject called subquantum physics.

Chapter 25

Introduction to the Background Sea

In the previous chapter, particles were introduced as clouds of photons and the concept of particle persistence was introduced. This suggests that a magic number of photons can randomly bounce off each other and maintain the structure of a particle. As the electron is a whirling cloud of photons held together by chance, photons will randomly escape the cloud.

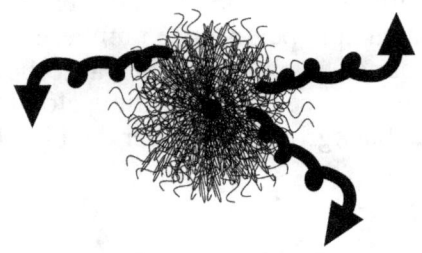

The phenomena of the right number of photons randomly maintaining a particle will keep most photons together as a group. If the particle is to continue to persist, there must be a way to replace those that escape. There are a large number of photons floating free throughout the universe. This is called the background sea of photons.

The Background Sea of Photons

The background sea of photons is constantly supplying extra electrons to photon clouds. The background sea is what clas-

sical scientists call the background electromagnetic radiation
left over from the big bang. In the theory presented in this
book, this sea is a source of photons that replace those that
randomly escape from particles. Thus, this background sea is
necessary for the existence of particles. The following picture
depicts many photons moving randomly in a variety of directions. In general, they hit nothing but just continue on flying.

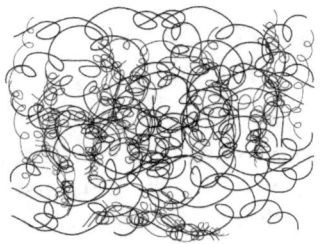

However, the photon particle clouds are immersed in this sea.
The photons escaping a cloud tend to fly into the background
sea and wander off forever. Then the cloud needs an extra
photon to keep its magic persistence number intact. A photon
from the background sea will simply be absorbed. The following shows photons escaping a particle cloud while other
photons from the background sea enter the cloud.

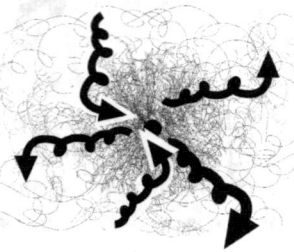

There is a balance between the background sea surrounding a
particle and the number of photons inside a particle. To support the existence and persistence of a particle, there is some
appropriate ratio of the densities of the background sea and
the density in the particle. In essence, the density of the back-

ground sea determines the magic probability number of a particle.

Notice that electrons and protons do not depend upon any set of photons for their existence. Randomly, an electron constantly allows photons to escape its grasp and is constantly accepting photons from the background sea of photons. Eventually, all photons in an electron and a proton will change but the identity of the particles remains the same.

Again, an important issue here is to realize that any force acting overall does not hold such a particle together. The photons in the cloud only react with direct collisions with other photons in the cloud or photons in the background sea of photons. These collisions are controlled by the Photon Rules of Interaction.

Spontaneous Appearance of Matter

While collisions of photons outside of particles are very rare, they do randomly collide. The collisions can randomly become ordered and manifest themselves as short lived particles. The following picture depicts two photons in the background photon sea randomly running into each other. The center of the black circle in the picture indicates where the collision occurs.

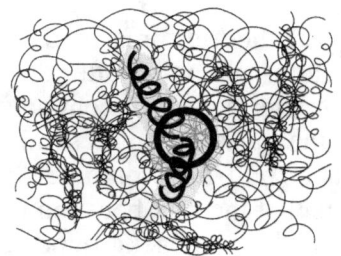

Then the two photons can bounce off each other for a time,

which is very short because the other properties needed for persistence are not present.

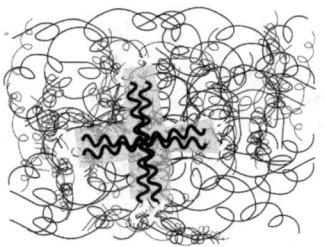

This is the source of the spontaneous appearance of matter. It is referred to as the appearance of virtual particles. This event occurs everywhere in the universe.[1]

More than likely, there are many more than two photons cooperating in the sudden ordering of motion. When studying the information available about the appearance of virtual particles, the information suggests that two particles are created when this event occurs. One is normal matter and the other is anti-matter. [2] The small particles attract each other and quickly annihilate each other leaving behind a burst of energy in the form of photons. This suggests that the background sea of photons is an even mix of right handed and left handed photons.

Within the scope of the theories presented here, the production of normal matter and anti-matter during the spontaneous creation of matter does not seem necessary. In this scope, any kind of small particle could suddenly appear and disappear quickly as its half-life would be small. However, perhaps this suggests that the mix between left and right-handed photons is equal in the background sea of photons. If a large number of left-handed photons suddenly group together, perhaps that leaves a large number of right-handed photons grouped that enter a dance of existence.

The conclusion here is that the appearance and disappearance of virtual particles consists of some number of photons bouncing off each other then returning to their chaotic roaming and appearing as free photons floating away from the incident.

The Dirac Negative Electron Sea

In 1928 Paul Dirac proposed that the universe is filled with electrons that contained negative energy. [3] He further proposed that we could not observe these negative energy electrons. In this theory, if one of these electrons gained enough energy it would become visible to us as an electron. The electron would leave a hole in the sea of negative energy electrons. This hole would appear as a positive electron or antimatter electron, a positron. This would explain the process of how an electron and positron could appear out of nothing.

This author has no idea what negative energy implies.

This was mentioned earlier in this book. Apparently, this process was observed near very heavy nuclei. If a photon with sufficient energy passed by a heavy nucleus, a positron and electron would spontaneously appear. [4]

Here is an explanation of this from the perspective the theory presented in this chapter. When a very high-energy photon passes near a heavy nucleus, the density of the background sea near the vicinity of the nuclei provides enough photon density to allow photons to cluster into particles, in this case, two electrons.

Perhaps Dirac's observation could be superimposed on the background sea. Perhaps the random action of the photons in the background sea could be visualized as a zero energy level of space. When a group of photons gather and move in an ordered way, that group could be regarded as an elevation of the energy of that group.

Conclusion

This chapter complements the previous chapter. In the previous chapter, we introduced the ordered motion of photons in the universe. In this chapter, we introduce the non-ordered motions of photons in the universe. Particles and the background sea of photons represent the two faces of energy in the universe, order and non-order.

Chapter 26

Quantum Weirdness Explained

At this point in the book, we have laid out an entirely different paradigm of universal structure. Usually when some new theory is presented, it is presented with the idea it resolves questions posed by some earlier theory. Or, perhaps, clarifies some factor a previous theory left out. For example, Niels Bohr, with the help of Rydberg equation and the Pauli Exclusion Principle, explained the spectral lines of hydrogen. However, that explanation could not handle atoms with more than one electron. The de Broglie matter wave function came along, extended that theory to any number of electrons, and was used by Schrödinger in his wave equation. This was a significant advancement in quantum theory.

Well, we have a new theory. The goal in this chapter is to review the unexplained or unusual factors of present day quantum theory and see if we can reassess what they mean in view of the new paradigms presented.

Heisenberg Uncertainty Principle

As a particle consists of a number of photons bouncing off each other, the center of such a particle would be the average position of all of the photons. Since all the photons are moving, that average position would be expected to change. The following picture illustrates this. The black dots in the center of the cloud of photons represent the average position of all the photons in the particle at different times.

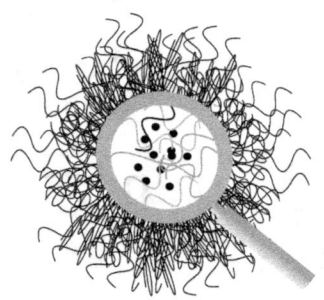

Why is this important? Proponents of hard ball physics like to assume the motion of something is the motion of its center. From that point of view, such a particle as shown would appear to be moving backward and forward while the whole is at rest or moving consistently in one direction. In addition, this would give rise to the uncertainty principle. If such a particle were moving through space, the center would appear to wobble. If such a particle were considered to be a solid ball, the entire ball would appear to move about with uncertainty.

This suggests an interesting point. When de Broglie introduced the concept of matter waves, the resulting wave equation was interpreted to be the probability of where the particle would be during its motion through space. Max Born proposed this. This hardball approach was that he particle moved in a straight line. Because the center of the particle is not the center of a hardball but the center of a number of photons constantly oscillating, the center of mass of the particle actually followed that wiggly line as it moved through space. The following picture actually represents the motion of the center of mass of atomic sized particles. It does not represent the potential of where the center of mass could be. It is where the center of mass is.

Of course, you can see why claiming the above wave is a probability curve would work. The math resembles the real motion of a real particle.

Consider the following picture that depicts the center of a particle moving through space. This is a picture of where the center of the particle should be at three different points in time.

From the foregoing, we understand that the center might be in front of those positions or behind them. The following picture depicts this. That is, because of the motion of all the photons in the particle, the center or average position of the particle may be in front of or behind the expected center of points.

This picture attempts to show a variety of forward and backward positions.

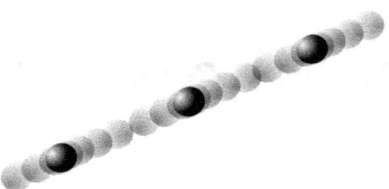

Of course, the center of particles may move up and down.

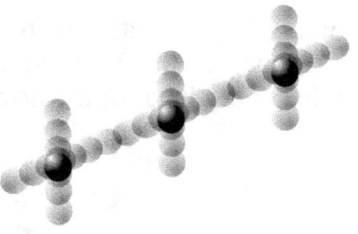

Moreover, the center might move side to side.

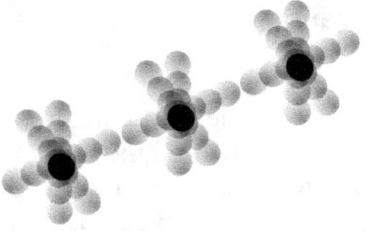

If we draw a line representing the probability of where the center might be, the lines would appear something as follows.

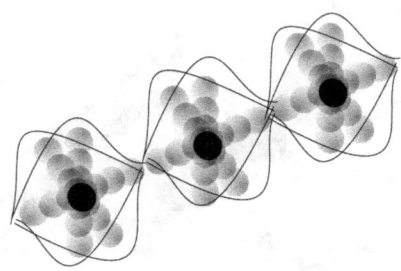

The point here is that this appears much like a wave function for the linear motion of a particle. With the understanding that a particle is a cloud of moving photons, we can understand that if we assume the particle is a hardball, the center of which would normally move in a straight line, we would be amazed to see the center at odd locations on its travel. With this inaccurate assumption, we would see uncertainty.

Particles as Waves

In the chapter titled "Matter Waves," we were introduced to de Broglie matter waves. Now that we have a different idea of particles, we need to revisit the idea particles display wave phenomena just as light. Let us review what we have seen a-bout particle waves.

The first important issue is single slit diffraction. If particles are aimed at some kind of barrier that consists of a single slit and they move at an appropriate speed, particles such as elec-

trons will pass through the slit and form a diffraction pattern
just as light does.

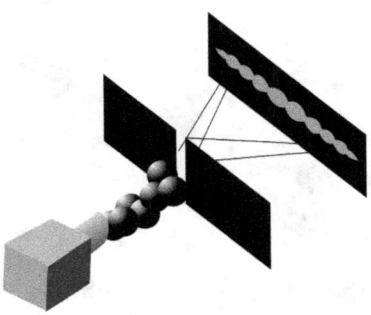

This experiment cannot be done at a scale we are accustomed
to. The slit barrier is not something we can make with a file
and a pair of tin snips. The barrier is some kind of natural
crystal that is thin and made with some complicated process.
The electron projector is also some kind of high tech device
capable of emitting electrons with the right speed and the
right direction. In addition, the target barrier is unusual and
the technology for taking pictures of the resulting diffraction
pattern is high tech. However, we can demonstrate that electrons can be diffracted just as photons. [1]

In a similar fashion, electrons can be subjected to double slit
experiments that display the double slit diffraction pattern.

This lends even more weight to the idea that electrons are waves.

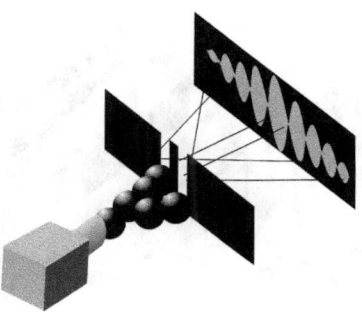

The weirdness continues. Light experiments demonstrated that single photons going through a double slit experiment continue to generate an interference pattern. The same occurs with electrons as well. Thus, the weirdness of electrons and particles keeps step with the weirdness of photons. The following depicts this. In the following picture, electrons shot at the slits must pass through a gate. The gate allows only one electron through the slits at a time. Even then, as time passes, the electron hits form the traditional interference pattern.

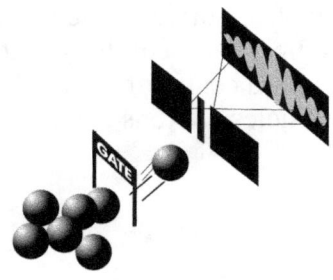

Then, to make the weirdness even weirder, if one slit is taped shut, the diffraction pattern appears. If the slit is opened, the interference pattern reappears.

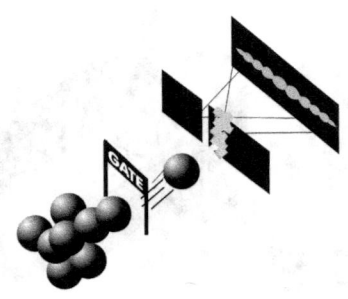

The goal here is to explain this weirdness with this new idea of how photons interact to form particles and how those particles react in the real world.

Single Slit

Let us take up the single slit experiment with this new kind of particle concept. Recollect what a particle looks like when it is moving. In the picture below, you can see two circles that represent the center of collisions in the particle. The one on the left is where the particle begins its most recent motion. The photons emanate from there, move out and return to another point to the right. This accomplishes the motion of this particle. There is no hard center of the particle. The center is simply where the particles photons constantly collide. The repetitive collisions and the angle of the photons to the right enable the motion of the particle.

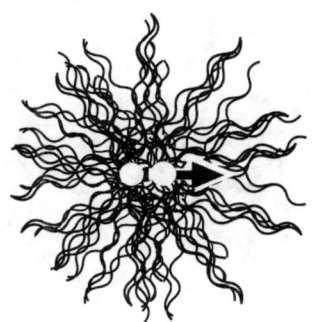

When a number of like particles move through a large slot, they pass with little incident. When they hit some barrier to the right, they will make a pattern similar to the slit they just passed through.

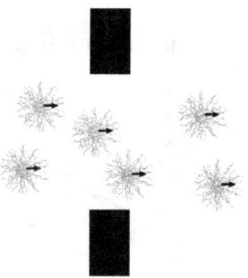

When the slit becomes smaller, the particles can no longer pass through the slit without event. Particles that strike the barrier straight on will bounce back and not pass through the slit. Those particles, however, that come close to the edge of the slit will interact with the edge of the slit. The circle in the following picture represents the area of particle collisions.

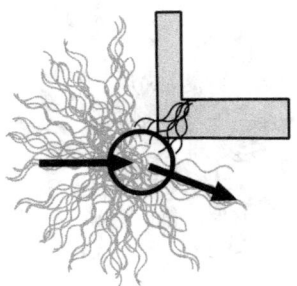

The photons not only encounter each other in the circle but also encounter photons emanating from the barrier. The particle reacts to this by bouncing off a bit and changing its direction of motion

The result is that the particles move through the slit at a variety of angles.

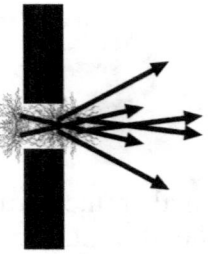

The next factor in this scenario is to recollect from the chapter titled "Introduction to Particles," the center of a particle is wobbling. We then get a situation that is very similar to single slit experiments for light. As a particle enters the slit, it also hits the edge of the slit. The reaction between the edge of the slit and the particle causes it to move at an angle through the slit. The particle continues moving through the slit wobbling as it moves.

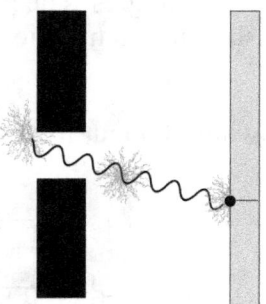

Eventually, the particle hits a barrier, the bar on the right. Normally, the center of the particle is considered the location of the particle. Thus, the dot represents where the particle hit the barrier. In these pictures, a horizontal line in the right bar indicates where the center of the particle hit. The picture below illustrates that several particles coming through the slit will hit the barrier at locations that are a varying distance apart.

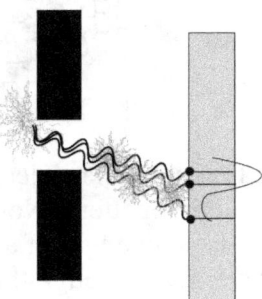

The curved line represents where collections of particles hit the barrier. The more the curve extends to the right, the more particles hit at a corresponding location. Bear in mind that the combination of the oscillating center and the particles hitting the corner of the slit; causes the particles to pass through the slit in the same phase. This is similar to light wave polarization. The particle matter waves are in phase and move toward the barrier at various angles. The pattern the particles make on the barrier is similar to a traditional diffraction pattern.

The following is a drawing of about forty matter waves passing through a single slit.

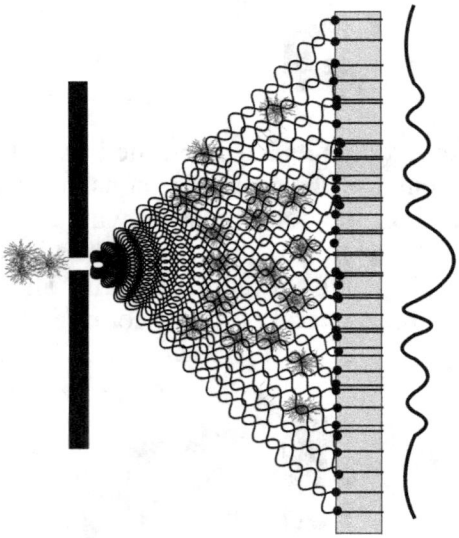

This is to demonstrate that many particles flowing through a slit will generate a diffraction pattern. Note that each wave flowing through the slit is at an angle of about five degrees from a neighboring wave. Even though the waves are equally spread, the change in distance traveled causes the middle of the particle to strike the right bar at a variety of locations. The horizontal lines mark those hits. You can see that the hits cluster in a pattern to produce the traditional diffraction pattern.

Double Slit

The double slit experiment is two single slit experiments done side by side. Because of this, the patterns made by each slit are out of phase just a bit. Thus, the two patterns superimpose to create a pattern that appears as a traditional interference pattern.

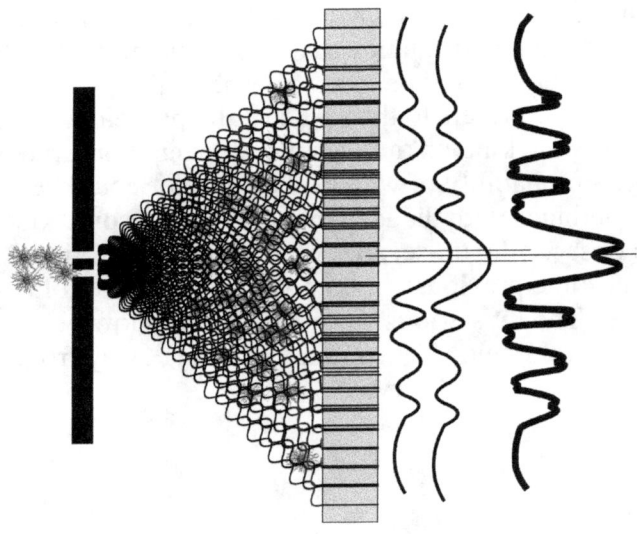

Each of the thin curved lines on the right of the picture represents a diffraction pattern coming out of each slit. They are superimposed to make the interference pattern shown as the thicker curved black line. A problem with traditional interference experiments is that when electrons are allowed to travel through the slits singly, the particle still hits the barrier and forms the traditional interference pattern. This is interpreted as unusual since there is only one particle. The question is asked how the electrons can interfere to produce the pattern. In the scenario presented, this is not a problem for there is no particle interference. The particle moves through

the slits and generates the pattern without interacting with
other particles.

What is a Collapsing Wave Function?

In the chapter titled "Quantum Weirdness," wave collapse
was presented. There, it was a function of wave spaces. The
Super Matrix theory proposes a different explanation.

To refresh your memory of what was presented, a wave function is a mathematical equation that describes the probability
of where a particle might be in space. The term *the collapse of
the wave function* refers to the idea that the particle while not
observed is some kind unknown stuff we cannot observe but
represents a probability of where "the particle" could be. The
concept continues with the idea that humans do some kind of
action to observe the un-seeable wave function and find a
hardball at some previously unknown location. The following
depicts this. The mysterious wave function is shown in the
left in the picture below. The human is facing away from the
function so he cannot observe it.

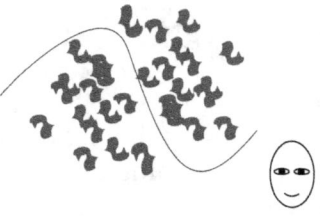

When we do something to observe the object, the wave function "collapses" and we observe a hard ball at a specific location.

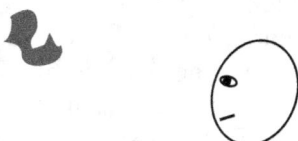

Because the hardball was probabilistically spread over the space of the wave function, it collapses at some unknown location, suddenly known, when we observe it.

This book proposes a different scenario. The wave function here is a mathematical equation that describes the motion of a cloud of photons. There is no hard ball. Hardball physics views a hard ball as being at the center of the mass of the particle. However, because a particle consists of many photons randomly whirling about each other, the center of mass or center of photons is constantly moving about in an unpredictable way. When one observes a particle, the task requires that the positions of all photons be averaged. The average position of all photons is the center and that is where the particle, as a hardball, is perceived to be. Observation is actually a mathematical act to locate something observers expect to find. Nothing collapses nor expands. For the particle, life just continues as normal.

How does the process of observation collapse the wave field or appear to? From the perspective of the Super Matrix theory, a particle is a cloud of flying photons. The, so-called, process of observation is the simple task of finding the center of the cloud. That would be the average position of each constituent photon. How is this accomplished? Consider a beam of electrons moving through an evacuated glass tube. The experimenter may measure the curvature of such a beam. In so doing, he selects a line that he perceives represents the middle of the beam. As such, he is "observing" the electrons. This observation consists of finding the average or center of the motion of the electrons in the beam. Consider electrons or some other particle slamming into a plate that causes a flash of light. The experimenter normally considers the center of the splash of light the spot the particle hit. That is, he finds the average of the light intensity and assumes that is the center of the particle interaction. Thus, observation consists of the estimation of the center of some event. Nothing collapses, whatever it is, it just keeps undulating.

The collapse is nothing more than the human deciding the particle must be the middle of everything. The human finds the middle of a particle that is changing all of the time.

Entanglement

Entanglement was presented in the chapter on "Quantum Weirdness." Let's review entanglement briefly. Recollect it is observed when some subatomic particle decays. In some of these events, two particles may be emitted. Such an event may yield two particles with opposite spin. In addition, they may be emitted so they move quickly away from each other. During their travel, they will be in some quantum state or states. The following depicts this event. The top figure shows a particle before the event. The next line shows the particle during the event. The bottom figure shows just after the event occurred.

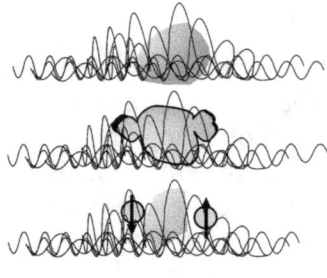

Thus, the particles are entangled and move far apart.

When far apart, one can measure some property of one of the entangled particles. According to this theory, the wave function will collapse at the time of that measurement.[2]

The idea is that when one of the particles is measured, the wave function collapses. The other particle will become real as well as its wave function has collapsed. Conservation laws require that one be spin up, the other be spin down when they exist together. The assumption is that they are up and down when this wave function collapses. Thus, the spin of the particle that was not measured will be discovered, as it must be different from the one measured. As the two particles can be far apart, the speed of communication of this spin state is apparently communicated instantaneously or at least faster than the speed of light.

The goal of this chapter is to consider the weirdness of present day quantum physics and find a bridge to the new paradigm presented in this book. This isn't attempted with Bell's Theorem or entanglement. The reason is that these concepts are based on Heisenberg's Uncertainty and the concept of a wave function being some unknown thing. Uncertainty has been shown to be a simple misunderstanding of what is actually going on. Likewise with wave functions that appear to collapse. Entanglement was based on something not understood. Thus, we will not consider a bridge of it to this new age paradigm.

Conclusion

The goal in this chapter has been to go over the common weirdness of present day quantum physics and show that they are logical consequences of the Super Matrix theory. Little math has been used and simple pictures have sufficed to present these explanations.

The old clothes have been dropped to the wayside and we can continue forward with new ones to see more about this new structure of the universe.

Chapter 27

Introduction to the Super Matrix

The first paradigm shift of interest here was the discovery that particles displayed wave phenomena just as light. We can thank de Broglie for that. The second paradigm shift was that photons are points that move in a helical pattern. The third was that the physical universe consists of a sea of photons. Particles, then, consist of clouds of photons whirling about each other. Could there be a bigger paradigm shift? In this chapter, we pursue the idea that photons are not part of the physical universe.

In this idea, the photon is shown to be a link between the physical universe and some other universe that spawned the physical universe. The physical universe possesses space, time, mass and energy. The other universe does not have these properties but spawned photons that support our universe. One may think of this idea as a bootstrap theory of the universe. Generally, a bootstrap theory refers to any method of reaching higher understanding by building on lower foundations. [1] The concept of the Super Matrix, the photon, the Photon Rules of Transmission, and the Photon Rules of Interaction; combine together to from a bootstrap theory. These four factors are a set of rules or conditions from which the existence of the universe unfolds.

There is another factor that some may find interesting. In an earlier chapter, comments were made about the universe being Euclidian or non-Euclidian. There, the idea was that we live in a universe that has superimposed a non-Euclidian space onto a Euclidian space. While this may be a mathematical problem, the universe is real. Part of this chapter is about how this superposition has taken place in the physical world.

In a Nutshell

Photons are artifacts from some non-physical universe. Photons have no substance but are a collection of data that is passed from one node of subspace to another node of subspace called the Super Matrix. This structure creates time and creates distance. Physical matter in our physical universe is manifested when photons cluster together as particles. The particles appear as matter. We do not observe photons. We observe the effects of photons on particles.

This is covering a lot of territory quickly. It bears repeating. The first point is that the physical universe was spawned from or is supported by some other universe. This theory attempts to answer the question, *What is matter?* In essence, the answer presented here is that matter is made of something other than space, time, mass, or energy. Since this is what our physical universe consists of, this something must be of some other universe. This statement appears dubious or silly. However, if pursued, it logically unfolds to what we see around us today. Somehow, this other universe made a three dimensional array of nodes. A photon is a bit of data that is passed from one node to another. The photon does not really exist as something physical. This book refers to this three dimensional array of nodes as subspace. A more interesting name is Super Matrix.

This chapter describes subspace/Super Matrix and further defines subquantum physics. Subquantum physics is about the behavior of photons in the Super Matrix.

Subspace and the Super Matrix

We need to describe the Super Matrix. What it consists of is somewhat beyond this book. We can only discuss how it appears to us in the real world. To make progress, an analogy is presented utilizing tiny computers to show how it might appear if we could see into the other side of reality. In this analogy, the Super Matrix consists of a three dimensional array of tiny computers.

It may appear as follows.

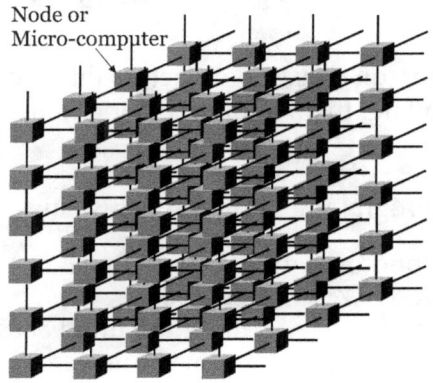

Node or Micro-computer

This Super Matrix is large and the physical universe exists within this three dimensional matrix. Each node is capable of processing a small amount of data. In this scenario, a photon is a set of data. Each data set or photon passes from node to node.

The following shows a photon passing downward in the three dimensional Super Matrix.

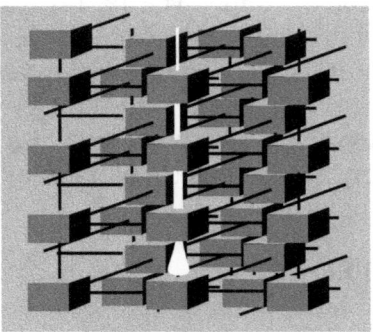

The data in a photon consists of vectors that indicate the direction it is moving, the photon's position in the Super Matrix, and so on.

Here is how a programmer might see this data

```
//photon data
int X,Y,Z;      //position
int x,y,z;      //direction unit vector
int l,m,n;      //radius unit vector
int r;          //radius
int i,j,k;      //direction accumulator
int d,e,f;      //motion unit vector
int counterA //collision counter
int counterB //jump counter
```

A photon moves when one node computer passes this data to another node computer. To be precise, one node computer copies the data from another node computer. The sending node computer then clears its memory. The receiving computer processes the information, selects one of six node computers (up, down, left, right, front, or back) to send the data on to. Then, the data is passed.

Each node has a bit of software or a program that processes the data. There are two parts to this program. One part is called the Photon Rules of Transmission. The other part is called the Photon Rules of Interaction. When a node gets a single data photon it executes the Photon Rules of Transmission. The node updates some information in the data set and determines which node the data should be passed to next. When a node gets two or more data photons, it executes the Photon Rules of Interaction. Again, the node updates some information in the data sets and determines where the data sets should be routed to next. In essence, these two subprograms describe how the universe functions.

Photon Rules of Transmission

These rules apply when the photon is moving through space without collision. Each photon in the Super Matrix must move into another node with all other photons. This creates time and is called a tick. During this transition, the photon must advance along a specific direction. The motion from one node to another represents a distance and is called a click. The photon must also advance some appropriate distance around the centerline of the helix. The rules define the math required to accomplish these two things. Thus, when the photon moves through the Super Matrix it traces out a helix. However, because the photon is moving through a rectangular array, it will not follow the helix pattern accurately. Therefore, the math must keep track of how much the position is in error so that for many transitions, the path will be corrected to keep the helix moving forward as accurately as possible.

Photon Rules of Interaction

When two or more photons attempt to enter a common node, the Photon Rules of Interaction are applied. These rules determine how the photons reflect off each other. Part of this calculation is to insure this collision reflects the motion of a particle as a whole. There is no communication between photons, other than that which occurs at collision time. The math must be consistent among all collisions so randomly the right result is produced. A factor of all this is that the photons are moving at the speed of light in a very small space. That is, they are clustering about in an area the size of an electron. Thus, many, many collisions occur and many collisions will not produce the appropriate result. However, the assumption is that due to the large number of collisions, the photons will share data and the particle as a whole will behave in a very logical manner. Although photon data sets move about somewhat randomly, the sheer number of motions support Conservation of Momentum, Mass, Energy and so on.

Conclusion

This chapter essentially introduces a bootstrap theory of the physical universe. This theory proposes that photons are bits of data that move from one node of a three dimensional array to another. Each node contains some kind of algorithm or program that determines which other node the data should be transferred to. These three factors are very small and concise. When these factors are implemented with a three dimensional array of an apparent infinite number of nodes, the physical universe in all of its complexity appears.

Part 8

Mass, Energy, Space, and Time

The purpose of this part is to identify with accuracy what mass, energy, space, and time are. This is to be done within the framework of the Super Matrix that was described in the previous chapter.

We will begin this task by considering the structure of particles and coming up with a tool to study a single photon particle within a larger particle. We will build an imaginary device called a particlescope. It will allow us to see inside a particle.

A goal is to describe each property of the universe: mass, energy, space, and time. Each is to be described with great accuracy. To demonstrate the practicality of this study, several of Einstein's theories will be developed from these primitive descriptions. In addition, a very physical description of kinetic energy and potential energy will be introduced and a description of precisely why the two are equivalent.

Right now, we do not have a tool that will peek inside the structure of particles to see how they tick. Developing one is our first task.

Chapter 28

Particlescope

This chapter is dedicated to the development of a device to peek inside particles to understand what makes them tick and click. It analyzes what is happening. Analyze means to cut apart and study. Unfortunately, we cannot simply catch a particle floating by and cut it apart to see how it ticks and clicks. First, we must come up with some method to isolate that which we wish to study. This method uses technology similar to the superposition of waves. Thus, we study that subject some more.

Fourier Series Analysis

When this author was taking a math class in high school, the teacher of that class spoke of the ability to make an equation of any curve conceivable. He spoke of an experiment performed when he was in college. He and his fellow students got a young lady to stand in front of a large sheet of white paper. A light was aimed at the woman so her shadow was cast against the paper.

The students traced her form on the paper getting a line that represented her torso and hips. Then the experiment consisted of determining an equation of that line. The teacher pointed out that the real fun part of the experiment was over.

Fourier Series Analysis was used. We need to talk about this kind of analysis so we can present the tool that will be used in a bit. The theory of Fourier Series Analysis is that any curve can be represented by the summation of some number of waves of varying amplitudes and frequencies. Consider a curve that might be from the above experiment.

The wave could be reproduced by adding the following sine
waves and many more like them.

The sum may appear as follows.

Realize that the curves summed need not be sine waves. They
could be square waves as well. The following illustrates a
number of square waves that might make up such a curve.

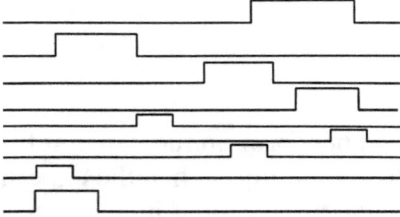

These square waves and a number of additional waves like
them could be added together to produce the same curve.

This introduces the primary concept that is being used in the
following presentations. That is, a combination of simple
structures can be combined in some appropriate sizes to produce some complex structure.

The process of examining some curve and finding a number

of waves that would result in such a curve is called Fourier
Series Analysis.

Applying Superposition to Our Problem

A particle has been shown to be a large number of photons
whirling about. A particle consists of many photon points
moving in a helical pattern. It may appear as follows.

Some number of equivalent particles that move in a square
pattern could represent this cloud. Here, the term equivalent
particle is as one of the square waves depicted in Fourier
Series Analysis that when added to other square waves became the curve under study. An equivalent particle consisting
of an equivalent photon would appear as follows.

Like the square waves that were added together to make the

curve of the woman above, many of these equivalent particles together may appear as follows

If enough of these square patterns of motion of various sizes were put together, we could come up with a group that appears as a cluster of helical structures.

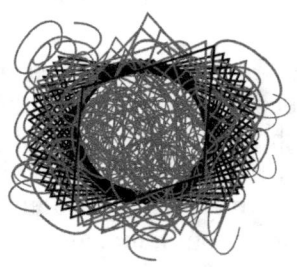

This being true, we can analyze one of these square patterns and generalize to all the constituent square patterns. That is, find out how one works and apply the same principles to all the equivalent particles in the group. The advantage of this is that, as the pattern is a square, the math to analyze it is much simpler than working with a cloud of helical patterns.

This is the device used to study the structure of particles.

The Equivalent Particle

Now, let us look at this equivalent particle to see how we will use it. The first observation is to realize that this square motion represents the smallest possible particle. It, essentially, consists of a single photon moving in a cloud of one. It demonstrates the two properties we have discussed earlier: persistence and motion.

Compare it to a motionless particle we discussed earlier.

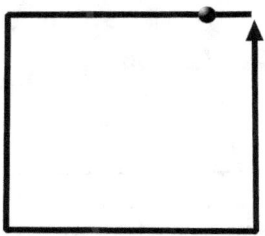

The photon constantly loops so it creates a particle hanging in space. The question is, how does the moving particle exist? Straight-line motion is understandable. However, how does the photon turn on those sharp corners? We assume that this particle is one within a large number of other particles. Photons in the other particles just happen to be in the right place at the right time for our test photon to bounce and make the sharp angles. This enables the photon to move in a repetitive pattern enabling it to exist.

Another factor is that, even though all the parts are moving, the motions in all directions add up to zero. Therefore, we see what appears to be a stationary object. Then realize that the point moving is a photon. Therefore, it is moving at the speed of light. One trip around the square is called a Cycle of Existence.

The next important aspect we wish to observe is the motion of the particle. What happens when a particle is moving? We have seen this in the chapter titled, "Introduction to Sub-quantum Physics." Here, however, there is only one equival-

ent particle to consider. The picture below depicts single photon motion that enables the particle to move.

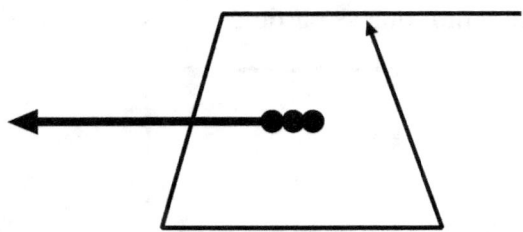

The black dots in the middle of the picture represent the middle of the moving particle. The large arrow pointing to the left
represents the velocity of the moving square pattern. As we
are only observing a single equivalent particle in this unit particle, we can see exactly how such a photon must move to enable the particle as a whole to move.

A Close Look at the Cycle of Existence
To understand the logic in this chapter, one must understand
how the photon particle changes during motion. Here, we
take a very close look at this.

First, consider what our goal is. The following picture shows
two identical Cycles of Existence. The particle on the left is
moving with some velocity to the left. The one on the right is
not moving.

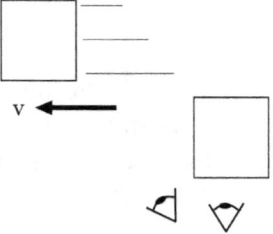

As each box or cycle consists of a photon moving in that box pattern, we can see exactly what the photon does to accomplish the motion of a particle as a whole. The eye symbols at the bottom of the picture represent points of observation.

The following depicts the motion of a photon in a moving particle with some detail. Note that both cycles are shown in the same frame of reference.

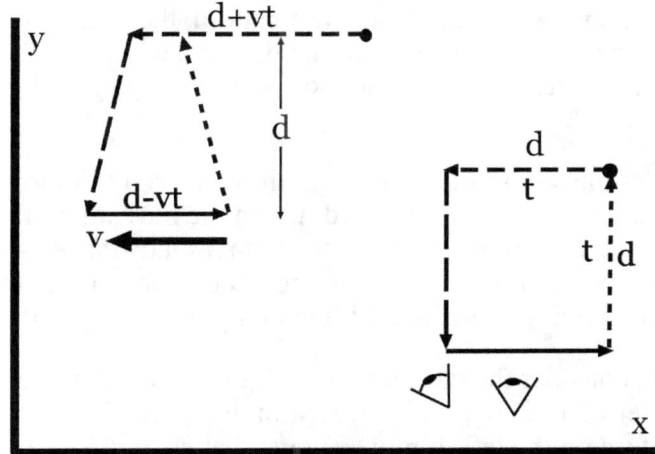

The pattern on the right depicts the particle not moving. The pattern on the left depicts the particle moving with some velocity v to the left. If both patterns were not moving, they would be identical. Note that the patterns depend on four basic motions. We can isolate these basic motions to get a complete idea of what is happening.

Consider the horizontal line made of medium sized dashes in both patterns. The dot in each represents the starting point of the photon. It moves to the left as depicted by the medium dashed line. It moves a distance d. That line is longer in the pattern on the left. This is to support the motion of the particle to the left. It is longer by vt or velocity times distance. Note that time t is the time the photon travels along one edge in the non-moving box.

Consider the line made of long dashes in both patterns. In the right pattern, the motion is straight down the distance d. In the pattern on the left, the motion is down and a slight angle to the left. This supports the motion of the particle as a whole to the left. Note that the long dashed line travels the distance d straight down. However, in the left pattern, the long dashed line is longer than d. The additional length is due to the motion of the particle. This is the most important part of this analysis.

Consider the solid line. In the pattern on the right, it moves a distance d to the right. The solid line in the moving particle moves to the right also. However, it moves a distance d-vt. This enables this phase of motion to enable the particle to continue moving to the left with velocity v.

Finally, consider the dashed line with the smallest dashes. In the pattern on the right, it moves straight up the distance d. In the pattern on the left, it moves upward at an angle. This serves the same purpose as the long dashed line. The line made with the shortest dashes moves straight up the distance d. However, it is longer than d as it angles to the left and supports another part of the motion of the particle to the left with velocity v.

Again, realize that both patterns are, as they would appear in the reference frame at rest relative to the page. The left pattern of course appears very unusual.

A critical point is that when the left pattern is observed from a reference frame moving with it, the pattern appears very normal. The following picture depicts this. It shows a reference frame around the left pattern that moves with it.

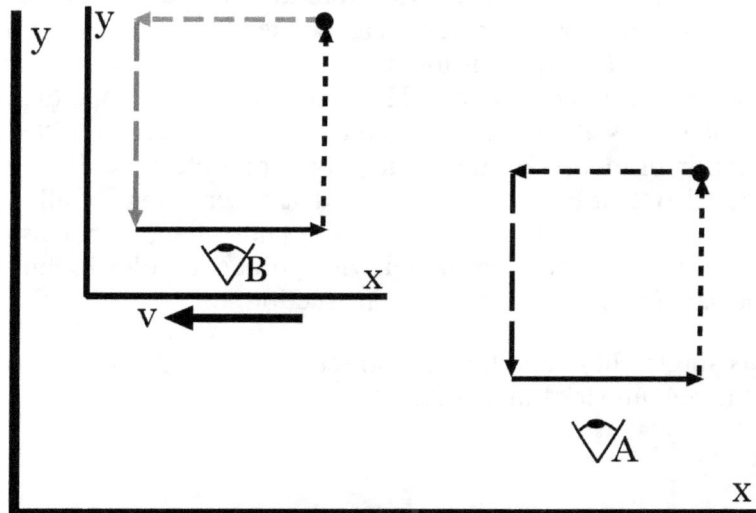

In the moving reference frame there is an observer B moving with that reference frame that observes the particle pattern as normal. Observer B sees his pattern exactly as observer A sees his pattern.

This scheme is derived from Einstein's first postulate of special relativity. That is, observations made in one reference frame will appear identical to the observations in other reference frames moving at constant velocities. Thus, the effort here is to be sure what A and B sees in their respective reference frame is the same. This also means what they see in the others reference frame is the same. So, when B looks into A's reference frame, he will see the same as A looking into his reference frame.

Conclusion

The primary factor to take away from this chapter is that the motion of photons in a particle can be viewed as some combination of many photons moving trough space as two-dimensional square patterns. The secondary issue is that the photon moves at an angle when moving perpendicular to the direction of motion of the single photon particle as a whole. Note, also, that Einstein's postulate that observations of all observers in the universe must see the same phenomenon as all other observers do from their viewpoint still holds in this unusual viewpoint of particle construction.

This single photon particle construct will be used extensively in the remainder of this book.

Chapter 29

Space and Time

This chapter continues with a discussion of space and time. The previous chapter established a tool we need for this task. Precise definitions will be offered for both of these. Then, the definitions will be used to develop elements of Special Relativity. These will include time dilation and Lorentz-FitzGerald contraction.

During this process, we will describe mintums. For time and distance, mintums are the smallest possible values for each. One would think the word quantum would be appropriate. That word is reserved for that defined by Planck's constant. Quantum, in that sense, is not the smallest amount of something but some unit of energy within particles to represent changes in energy levels. For distance and time, the term mintum is used to denote the smallest possible value.

This chapter attempts to show the underlying structure of space and time and clarify what they mean. The preceding chapters have purported to explain the structure of our environment. If those explanations are accurate then we should be able to define what time and distance is.

Time
As mentioned earlier, there has been no explanation of what time is. Standards are used to compare against some unknown time to measure it. Here we define time as change in position of a photon as it travels from one node to another node in the

Super Matrix. The white comet in the following picture represents a photon jumping from one node to another.

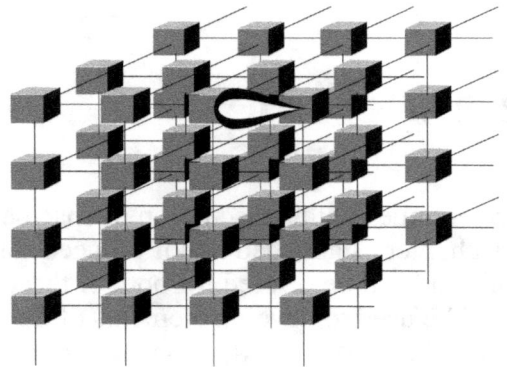

This would be a mintum of time. That is, there is no smaller time in existence. It is called, in this book, one tick. All ticks are defined as equal to each other.

Space
Space consists of the three dimensions of distance. These are up and down, left and right, and forward and backward. Other than direction, each of these exhibits the same properties.

Like time, distance has no clear definition. Applications that require it use operational definitions that specify the units of measurement that quantify distance. In the philosophy introduced here, the unit of distance consists of the photon moving from one node to another in the Super Matrix. An assumption here is that the Super Matrix is homogeneous. That is, the distance from node to node is the same throughout the matrix. This distance is the shortest distance a photon can travel. This then is a mintum of distance. That is, this distance cannot be subdivided. This distance is called a click so the unit of distance is a click.

Also, note that the photon can move in six ways making the three dimensions. These are up, down, left, right, forward, and backward. Each of these is called a mintum. These minta have no other direction other than these six vector unit directions. Likewise, any distance that exists is some set of these six minta.

Observations of Space and Time

Part of the theory here is that all nodes act at the same time or appear to act at the same time. Thus, every photon moves through the Super Matrix one click per tick, the speed of light.

Something of note is the high coordination between time and distance. A photon is always moving. This means, from a photon perspective, the time of travel is always equal to the distance traveled. Perhaps this does not have significance but it is interesting. A factor is that observers of particles observe time as advancing while the distance a particle moves can be zero or almost any value shy of how far a photon can move. Since a particle consists of photons circling about, the sum of the distances traveled, of the particle as a whole, can add up to zero. In general, however, time appears as a scalar. That is, it is a number that represents something without direction. Distance appears as a vector. Adding time always increases. Adding distance takes direction into consideration. Thus, adding many distance minta can result in zero.

Note that, because a particle consists of photons whirling about, a particle cannot move faster than any of its constituent parts. This implies that particles cannot travel faster than the speed of light.

Those that are astute may have questioned that the motions of a single photon is represented as a two dimensional square. Such a person might ask why there is not some basic motion in three dimensions. A great deal of time was spent during the development of these theories seeking such a structure. None could be found that worked. The two dimensional square pattern fit many necessary demands of the concepts introduced in this book. This perspective simply flowed forward into the principles presented.

New Age Special Relativity

Measuring the smallest distance and time is far beyond the capabilities of this work. That is, this author sees no way to prove these ideas are accurate. In an effort to present something that demonstrates these ideas are at least plausible, the author has applied these concepts to developing equations related to special relativity. There are four important relations in special relativity.

- Time Dilation
- Length Contraction
- Relativistic Mass
- $E = mc^2$

Time dilation happens when some object moves very fast relative to an observer. Time on the object, relative to the observer will slow down.

Length contraction also occurs when some object moves very fast relative to an observer. The length will become shorter relative to the observer.

Likewise, relativistic mass occurs when some object moves very fast relative to an observer. The fast moving object will increase in mass relative to the observer.

The final relationship is the relationship between mass and energy.

$$E = mc^2$$

All of these are developed in this book using two concepts. One is that the photon is data that moves from node to node in the Super Matrix. The second is the concept that particles consist of photons whirling about each other like clouds. Relativistic mass is discussed in the chapter titled, "Mass." The energy-mass relation is discussed in the chapter titled, "Energy." Time dilation and length contraction are presented in the appendix titled, "New Age Special Relativity."

The development in appendix, "New Age Special Relativity," uses the tools established in this part of the book. Here is a brief description of the approach used. The main object of study is the single photon particle.

It is observed while not moving and while moving. Many details are observed as in the following picture.

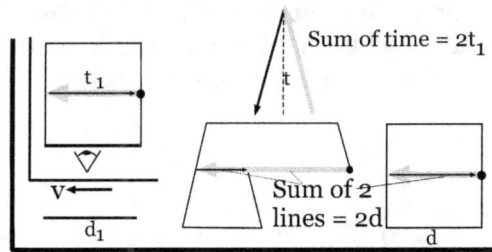

You will notice that the non-moving single photon particle is

compared to a particle moving relative to a non-moving
observer. The following picture isolates the important motion.

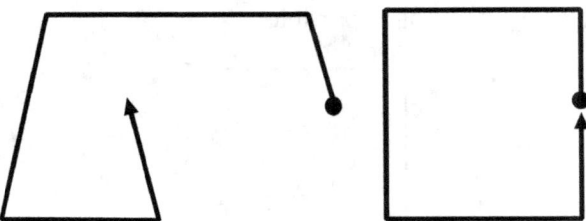

The photon motion on the right is the non-moving single photon particle. The image on the left is the moving single photon
particle. The black dot represents the photon in each at the beginning of the Cycle of Existence. The important factor of this
display is that the photon on the left travels further than the
photon on the right. This enables the particle on the left to
move to the left. Let us simplify this some more.

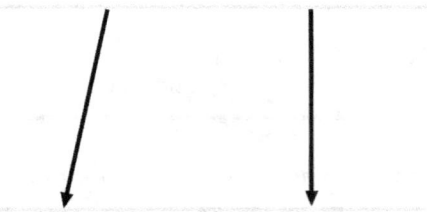

These two lines cut from the above picture represent the
downward motion of the photon from both images. As the
photons are traveling at the speed of light c, the photon on the
left will take longer to travel that distance. This difference is
responsible for time dilation and length contraction.

Consider time dilation. As all matter is made of photons, the
pendulum of a clock will of course be made of photons. The
pendulum swinging back and forth measures time for us. If
such a clock is moving past us at great speed, the photons in

the pendulum will need to move some extra distance to keep the correct form of the clock. All the photons in the pendulum will need to move at that angle and travel a little further. Therefore, the pendulum will appear to move slower relative to an observer not moving. This is the heart of time dilation, length contraction, and relativistic mass.

This is the essence of what is presented in the appendix, "New Age Special Relativity."

Conclusion

The point of this chapter is to show that time and distance are defined by node-to-node jumps in the Super Matrix. Understandably, these minta are so small that direct measurement of them is virtually impossible. We have used these concepts to develop significant parts of the theory of special relativity in the hope of showing some relevancy.

Chapter 30

Mass

To the average person, mass is something that feels heavy. He or she will tend to describe mass as something that has weight. Those in science have a very mechanical way of dealing with mass. Mass is defined as that property which opposes a change in velocity. That is, a more massive object will take more force to move than a lighter object. Beyond that, mass is unexplained. Therefore, it is dealt with in a manner similar to time. Something is selected that many have access to. Some standard is set for its size and temperature. That object is selected as a unit of mass. In the cgs or centimeter, gram, second system of units: that object would be a cube of fluid water 1 centimeter to an edge and at 0 degrees Celsius.

Note that in the metric system (using cgs units) the unit of mass is called a gram. In the English system, the unit of mass is called a slug. People that are accustomed to the English system usually refer to the mass of something as pounds. This is inaccurate as pounds are a measure of force. A pound is actually a unit of force a mass of one slug causes at sea level.

This chapter attempts to go below this description and ascertain the nature of mass.

We have done well with the definition of space and time. Can we continue the effort and describe mass in the same manner to see if the concepts presented thus far will continue to hold water? So, let us dig deeper.

Mass and the Super Matrix

So, what is the mass of a particle? With all that has gone before, this may seem odd for we have described a particle as a whirling cloud of photons. One could easily ask the question, *Why would a bunch of light resemble anything like a block of lead?* If the theory presented here is accurate, there should be an explanation of why clouds of photons have mass.

To repeat, mass is defined in physics as that property of matter that opposes change of velocity. Bear in mind that the velocity of a particle might be zero. So, in that case, mass would oppose that which attempts to move such an object. This would also apply to a rock flying through space. The mass of the rock opposes anything that attempts to speed, slow, or change the direction of motion of the rock.

In the theories presented here, a particle formed out of photons can appear to hang in one position in space. As demonstrated, the sum of the velocities of all of the whirling photons add up to zero. That is, the particle is not moving. For such a particle to move, some kind of mechanism would be required to change the paths of motion of all photons in such a cloud so it would appear to move. Without such an external device, the cloud will remain in a fixed position. Thus, a photon particle appears to resist a change of position and therefore has mass.

Consider a Single Photon Particle

Consider an equivalent particle that consists of an equivalent photon moving in a square. As before, the picture here rep-

resents a particle in one Cycle of Existence. It essentially is a
particle of the smallest kind.

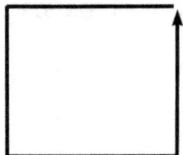

As the description of mass requires some attempt be made to
move this representative stationary particle, consider how this
might happen. The following picture depicts this. Start with a
composite photon moving in the square pattern. When the
photon is at the lower right corner of its travel, some force or
action, not defined here, causes the direction of the photon to
change.

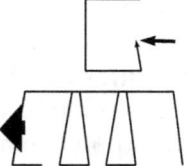

Let us assume the direction changes by ten degrees. This is
shown in the upper image of the picture. The lower image
shows the change affecting the the vertical motion of the photon. It always angles to the left ten degrees as the Cycle of
Existence continues to beat out its path through the universe.
If the Cycle of Existence is small, the change in direction that
occurred at the lower right corner will not cause the entire
cycle to change much.

Consider a photon with a larger Cycle of Existence but with
the same force applied causing the photon to fly up at an an-

gle of ten degrees again. The size of the particle is larger and the force will cause a bigger movement in the particle.

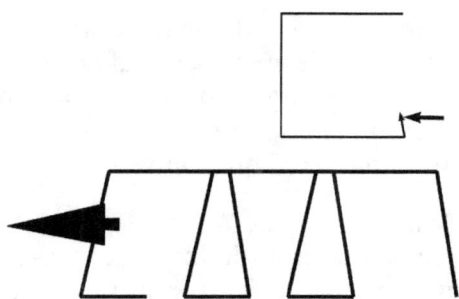

Thus, smaller particles resist change more than big particles. Hence, particles with less area in the Cycle of Existence appear to have more mass.

The point of this is that the resistance to change depends upon the inverse of the area of the Cycle of Existence. That is,

$$1 \text{ mick} = \frac{1}{\text{Area}}$$

Here Area is the area traced out by the Cycle of Existence. We can use this as a definition of mass. If the Cycle of Existence is one click square, the area is 1 square click.

$$A = 1 \text{ click}^2$$

We will call this unit of mass a mick so all of our new units end with "ck." Then,

$$1 \text{ mick} = \frac{1}{\text{click}^2}$$

An Odd Thing

The odd thing is that mass here is defined as the inverse of the area. This means that the smallest area we can have for a single Cycle of Existence is as large as the smallest particle can have. For this reason, the definition of a unit of mass is called a maxtum.

When this was being developed, the hope was to locate a quantum of mass. The unit of mass found through these methods turned out to be very large. This author does not know how to cope with this. The solution was to give it a name to differentiate it from the word quantum.

Relativistic Mass

Theoretically, we have a clear definition of mass. If this is true, can we develop relativistic mass as in special relativity? Consider the two Cycles of Existence below. The top pattern represents our smallest particle at rest. The lower represents a smallest particle moving with some velocity to the left. The labels indicate the length of the edges and the masses are named in the middle.

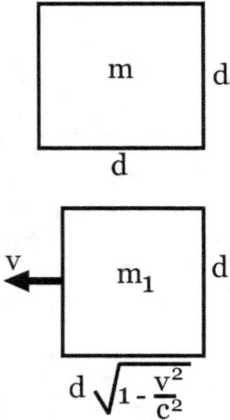

At rest, both patterns are the same. Both would have d as the length of its edges and have mass m. The lower one is moving so it has a velocity and an edge that is a distance defined by the Lorentz-FitzGerald contraction.

Let us calculate the mass for the stationary particle.

$$m = \frac{1}{A}$$

$$A = d\,d = d^2$$

$$m = \frac{1}{d^2}$$

Since mass is the reciprocal of A the area and the area is d squared, the mass m is one divided by d squared.

Calculate the mass for the moving pattern. As before, the mass is the reciprocal of the area.

$$m_1 = \frac{1}{A_1}$$

The area here is a bit different because the pattern is moving.

$$A_1 = d\,d\sqrt{1 - \frac{v^2}{c^2}} = d^2\sqrt{1 - \frac{v^2}{c^2}}$$

Substitute this value for in the equation for mass and we get the following.

$$m_1 = \frac{1}{d^2\sqrt{1 - \frac{v^2}{c^2}}}$$

Simplify it a bit isolating one divided by d squared.

$$m_1 = \frac{1}{d^2}\frac{1}{\sqrt{1 - \frac{v^2}{c^2}}}$$

We have seen that one divided by d squared is m. Therefore, we get a relativistic relationship.

$$m_1 = m\frac{1}{\sqrt{1 - \frac{v^2}{c^2}}}$$

This is relativistic mass as defined in special relativity.

Conclusion

We now have a definition for mass. It depends upon the area within the Cycle of Existence. In addition, the mass is the reciprocal of this area. This seems a strange definition. When tested to see if it logically satisfies another aspect of mass, that of special relativity, it seems to fit quite well.

Let us see if this definition of mass works as we discuss energy next.

Chapter 31

Energy

Our goal in this chapter is to define, with precision, what energy is. If you read the popular definitions of energy you come up with some statement that energy is the ability to do work. This does not help us for work is defined as the amount of energy expanded during some length of time. This is a circular definition. That is, it is like defining the word cat by saying, "A cat is a cat." Then these "definitions" of energy continue to talk about the various kinds of energy. Apparently, if a number of examples of energy are discussed, you will get an understanding by association. That, also, does not help us here.

We are after a definition or explanation of energy that can be used mathematically.

Our first goal is to define the essence of energy. This has been done with distance, time and mass. Once the essence of energy is defined, the plan is to see how it relates to a formula from special relativity. Then, we will show how the definition presented can be used to derive the accepted equations for kinetic energy and potential energy.

Here are the equations we will derive using the basic structures we presently have. The first is the relationship Einstein developed between mass and energy.

$$E = mc^2$$

Next, we will develop a formula taught in basic physics classes, kinetic energy. Picture a block of wood moving past your face. The energy of motion is expressed with the following equation.

$$E = \frac{1}{2}mv^2$$

E is the energy, m is the mass of the block, and v is the velocity of the block as it moves past your face. We will study this in detail.

We can hold the same block of wood ten inches above some tabletop. There is some potential energy from the block of wood relative to the tabletop. This is expressed by the following equation.

$$PE = FD$$

Here, PE is the potential energy, F is the force applied to move the block up from the tabletop and D is the distance moved while experiencing that force. In this case, the force is due to gravity.

The Essence of Energy

We begin by reviewing what we have developed so far. Consider distance. Distance is defined as some set of unit distances a photon travels from one node to another node in the Super Matrix. A unit of distance appears as a white comet in the representation of the Super Matrix shown below.

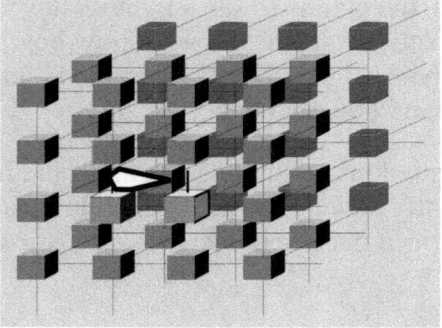

Time is defined as some set of time units spanned when a

photon travels from one node to another node in the Super Matrix. Below, one unit of time is depicted as a gray comet. That is, the smallest unit of time is the time it takes a photon to travel one click from one node to a connected node.

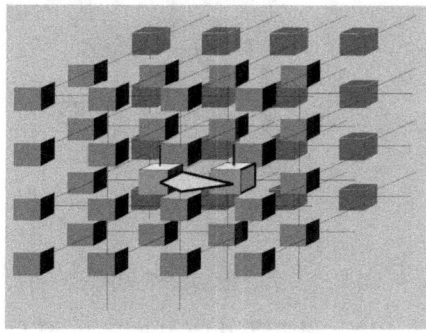

Mass is defined as the reciprocal of the area traced by a photon that moves in a closed pattern.

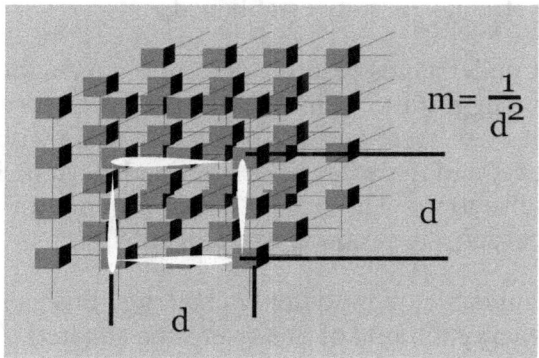

$$m = \frac{1}{d^2}$$

The goal is to add a definition of energy that adds to these just presented. After much trial and error at guessing what energy could be, a conclusion is that energy is the reciprocal of the area traced by a photon using the time elements as time vectors. The following depicts this with gray comets.

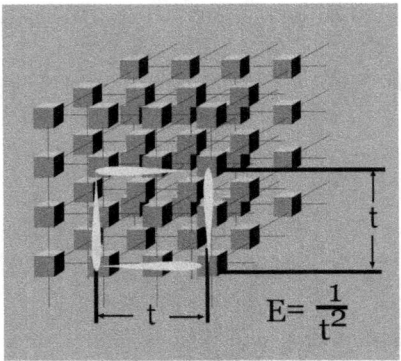

$$E= \frac{1}{t^2}$$

A unit of energy, the reciprocal of one tick squared, is called a maxtum as this unit value is as large as a single photon can have. Here, we apply the name eick to this unit to continue the naming strategy already established.

This author understands that this does not make a lot of sense. Two factors suggest the reciprocal of one tick squared is appropriate. First, there was not much else to do with the way time, distance, and mass have been defined. That is, it was something else to try. The second is that moving forward with this concept seems to produce workable results.

The knowledgeable might quickly challenge this as it expresses energy in units of one over time squared. Actually, this is fitting. To show this, let's begin with the standard mathematical form of energy.

$$E = \frac{1}{2} mv^2$$

Consider the expressions for mass and velocity.

$$m = \frac{1}{d^2} \quad \text{and} \quad v = \frac{d}{t}$$

Insert these into the energy equation for E to get the following.

$$E = \frac{1}{2} \frac{1}{d^2} \frac{d^2}{t^2} \quad \text{or} \quad E = \frac{1}{2} \frac{1}{t^2}$$

Thus, the units of this definition of energy seems consistent with what has gone before. Let us see how this unfolds in other ways.

Using the New Definition of Energy
Begin with a photon moving in a square pattern.

Now, interpret the area, as just suggested, as a square of time vectors. We get the following.

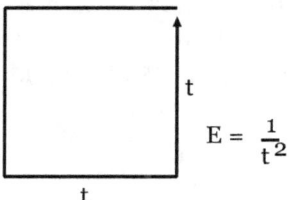

$$E = \frac{1}{t^2}$$

This would be the new definition of energy. The goal is to change this relation to one using distance instead of time. Convert t to d using the relationship c, the speed of light. We solve for t.

$$c = \frac{d}{t} \quad \text{or} \quad t = \frac{d}{c}$$

Replace t in the new equation for energy.

$$E = \frac{c^2}{d^2} \quad \text{or} \quad E = \frac{1}{d^2} c^2$$

Recollect the definition for mass.

$$m = \frac{1}{d^2}$$

Then plug that value for m into the equation we just got for E.

$$E = mc^2$$

This is the well known energy-matter relationship developed by Einstein.

Although the initial idea at first appeared quite strange, this result is quite startling. Next, we are going to take this strange definition of energy and derive the equations for kinetic energy and potential energy. We will apply the use of time as a vector to our tasks below. Remember, calculate energy using time vectors then convert time to distance vectors to get energy expressed in terms of area as a function of distance squared. The speed of light is used as kind of a catalyst in this conversion.

Kinetic Energy

Here is the approach that will be used to calculate kinetic energy. Consider the following picture.

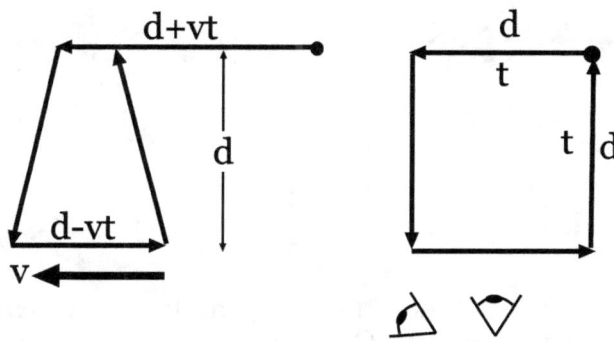

The right image is of a unit photon particle not moving. The image on the left is of a unit photon particle moving. To the observers in the frame of reference with the non-moving particle, the vertical components of the moving particle will be seen to angle to the left as they move up and down. Because the time area of the particle on the left is smaller, it has more energy. The approach, then, is to calculate the energy in each particle and subtract them to get the energy of motion of the moving particle.

The following picture represents this a bit more accurately.

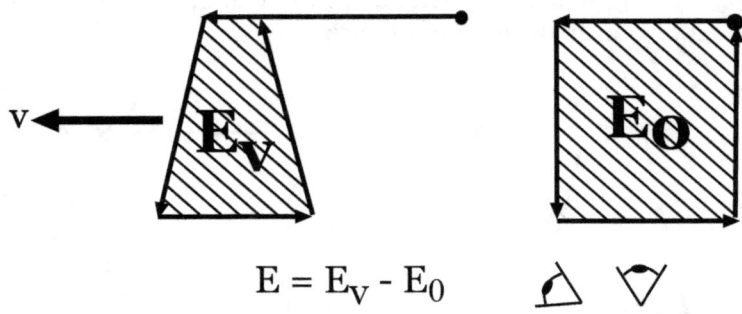

$$E = E_V - E_0$$

Due to the motion of the left particle, the area is smaller. This means it has more energy. Our goal is to calculate the area of both particles using time vectors, convert that to a distance perspective and find this difference. The difference will be the standard equation of kinetic energy.

$$E = \frac{1}{2} mv^2$$

The development is complicated. It is presented in the appendix titled, "Kinetic Energy."

An important point of note here is that kinetic energy is the change of area from one state of motion to another. In other words, energy is area. This point is reinforced as we study potential energy.

Potential Energy

Potential energy in classic physics is defined as:

$$PE = F D$$

PE is potential energy, F is the force applied to the particle as it is moved the distance D against that force. Consider the following picture.

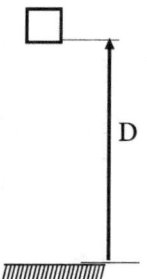

This illustration shows a photon particle some distance D from some supporting surface. The potential energy relative to the surface is due to the force of gravity, here F, and the distance D between the surface and the particle.

The goal is to derive the equation of potential energy by visualizing the particle move from the surface of the tabletop to the position D above the surface. While doing this, we trace the motion of the photon as its particle moves upward. From what has gone before, we know that the particle will move as depicted in the following picture.

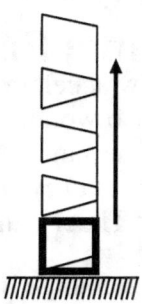

As per our previous discussions, the change of energy will be due to this motion. As the particle moves upward, each Cycle

of Existence will cut a smaller area than when the particle is
not moving. The following depicts the change in each cycle
as the particle moves upward.

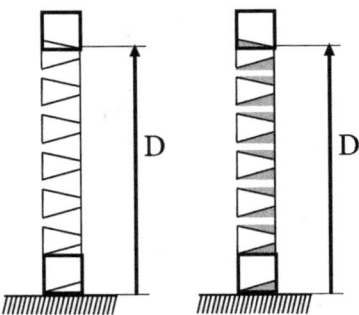

The image on the left depicts the path of the photon as the
particle moves upward. The image on the right shows the
change in area of each Cycle of Existence in gray. The approach to developing the formula for potential energy is to
sum each of these gray areas. This development is presented
in an appendix titled, "Potential Energy." The result of this
development is

$$PE = F D$$

This demonstrates again that energy consists of the area photons carve out in space as they move.

Kinetic and Potential Energy

Early in college, when I saw acceleration and gravity producing the same effect, I also wondered why kinetic energy
and potential energy were the same. My question was, *How
can something moving have energy equal to something not
moving?* The Super Matrix Theory answers the question.

Energy change consists of a change in area the photon sur-
rounds as it beats out its path in the Cycle of Existence. A
particle moving with some velocity will have the same kinetic
energy as a block motionless some distance above a table if
the area to produce that change is the same in both items. In

the picture below, the horizontal motion patterns represent kinetic energy. The square particle starts from zero and accelerates to the left until it reaches the velocity v. The gray bars represent the change in the area of the Cycle of Existence as the particle accelerated.

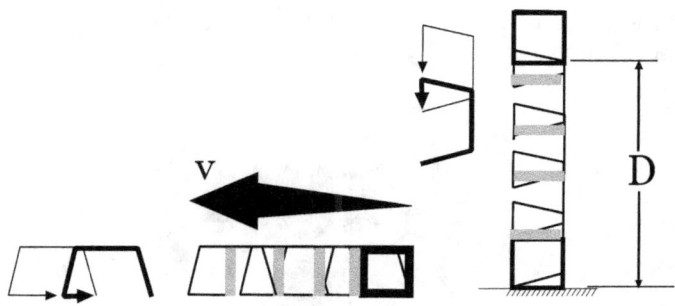

The vertical motion patterns rising the distance D represent a potential energy change. The particle begins at the bottom and moves upward against the force of gravity. Each gray bar represents the change in the area of the Cycle of Existence as the particle moves up distance D. If the sum of the gray bars in each is the same, the kinetic energy in the left display is the same as the potential energy in the right display. Therefore, the answer to the question is that kinetic energy is the same as potential energy because both depend on the change of area in the Cycle of Existence as the energy changes in both systems.

Photon Energy

Now we can address a very interesting question. How does a photon contain energy? In addition, why does the energy increase in a photon when its wavelength gets smaller? Here is the answer.

As described in this chapter, the energy is the reciprocal of the area in time cut out by the motion of the photon. As the components of particles are photons that travel as a helix, we can see how energy is manifested in the photon. It also explains why the energy in a photon is higher when its wavelength is shorter.

$$E = \frac{1}{A}$$

In this equation, E is energy and A is area. This says that the larger the area of motion, the smaller the energy. Consider the motion of a photon.

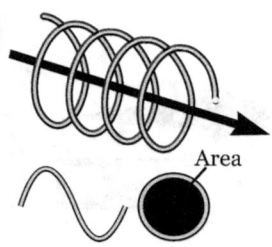

Area

Essentially, a helix is a circle moving through space. From what has gone before the energy in a photon is dependent on the inverse of the area in such a circle. Part of this theory is that the amplitude of the helix is proportional to the wavelength. Therefore, the shorter the wavelength of a photon, the smaller this area is. Thus, shorter means more energy.

Conclusion

The most important bit of information to carry away from this chapter is that energy is the area in space the photon describes as it spins through the Super Matrix. This chapter has backed up this concept by showing how it leads to the classic equations of physics, Einstein's mass-energy relationship, kinetic energy, and potential energy.

Part 9

Universal Gas

The view of this book is that existence consists of a vast sea of photons that interact with each other. One might be inclined to refer to this as a Universal Gas. Consider thermodynamics as an analogy. The laws of thermodynamics evolve from the study of the random action of molecules. [1] In a similar fashion, this book views the physical laws around us to be a study of the random action of photons. The molecules in a gas go their merry way unless they crash into each other. The laws of thermodynamics emerge due to those physical collisions of gas molecules with each other and the sides of some container. Likewise, one may consider the photons that fill the universe as a gas of photons. This part of the book presents some more details of photon interaction and photon transmission. Some attention is devoted to electron energy levels, orbits, and chemical bonding. This part ends with a discussion of the four forces: gravity, electromagnetic, strong, and weak.

Chapter 32

Particle Dynamics

We have gone over some of this before. Here, we fill in some of the finer details of particle persistence and motion. Perhaps the most important aspect of the material presented here is that the nodes in the Super Matrix are some kind of calculating device. Photons are sets of data that are passed from node to node. The node must analyze the data in the photon and make a decision which node gets the data next. This process enables a particle to persist, remain stationary, and move.

Persistance
Given that a particle consists of a group of photons interacting with each other to maintain existence, we need to determine how the photons persist as a group. Here is one scenario. Photons bounce around a spot in space. The following picture illustrates this.

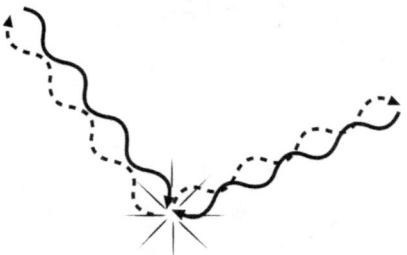

This depicts two photons bouncing off each other. The waves formed by the dashed lines represent each photon moving away from a collision point. The solid lines represent the two photons traveling back to the collision point. The philosophy here is that when two photons collide, they bounce away from each other. Note that the collisions are not reflective like two billiard balls bouncing off each other. Here, when two photons collide, they reverse their direction of motion. At the time of collision, they are programmed to move directly away from the collision point some distance. Once they move that distance, they reverse direction and travel back the way they came. If nothing else changes the return motion, the two photons collide, then continue the back and forth motion. This is the primary device to maintain the persistence of a particle.

The following picture depicts eight photons going through this dance. The short dashes represent inbound photons while the solid lines represent the outbound path.

Note there is nothing in the middle of the particle.

The picture below represents many photons bouncing off each other. The distance they travel is chaotic. Some travel near and some travel far. However, due to the large number of

photons participating in the existence of the particle, it will
appear spherical.

Again, note there is nothing in the exact center of the particle.

Recollect the magic number presented in "Introduction to
Subquantum Physics." If a particle contains this magic number of photons bouncing about in its center, it will continue to
persist. The magic thing about this number is that particles
with photon counts slightly above or below this number will
rapidly vent or absorb photons to enable the particle to persist.

Motion
Once we have established that particles can persist, we can
consider how they move. Keep in mind that one of the rules is
that photons only interact when they collide. Thus, the motion
of a particle is executed at the time the photons collide. To get
our mental hands around this, let us look at a particle not
moving.

Stationary Particle

The following is a particle with four photons in it. The particle is not moving.

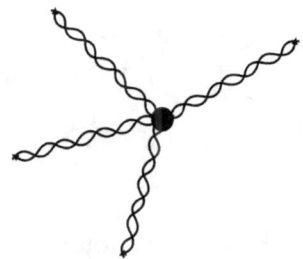

The black disk suggests where the center of the particle is. As before, we note the particle persists because the photons are bouncing back and forth off each other.

Moving Particle

Motion is accomplished when a photon is sent out of the particle center at an angle relative to a non-moving particle.

In this image, a photon has been programmed to go out so far then turn and return to a collision point. The small circle represents the spot from which the photon was sent. The large circle represents the final destination. Motion programming directs the returning photon to a different collision point. This enables the particle as a whole to move to the left.

This motion is depicted in the following picture.

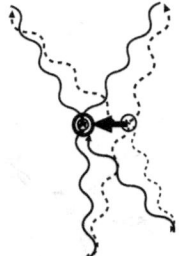

Here the dashed lines moving away from the collision point angle a bit to the left. When the photons return on the solid path, they angle again to the left so they all collide at a different point to the left. This produces an overall motion of the particle.

Particle Motion Vector

How do the bouncing photons know how to travel some distance away and then return to the center of the particle to bounce away again? The answer is a bit complicated. It is complicated because photons have no effect on each other unless they collide. Then, any manipulation of photon motions must be determined at that time. The solution is that each photon carries a great deal of data with it to coordinate inter-photon activities.

A primary task of this data is to establish the direction of a photon away from the center of its particle, reverse direction at some distance traveled, and then to return to the center of that particle. Let us look at this process closer. First, let us review some data that the photon might carry.

> Out Count
> In Count
> Out Vector
> In Vector
> Particle Direction Vectors
> Particle Direction Amplitude

Home collision count
Home count

Recollect that a node is an intersection in the Super Matrix. The Rules of Photon Transmission governs the motion of a photon as it jumps from node to node. The data in a photon contains several node counts. Each of these jumps or transitions may increment or decrement a node transition counter. The Out Count is the number of node transitions the photon makes before reversing directions. The In Count is the number of node transitions the photon makes before expecting to be back in the center of the particle.

A number of the photon parameters control which node a photon will be passed to. The vector values control this. The Out Vector is the direction of the photon motion when moving away from the center of the particle. The In Vector is the direction of the photon motion when moving back toward the center of the particle. The vectors are used to determine which node the photon will be passed to: the up, down, left, right, back, or front. Other parameters deal with the overall direction of the particle.

The concept of the vector here is perhaps not clear. So let's go over it a bit. As has been explained, the photon is a bit of data that travels from node to node. The following picture represents the Super Matrix. We will consider it from a two dimensional point of view. The photon, the dot, is the photon

residing in a node of this rectangular array. The arrow or
vector represents the motion the photon is to follow.

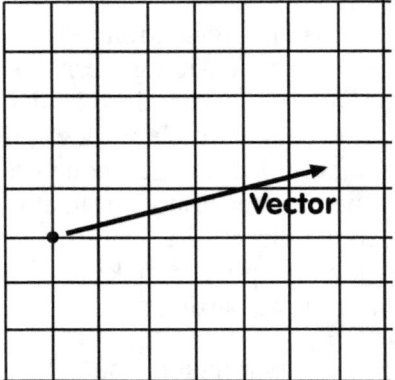

The photon cannot follow the line of the vector for the photon
can only go to up, down, left, right, front, and back nodes.
The following picture shows these possibilities. The possibilities are the dots one jump from the node containing the
photon.

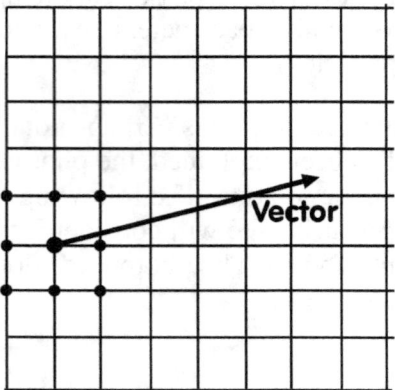

The node must perform some calculations to determine which
node the photon will be passed to. The goal is to select the
node that is the closest fit to the line the vector is indicating.

The next node is the one to the right as shown in the following picture.

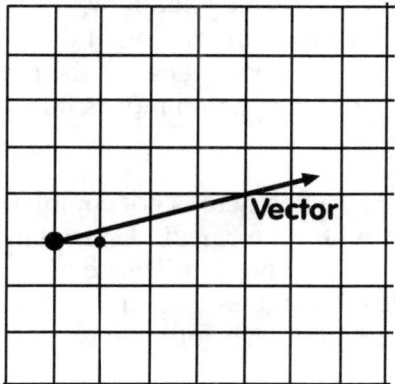

As the photon continues to move from node to node, each node looks at the vector data and makes a mathematical decision about which node to pass the photon to next. The following picture shows six jumps from the original position shown above.

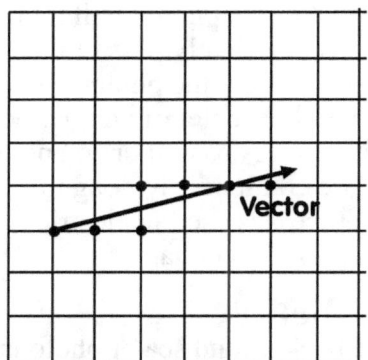

In this example, the data for this vector would be:

```
int X,Y,Z //Base of the vector
int x,y,z //Direction of the vector
```

The base of the vector is where the tail of the arrow resides in the rectangular matrix. As the photon moves forward, this is changed to reflect the forward motion of the photon. The direction of the vector does not change as the photon moves forward. It is the same as the photon moves forward with no change in direction.

The information presented here is not complete. It is clear to someone familiar with vector math. Understanding the details of this is not necessary. The significance of all this is to indicate that the node manipulates data in the photon. This enables the photon to move in a common way with all other photons.

The theory here is that when two photons collide, they share their motion vectors coming to an average value. Then, they adjust the motion vector of each so the two particles will return and collide again.

Note, there is a probability a photon will return and not collide. Then, the photon will likely pass through the center of the particle untouched. Then, the photon can move some distance and collide with the center of another nearby particle. In this way, the information is passed on to another particle. Since all particles are continuously dong this, the particles can interact. This process is used in electric force. This subject will be taken up in a later chapter.

Background Sea

The concept of the background sea of photons was introduced in "Introduction to Subquantum Physics." In this theory, all of space is occupied by photons randomly moving around. The following picture represents this random motion.

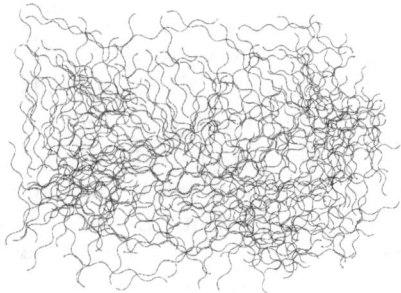

This is in contrast with photons that participate in the persistence of a particle.

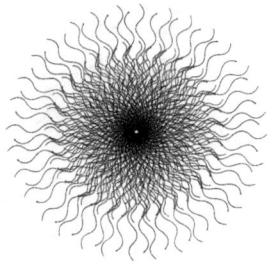

The photons in a particle behave in a very ordered way. The photons in the background sea move about randomly.

Background Sea to Particle Ratio
The background sea supports the existence and persistence of particles. In order for a particle to exist, the ratio of density of the sea to the density of the particle must be some specific value. Particles are constantly releasing and acquiring photons from the background sea. Let us name these actions and look at them a bit closer.

Venting
Venting refers to a particle process in which a normal photon leaves the group that supports the particle. There are several reasons for this to occur. As the photons randomly move in and out of a persisting particle, some will leave the particle as the particle has moved in an unpredictable way. Something outside of the particle can capture a photon so it cannot return. Then, some photon that is returning to the center may not encounter a photon to continue the bounce pattern. It can fly by the center as if knocked out. The following picture illustrates photons venting from a particle.

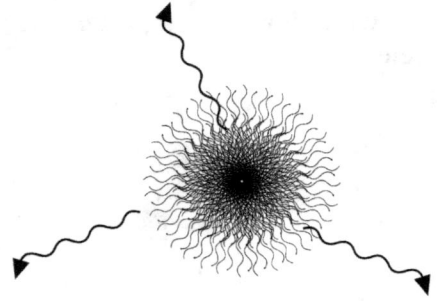

Absorbing
As photons are constantly venting from a particle, the number of photons in a particle can fall below the magic number. Then, photons from the surrounding background sea can slip into the particle. That is, they are absorbed. The black and

white wavy lines entering the particle in the following picture depict this process.

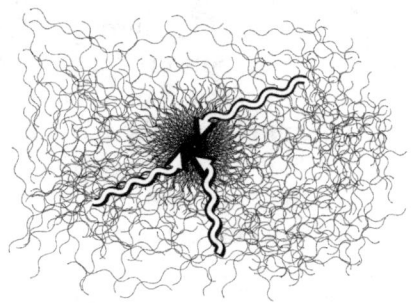

Particle Balancing
Particle balancing is the name of the continuous balancing process of venting and absorption. When a particle loses a photon, the particle would quickly absorb photons from the background sea to keep the magic number in the particle correct. If some extra photons from the background sea manage to get into a particle, the particle would quickly vent photons to keep the magic number in the particle correct.

Virtual Background Sea
As mentioned, the existence and persistence of a particle depends upon the background sea of photons. The ratio of the background sea density to particle density determines the magic number of a particle. Most proton-neutron clusters are immersed in an electron particle or electron particle cloud. The proton-neutron cloud then sees the electron cloud as its background sea. Then, it is a virtual background sea, as it is not the chaotic background sea that permeates universe subspace. Then the ratio between these two determines the magic number of the proton-neutron cloud.

This would apply to any particle cloud immersed in another particle cloud.

Background Sea Friction

There is an interesting consequence of this photon structure of
particles. When a particle is moving through the background
sea of photons, the aggregate motion of the photons in the
particle is different from the aggregate motion of photons in
the background sea. The photons in the particle are moving as
a group. The photons in the sea, as a group, are not moving.
Yet, to support the existence and persistence of a particle,
photons are exchanged between the particle and surrounding
sea. The result of this exchange is that the particle motion will
slow as the particle continues to move. That is, the slower
photons absorbed will slow the aggregate motion of the
whole of the particle. This is called Background Photon Sea
Friction.

At what we consider normal speeds, this friction is not noticed. The speed of the photons moving about is vastly higher
than the aggregate motion of the photons in a particle. Thus,
the friction is virtually non-existent. However, it has an effect
in extreme areas. This principle would suggest that heavenly
bodies would have an upper limit to their speed through free
space. For example, if a galaxy were flashing along at half the
speed of light, it would slow down due to this effect. Another
area is in high-energy particle accelerators.

Consider what happens in an accelerator. Particles are
accelerated near the speed of light to smash into other particles moving near the speed of light. As these particles are
accelerated at these high speeds, they will encounter background sea friction. As mentioned, photons are exchanged between the accelerated particle and the background sea. The
photons vented from the particle will have more energy than
those in the background sea. Because of this difference, outside observers will see the venting as energy flashing out of
the accelerating particle. This is observed in accelerators. As
objects such as protons are accelerated in super-colliders,
they shed enormous amounts of energy. The collider must
pump more and more energy into the photon to get it to keep

moving and move faster. The energy shed from the accelerated particle is the photons that have been vented during the
vent/absorb dance.

Conclusion

This chapter has reviewed the process of persistence and motion a bit more than was done in the particle introduction. The
newest factor introduced here is the idea that the photon carries data with it that is used to control where a photon goes
during collisions and normal non-collision transitions from
node to node. The nodes use this data to process which node
the photon will be transferred to next. The real difficult aspect
of this to grasp is that very large numbers of photons are
bouncing back and forth at the speed of light. This large number of collisions and the speed enable this data to be moved
through all the photons participating in the cloud. As particles
normally move much slower than the speed of light, the data
transferred in the middle of the photon can be quite complete.

Chapter 33

Energy Levels, Orbits, and Bonding

With the present understanding of atomic physics of today, energy levels are dependent upon the orbit of an electron around the nucleus. Then, bonding is thought to be an electron orbiting around two nuclei. In this present scheme, an electron would change orbits as the energy in the electron changed energy level. This is an awkward picture from the perspective of quantum physics.

In quantum mechanics, waves moving around the nucleus as clouds do not seem to be orbiting much less changing some kind of orbit. The concept of energy levels is confusing. Is the energy of one of these wave particles dependent on the wavelength? In traditional quantum physics, the precise position of the electron is only detected when observed by a human. However, the structure of the atom depends on this knowledge to exist. However, no human is involved with the observation. How do the many atoms exist without this aid?

This chapter proposes that the electron is a particle made up of a cluster of photons bouncing off each other. Some combinations of these photons persist as a particle. Photons can join the cluster and increase the energy in the cluster. This would represent a different energy level. The idea is that that combination lasts for a bit of time. That combination of photons would then decay, throwing off a photon so the cluster would return to a stable combination.

This chapter examines this and how this cloud of photons orbits a nucleus.

We also take a quick look at something called the Virtual Background Photon Sea.

Energy Levels

The structure of particles depends upon chaos and chance. There are four factors supporting the chaotic existence of a particle.

- Number of photons in the particle.
- Combined radius of all photon helix.
- Number of stable patterns for that number of photons.
- Density of the background sea.

We have defined a magic number for the existence of particles. A magic number of photons in a particle enables the photon bounces to continue so the particle persists.

The radius of a photon helix determines the wavelength of the photon and, hence, its energy. The total energy of photons participating in a particle that exists must also fall within a distinct range of values. As energy depends upon the radius of helical motion, the combination of all photon radii must also fall within some specific range of values.

For any set of photons, there are combinations that bounce such that they continue bouncing together and maintain integrity as a group. Then, there are combinations that do not persist but decay. The chance the particle will persist is the ratio of stable to non-stable combinations.

The density of the background sea also determines the stability of particles. If the background sea can supply extra photons when a normally stable combination begins to decay, the particle can continue to persist even though it falters.

To grasp the effect of these factors on particle existence, consider a hypothetical example of a combination of these factors.

4 million photons
Each photon has a radius of 1000 clicks.

400 trillion patterns of this combination may be stable. 1 million are unstable. That is, the magic ratio of existence would be 400 trillion / 1 million. The magic ratio is defined as a ratio of stable combinations to unstable combinations.

This may be a single electron at its lowest energy state. Such a particle would continue to persist for a long time as almost any combination it fell into would be a pattern that supported persistence. However, some combinations do not work. Should one of these occur, the particle would decay. A photon would be vented. That smaller number might not support persistence. However, a photon from the background sea of photons could enter the particle bringing the magic number to a level of persistence.

Chances are that particles are always absorbing and venting photons from and to the background sea. The existence of the particle would not depend on any particular set of photons.

Enter a Stray Photon
Now, let's suppose we add a photon that has a radius of 900 clicks. Then we have the following combination.

4 million and 1 photons
Each photon has a radii of 1000 clicks and one has a radius of 900 clicks.

Possibly, 50 thousand of this combination is stable.

As the possibility of existence is very low, the particle would tend to decay quickly. As it decayed, one photon would leave, vented. When that one leaves, the remaining pattern of movement would be in one of the 400 trillion stable patterns. The particle would then continue to exist.

Enter a High Energy Photon

To continue this example, suppose a very high-energy photon with a radius of 150 clicks were to enter the particle. Then we might have a combination that has a higher probability of persistence.

4 million and 1 photons
Each photon has a radius of 1000 clicks and one has a radius of 150 clicks. 200 trillion patterns of this combination may be stable.

The result would be a particle that is likely to last for a long time. The number of possible combinations is near the lowest energy state of the particle.

However, if this combination were to absorb a stray photon, the result might be a very unstable particle. The result could likely be the venting of a high-energy photon so the particle as a whole would return to the lowest energy state. Also, as the magic ratio of existence would be 200 trillion / 1 million, that state of the electron would have a shorter half-life than an electron at its lowest and stable state. Such an electron could randomly vent a photon to bring it to the lowest level.

This scenario could explain how electrons pick up photons and move to a higher energy level then shed a photon and drop back to a base energy level. It also explains why electrons will not raise to any energy state. The electron will only absorb photons that increase the magic number and magic ratio to some stable state. Other photons may be absorbed by an electron cloud but be vented immediately as if the photon were never a part of that particle.

Energy Level Map

The following graph is an attempt to show a relation between somewhat stable photon-energy combinations.

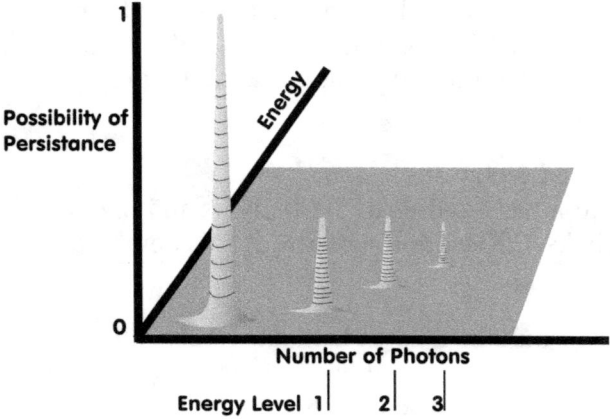

Note that this graph has three axes. The vertical axis represents the possibility a particle will exist. Note there is a 1 at the top of the line. If the possibility of existence is 1, the particle is guaranteed to exist. The first peak in the graph falls short of 1. This means that the particle will probably exist at that combination but there is a possibility it could decay. The horizontal axis represents the number of photons in the particle. The largest peak is at its lowest energy level. As the number of photons increases, the graph shows where the possibility of some stable combinations would be. The purpose of this graph is to show the effect of the number of photons vs. the amount of energy in the particle. Thus, the axis into the page represents the amount of energy in the particle. This particular graph shows that there are three more-or-less stable combination levels above the lowest level or highest probability level of persistence.

Orbits

Here we address the question of how electrons might orbit the nucleus. In the traditional classic physics model, the electron is a small hardball that spins like a planet around the nucleus. In present day quantum physics model, an electron is a probability cloud of which we don't know where the electron is until we put our finger in the cloud to stop the electron so we can see it. In the view presented in this book, an electron is a cloud of photons that are bouncing back and forth.

Seeing what it is and how it works is still difficult. The cloud is very big compared to the size of a photon, which makes it difficult to see how many clouds would fit around a nucleus without interfering with each other. The purpose of this section is to show how photon clouds can coexist with other photon clouds comfortably and form atoms.

Particle Proximity
Again, we extend what has gone before to consider some workable possibility. From what we have seen before, the photons from one particle will not interfere with the photons in another particle even if they seem to overlap almost 100%. The reason is that the clouds maintain existence by collisions at their very center.

Consider the following two particles.

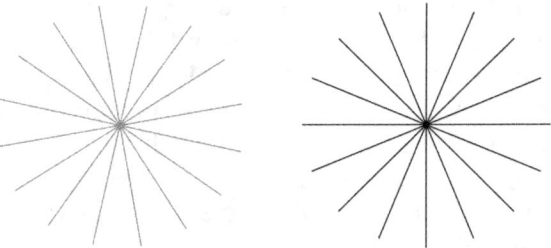

There is nothing in the middle of the particle. That is where the collisions occur that maintain the existence of the particle.

When the two particles are very close they appear to overlap as is demonstrated in the following picture.

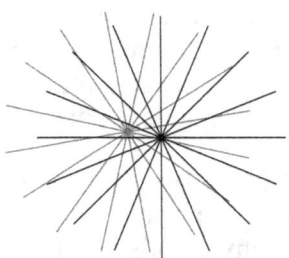

Note, however, the centers are not on top of each other. The photons bouncing back and forth do not interact. Realize, photons are points that do not have any dimension. Photon collisions are very rare. As long as the centers are not on top of each other, both particles continue to coexist as if the other were not there. Of course, if these were the only two particles as electrons in the experiment, they would react with each other. We assume the two electrons are together with two protons in the middle of an atom. The positive protons would be keeping the electrons together.

Consider something else. If the centers were on top of each other, the particle would probably decay, as the number for two particles would be bouncing in and out. That is, the magic number of would not be suitable for the existence of such a particle. As long as the two centers are not near each other, they would seem to exist with each other just fine. This is essentially an explanation of the Pauli Exclusion Principle. That principle states that two particles cannot exist in the same time and place. [1]

Moving Particles

Right now, we have established that particles, specifically electrons, can almost overlap each other and maintain an identity. Let's advance our understanding one step more. Recollect that, per previous discussions, an electron oscillates as it moves through space. This emerged while discussing de

Broglie wave functions. The following picture is to remind us of that discussion. The top image shows an electron being shot out of a gun. The wavy line represents the path of the electron. The middle of the electron is actually tracing out that curvy line of motion.

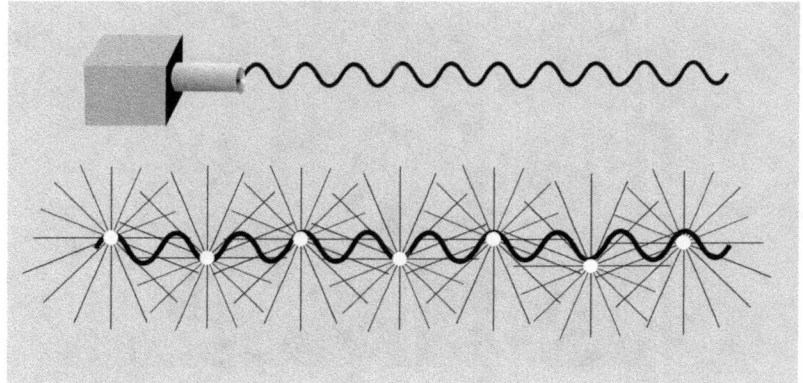

Be careful to note that this line represents the center of some particle moving through space. The image at the bottom of the picture adds the bouncing photons that make up a particle. Seven particles appear in this picture to represent a number of particles shot from the gun. White dots have been placed in the middle of each particle to represent the center of each. Note that the lines representing the particles moving to the right are in what appears to be a straight line. That is, an observer will view the electrons moving in a straight line. The centers however are moving up and down due to the random motion of the photons bouncing around. However, as traditional physics treats the center of mass of a particle as the position of the particle, that perspective suggests that the particles are oscillating as they move through space.

This is being presented now to show that two particles such as neutrons could move along the same line without interfering with each other. This is demonstrated in the following picture. The top line represents a neutral particle moving to the right.

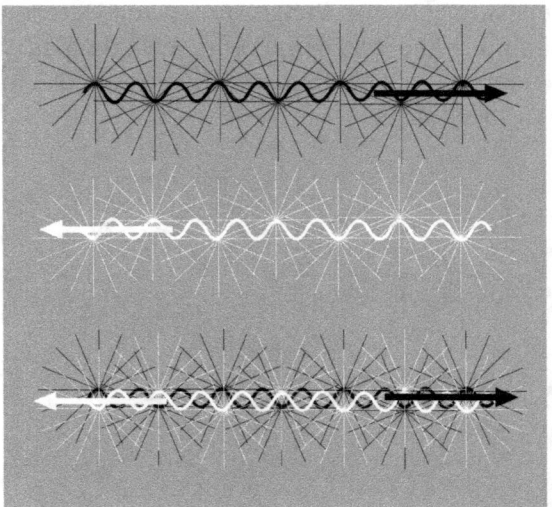

The middle line represents a neutral particle moving to the left. It is 180 degrees out of phase from the top line. That is, when the top line goes up, the middle line goes down. Both of these are placed on top of each other in the bottom image. The particles seem to go right through each other. Their up and down motion enables them to miss each other even though their center line of direction falls on top of each other.

Circular Motion

Our next task is to apply this understanding to electrons moving in a circular pattern around in an atom. Let's consider one electron moving about a proton forming a hydrogen atom. The upper left curved line in the picture that follows represents one wave of motion of the center of an electron. That wave is moving in a straight line. In an atom, the electron is not moving in a straight line. It is moving in a circle around a proton.

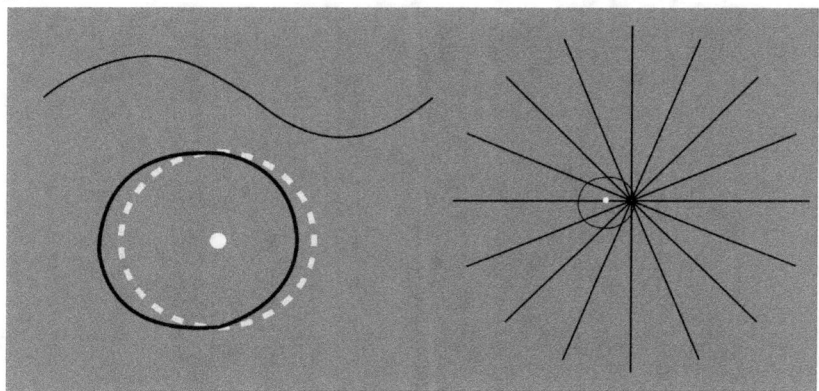

The white dot on the left represents the proton. The dashed white line is a circle around the proton. The black line represents the wave moving in a circle around the proton. That black line is the center of the electron as it moves. Remember, there is nothing there. That is where the photons of the electron collide as the electron cloud moves. The image in the right of the picture is to suggest the photon cloud moving around the proton.

To extend this idea, add another electron to the mix. In the following picture, the upper left curves represent the motion of two particles passing each other. Each curve represents one wavelength of each electron. In the straight line, the particles can move past each other following the same helical direction vector. The two offset circles lower in the picture represent the two wavelengths moving in a circular pattern. Two par-

ticles or two wavelengths can move in this pattern without
interrupting each other.

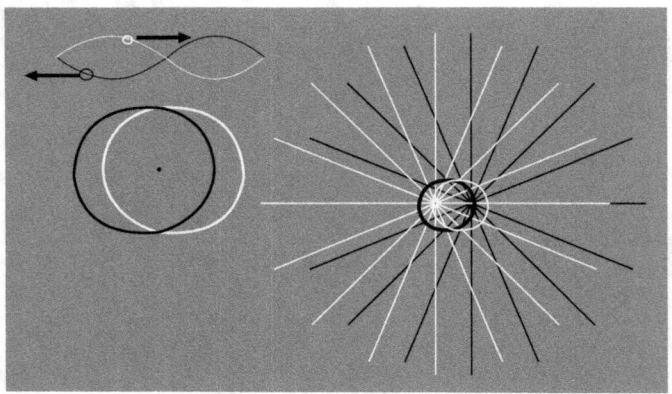

The pattern on the right illustrates how two electrons may
appear orbiting a helium nucleus.

The smallest orbit allowed is that of one wavelength. The
wave nature of the particle enables two particles to move in
that one wave cycle orbit.

Two Wavelengths

Next, we consider an orbit that consists of two wavelengths.
The picture below suggests how this might appear. The top
image in the following picture shows two cycles of wavelength moving in a straight line. The image below shows
what that wave would be like moving in a circle. The white
doted line represents the centerline of motion of the wave.

The black line represents the motion of a particle wave following the circular path.

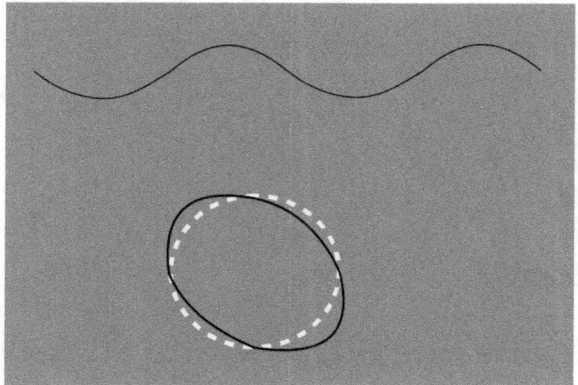

From the forgoing, we have seen that each possible wave
cycle in the wave can support a particle in that wave. The following depicts the possible cycles in the wave. The top
shows that there are two formed by splitting the two cycle
wave into two parts.

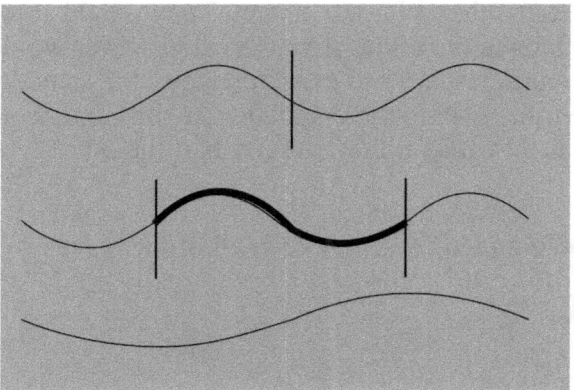

The middle line shows that there is a complete wave cycle in
the middle of the two-cycle wave. Finally, a single cycle
wave is shown that spans the two-cycle wavelength.

The following picture shows the possible particles a two-wavelength motion can support. There are 8. The two cycles can be broken into 6 distinct particles while the single wave that extends two wave lengths generates 2.

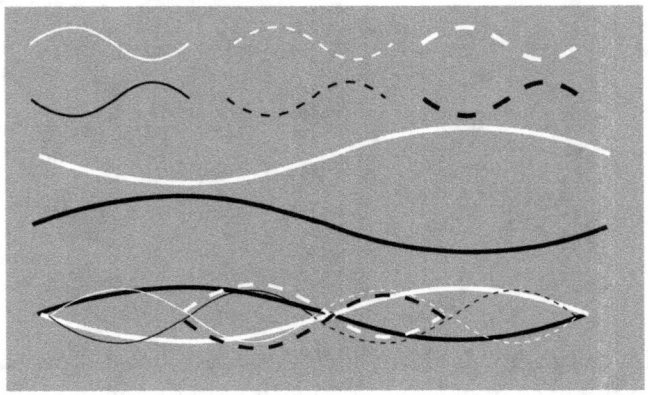

The bottom of the picture shows all of these superimposed on the same line of particle direction. The point of this is that neutral particles traveling on each of these 8 waves could move by each other without contact. They would slip by just like two ships in the night. An external observer would see that the 8 particles seem to exist at exactly the same location and same time. Due to the oscillation of the centers of these particles as they move, they coexist as if the others were not there.

The following picture shows what all of the waves would

look like moving in a circle. The lowest level of one wavelength containing two particles is shown in the middle as well.

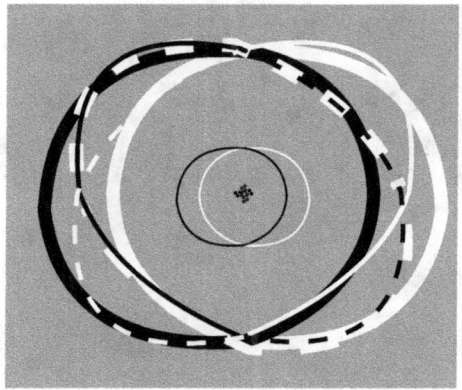

Bear in mind that these lines represent the middle of particles
where nothing exists other than collisions of the photons.
That space is extremely small.

The following picture attempts to display these lines with
photons bouncing in and out of the center points of the particle clouds. With 2 electrons in the lowest orbit and 8 in the

second orbit, the following is an attempt to illustrate a neon atom.

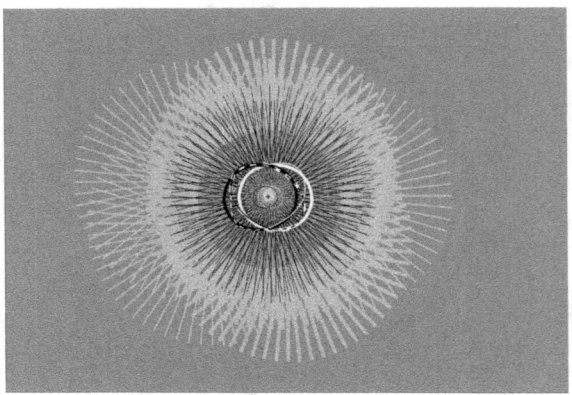

Virtual Background Sea
While viewing some of these images, you may have noticed that clouds of photons exist within other clouds of photons. The following picture depicts this and a ramification of what this could imply.

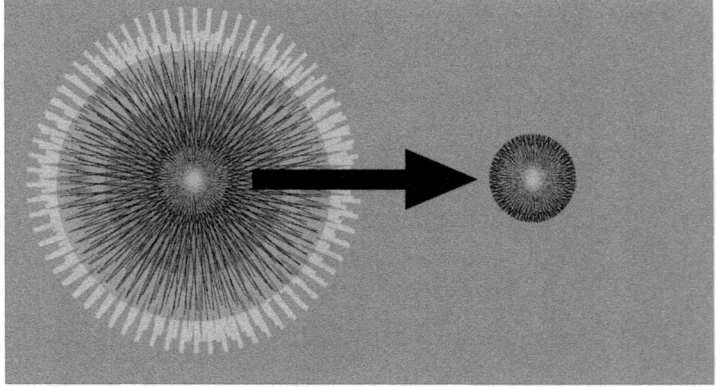

This picture shows an inner cloud that has left a larger, more complex particle. This could be due to a high-energy collision experiment. While this smaller photon cloud was inside the complex cloud, the complex cloud appeared as a background sea to the smaller particle. This virtual background sea appears significantly denser than the normal background sea. Therefore, the magic number of existence is different for the smaller particle when immersed in the larger. The theory in this book is that the magic number increases when in a more dense background sea.

This is being raised to point out a phenomenon that could be very interesting. Since the magic number is higher inside than outside, do the mass and/or energy of the particle change when moving outside of the larger particle? This would suggest that when a small particle leaves the middle of a larger particle, some energy and photons might suddenly be vented to maintain the structure of the, now nude, smaller particle.

Bonding

Our next task is to discuss how electrons can circle two nuclei at the same time to bond two atoms together to make molecules that are more complex.

Let's pick up with the electron configuration we have been working with. The following picture is a different way to look at that configuration. There are two primary orbits in this situation. The smallest supports one wavelength of electron motion and the other supports two wavelengths of motion. The segments in the following picture represent the motion of an electron. The inner band has two segments representing two electrons. The higher band has 8 segments representing 8 electrons. This configuration is the same as was presented in the previous section. This configuration represents a neon atom. One would say that both bands are full. There would be 10 protons in the nucleus so the neon atom is balanced.

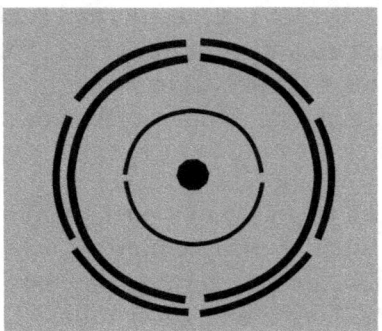

Neon is an inert element meaning that it does not react chemically with other elements. A chemist would say that the bands are full and have no openings for electrons from other atoms.

To make this more interesting, let's consider a configuration with two less electrons. That would be oxygen. The following is a picture of this using this unusual way of depicting the electrons in an atom. There are still two bands but the outer band only has 6 electrons in it.

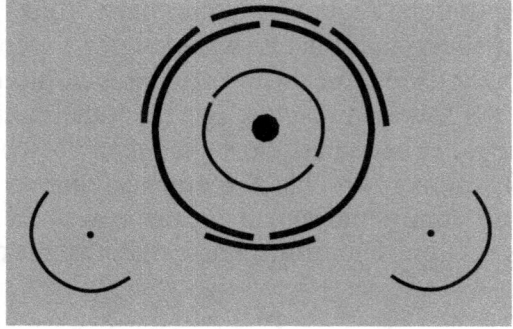

The figures to the left and right of the oxygen atom represent hydrogen atoms. They each have one electron that is represented by a segment going around half way to represent one of the two possible waves allowed in the lower band.

Since there are two empty slots in the oxygen atom, the electrons from the hydrogen atoms can slip into them forming H2O or water.

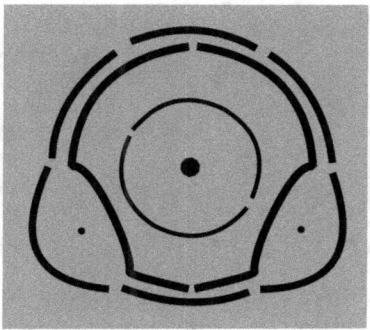

Keep in mind that the lines represent the centers of the photon clouds that are the electrons whirling around the nucleus.

The following is an attempt to represent a molecule of hydrogen fluoride. Fluoride has 7 electrons. The attempt in this picture is to show that the nucleus from both atoms are close together. Then the motions of all electrons spin around what appears to them as a single nucleus. The point here is that the electron from the hydrogen does not travel far from the other electrons to participate in the spinning electrons. That extra electron is very close to the orbits of the other elec-

trons. Again, bear in mind that the lines here represent the
middle of the photon clouds.

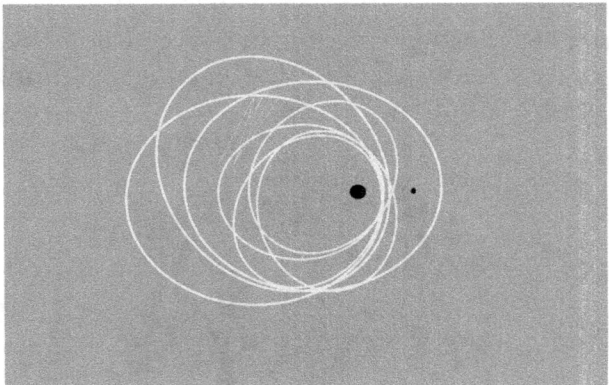

The following picture is an attempt to insert the photons into
the image to clarify how the entire molecule might appear.

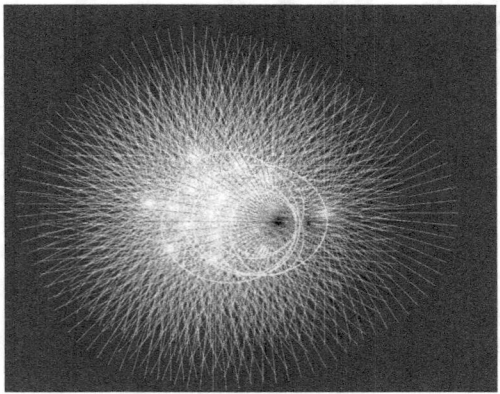

Note the white spots in the picture. These represent the centers of the electrons. They are to follow the white lines in the
picture. The white lines are drawn to suggest the path of the
bright spots. The bright spots in this picture were not inserted
into the picture. They are formed by the intersection of the
lines representing photons moving in and out of the particle.
This simulates what the center of a particle would look like.

There is nothing there except the collisions of the photons that make up the particle. Again, keep in mind that the center where the photons collide is very, very small.

Conclusion

The significant factor to take away from this chapter is the ease with which one particle can pass through another. In general, photons do not interact because they can only do so because they must collide in exactly the same space coordinates at exactly the same time. This can only occur when there are a very large number of photons present in a very small space. This happens near the center of a particle where the photons collide to maintain the existence of the particle.

The other important factor is that particles wobble as they move through space. The equation describing the motion through space is not some kind of probability. The equation describes exactly what the center of the particle is doing.

Chapter 34

The Four Forces

As science has progressed over the last thousands of years, we have come to observe four forces. They are:

Gravity
Electromagnetic Force
Strong Force
Weak Force

Let's briefly review the forces and see where they have been observed.

Gravity is the most apparent. That force keeps our feet planted on earth. It keeps our computer on a table and keeps our dishes on a shelf or on the dinner table. We know when we drop an apple it will fall down. That is due to the force called gravity. Newton pointed out that the force of gravity occurs between any two masses. In the case of the apple, the force is generated by the mass of the apple and the mass of earth.

The next most apparent force is electromagnetic. Its presence is not normally visible. Usually we observe it when playing with magnets. We see that magnets attract or repel other magnets. We also observe its effects when a small experiment is done when we comb our hair then use it to attract some bits of paper lying on a table. We are told that this is static electricity. When we take a science class, we learn that the force between magnets is caused by something similar to the static electricity in a comb brushed through our hair. The electromagnetic force is caused by photons moving between electrons and protons. [1]

In advanced classes, we learn about electromagnetic radiation. It is the light we see and that which radios send through the air for entertainment.

Strong force is not apparent at all. The study of it only appeared when scientists began to study subatomic particles. Strong force holds protons and neutrons together in the nucleus of an atom. The Standard Model proposes that small particles called mesons are responsible for the strong force. They are transferred between hadrons (heavy particles) and produce the strong force binding the nucleus together. [2] In this scenario, the mesons spin around nucleonic particles as electrons spin around nuclei. The mesons are capable of spinning around several nucleons holding them together as do electrons hold the form of an atom.

Weak force is even more obscure. Scientists observe it during nuclear reactions. It is revealed when particles decay and give off radiation. [3]

The primary point of this chapter is that these forces are due to a common thing. That thing is the random interaction of photons. As demonstrated in the previous chapter, photons move from one point in the Super Matrix to another point. A collision occurs when two photons attempt to enter the same point, node, or intersection in the Super Matrix. The forces are due to these collisions.

During a collision, two or more photons colliding do not simply bounce off each other and go their separate ways. The photons have acquired data during their travels throughout the matrix. Each node or intersection performs some logical calculations that follow the Photon Rules of Interaction to determine the fate of the colliding photons.

Those rules along with the Rules of Transmission govern photon motion. These rules are a bit more complex than angle of incidence equals angle of reflection. For example, when a

photon is flying free without collisions it may be recording
how many node transitions it has traveled through since
experiencing a collision. In addition, a photon may change direction if some internal photon counter keeping track of transitions is decremented to zero. Also, during collisions, one
photon can transfer data to other photons.

Electromagnetic Force

Electromagnetic forces are due to the exchange of photons. In
general, like particles repel and unlike particles attract. One
electron will repel another electron. Likewise, one proton will
repel another proton. On the other hand, an electron and a
proton will attract each other. During the exchange of photons, particles emit photons that travel to another particle that
collide with photons in the middle of that particle and cause
the hit photon to veer one direction or the other.

In a Nutshell

As we have seen, particles consist of photons. We have also
been exposed to the idea that photons are left-handed and
right-handed. In the scenario presented here, electrons and
protons consist of a single kind of handedness. For the sake of
argument, we can say electrons consist of all right-hand photons. In a similar fashion, protons will consist of all left-hand
photons. As we have seen, motion in a particle consists of
photons bouncing in and out of a particle at an angle in the direction of motion of the particle. A force, then, would consist
of some action that would be capable of moving a particle
from a rest position to some motion.

The basic element in the force is a collision between two photons. If two photons of like handedness collide, the photons
rebound from the collision by angling away from the incoming photon. If the handedness of two photons is different, the
photons rebound an angle toward the incoming photon. The

logic in the Super Matrix nodes manages this property. The following picture shows this action.

In the collision on the left, both photons are right handed. The photon heading downward collides with the photon moving horizontally. The collision point is circled. After the collision, the photon moving vertically moves away from the impending colliding photon. This contributes to the motion of the particle to the left or a repulsive force. In the right image, two photons collide that have different handedness. Then, the upper photon rebounds toward the impending photon.

Repulsion and Attraction

Because two protons contain like handedness, the photons each emit like-handed photons causing the protons to repel each other as the transfer of photons occur between them. Likewise, the same occurs with two electrons. As each emits like handed photons, electrons repel each other when photons are exchanged. With the case of an electron and a proton, the exchange of photons causes the two particles to attract each other.

The following picture demonstrates repulsion. Both particles are emitting photons. The photons only collide in the high-

density centers of the particles. The collision causes the constituent photons to angle away from the impending photon.

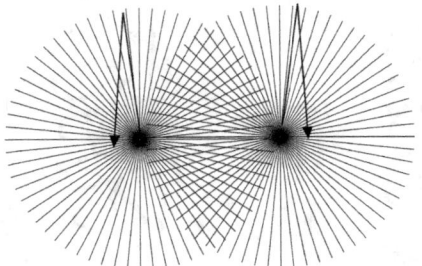

After the collision, the two particles are moving away from
each other.

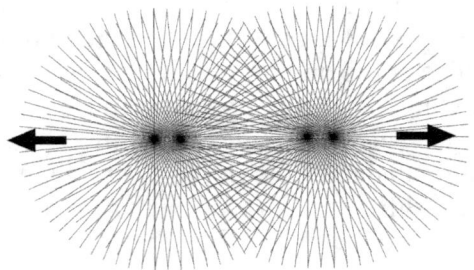

Attraction is similar but opposite. The following shows two
unlike particles. There the impending photon causes the
bouncing photons to angle toward the impending photon.

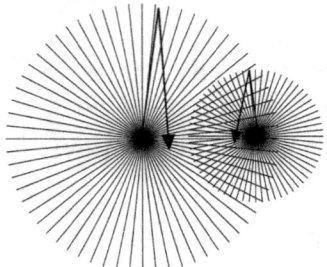

This causes the particles to move toward each other.

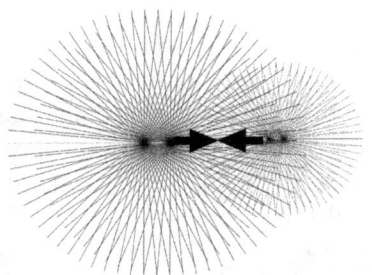

Neutral Forces

Particles, such as neutrons, have no charge and therefore do not exhibit electromagnetic force with electrons, protons or other neutrons. In the theory presented here, neutrons consist of left and right-handed photons. Although emitted photons are causing target photons to angle toward and away from impending photons, they balance each other.

Clouse Encounters

You will note that in the forgoing explanations of particle interaction via photons, the photons involved were vented. This is not necessary for particle interaction. The photons going through the Out/In Cycle can participate in the electric force phenomena as well. Then a photon, on its outward journey, can visit a stranger particle and interact with the stranger causing a repulsive or attractive force. These photons are called virtual photons. They appear as virtual photons near radio antennae or when particles are close.

Strong Force

The strong force is the manifestation of particle persistence. That is, the chaotic interaction of photons randomly clustering together gives the appearance of some force holding particles together. In the case of a single particle such as a proton, this chaotic force appears as a strong force that holds it all together. In this book, this has been referred to as persistence. In the case of the nucleus, this chaos force appears as a strong force holding the nucleus together.

Here, the nucleus is considered to be a large single particle. That is, the particle contains the number of photons for each proton and neutron that make up some nucleus. The nucleus would appear as a drop of dense photon fluid with a very small center that is empty. Many collisions occur there. Perhaps photons are colliding three or four at a time.

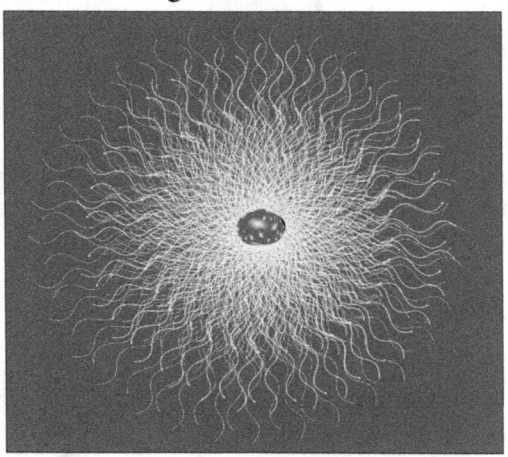

When we hit the nucleus with a high-energy exterior particle, the nucleus cloud is split into parts. The parts split out in random numbers to form many particles and possibly many elementary particles. Each of those sub-particles has a separate magic number.

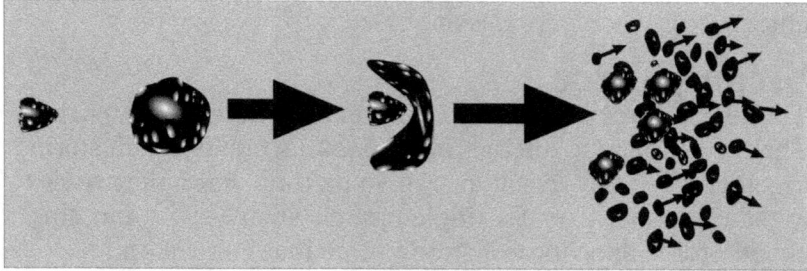

Thus, when an atom is divided we see a number of separate

particles form because of the apparent catastrophe. Consider the following analogy. The nucleus is but a single drop of photon/neutron fluid that splits up into various sized droplets when hit with a high-energy particle. Each droplet size has its own magic number and, hence, properties.

Because the big drop of photon/neutron fluid before the collision was embedded inside an electron cloud, the magic number of existence of this massive particle is considerably different from the separate protons and neutrons in the particle. That is, the massive particle, a nucleus, would perceive the electron cloud as a very dense background sea.

In experiments that inject high-energy particles into the middle of this massive particle, the nucleus is ripped out of its electronic cloud and split apart. The photon droplets then gather into their magic numbers established by the normal background sea.

The splitting is random, however, so differing numbers of photons may attempt to group differently forming a wide variety of particle types. Energy introduced into the nucleus due to the collision could force unusual groups of photons away from the collision at high velocity. The unusual number of photons that group together could prevent them from forming protons and neutrons. Instead, a number of smaller particles could form with a wide range of properties.

At this point, you must realize that this book does not agree with the concept that mesons are responsible for the force holding the nucleus together. In this book, mesons are the result of the nucleus proton/neutron fluid drop leaving the high density of the virtual background sea and entering the normal background sea of photons. The more dense virtual background sea causes the magic number of the nucleus particle to have a higher magic number than when the nucleus particle is in the normal background sea. In the normal background sea the magic number of particles making up the nucleus particle is lower. Thus, if you sum the photons needed for protons and

neutrons outside the middle of an atom, you will see there are
more photons than necessary. The theory here is that those
extra photons are enough to make mesons, which happen to
form after the catastrophic collision in high energy colliders.

The primary conclusion here is that the chaotic action of photons causes particles to cluster together and to persist as a particle cluster in the middle of an atom. This chaotic action is
referred to as the strong force.

Gravity

The philosophy presented here is that gravity is the result of
shifts in the background sea. Vented photons fly straight
away from their home particle. While they are part of the random background sea, the vented photons move away from the
home particle in all directions. Eventually, these photons will
collide with another photon. That creates a higher density
background at that point. As the background sea continues its
random dance, the collisions will chaotically cause a shift of
this higher density volume to a lower density volume of
space. The background shift tends to move particles immersed in the sea. This appears as gravity as the shift moves
the immersed particles toward the original home particle that
caused the shift in the beginning.

The following series of pictures demonstrate this process. In
the picture below there is a few particles shown on the right to
represent a massive number of particles. The wavy lines on
the left represent the background sea. They are to illustrate
the photons are moving randomly and in no particular di-

rection. There is a particle in the middle of the randomly moving photons.

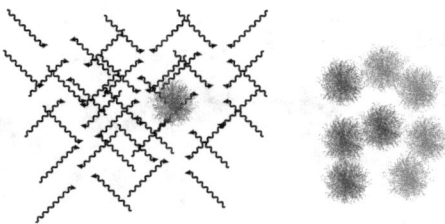

The massive number of particles on the right vents photons as all persistant particles do.

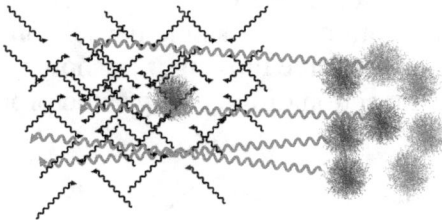

The vented photons will eventually collide with background photons far to the left of this picture. There they stop and become one of the photons in the background sea. At that point, the background sea becomes denser.

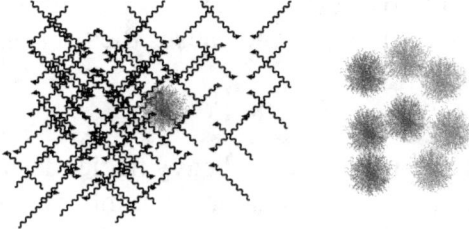

Through random collisions, as thermodynamic molecules in a gas, the sea shifts to balance the density of the photon gas. The wavy lines below are shown to move as a group in the direction of the mass of particles that initially vented photons to the left.

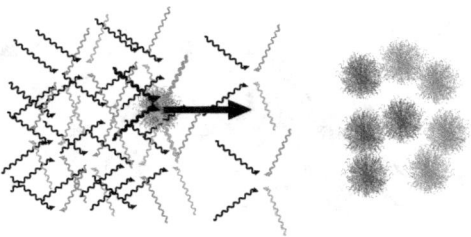

As the background sea is shifting, the particle in the middle of it will vent and absorb photons. The photons absorbed will be moving in the direction of the mass of particles. The tendency is translated into motion of the single particle on the left. The particle experiences a force moving it as shown in the picture. This represents the action of the background producing gravity.

There is another way to view the force of gravity. The particle above is immersed in a shifting body of background photons. Because the photons are moving at the speed of light, the shifting mass occurs quite rapidly. In a sense, the particle is subjected to Background Photon Sea Friction. That is, the particle is dragged along with the natural friction that occurs between a moving particle and the background sea. Here, however, the sea is moving and causes the friction between the particle and the sea.

Weak Force
Weak force is manifested when heavy nuclei decay. The reason is that heavy nuclei have a magic number lower than stable nuclei. Then, the photons can occasionally oscillate in a non-persistent pattern and the particle decays. In this kind of decay, the heavy nucleus emits an electron and a neutrino. Based on what has been presented in this chapter, a specific number of photons and specific amount of energy are bundled together inside the nucleus and pushed out of the atom. That number and amount are determined by the nucleus magic

number inside the atom. When the particle emerges from the inside of the atom, it leaves the virtual background sea of the other nuclear components and the electrons. When the particle hits the outside world, it sees a different magic number. This magic number is lower. This means the ejected particle has too many photons so it vents them immediately. However, the ejected particle has used up some energy during its travel out of the atom. The result is that the new particle vents photons with no energy.

This has an interesting consequence. Understanding that energy is the inverse of the area the circle of the helix cuts out in space, the meaning would be that the helical motions of the photons set free have no energy or have a radius of infinity. As the mass of a particle depends upon the inverse of the area it cuts as it moves through space, the implication is that the photons also form a particle with no mass. The following picture depicts this situation.

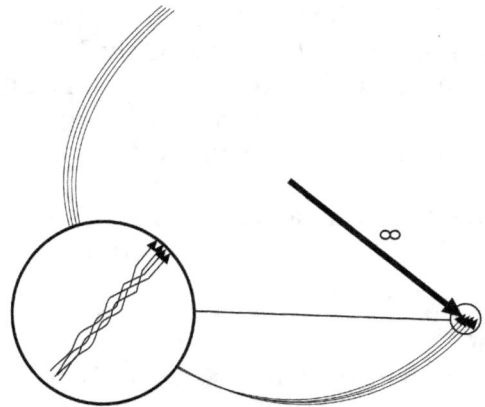

The radius may be so large the mathematical computations to calculate photon motion in the Super Matrix break down. Compare this to the use of computers today. The memory cells inside a computer can hold numbers that are only so large. When numbers in such programs get that size, the mathematical capability suffers. Those that program computers have ways that get around the problem. For example, aircraft guidance systems have trouble calculating sine and cosine functions when directly over the North Pole. Programmers must do the calculations as if the airplane were at the equator and then translate the results to the North Pole.

The point here is that numbers used in the collision algorithms can get so large that the results are abnormal. This is the area of weak force. In this realm, interactions produce different results. A theory here is that this area manifests neutrinos. In this theory, neutrinos are clusters of photons that have essentially infinite or very large helical radii. This creates the appearance that such photons are moving in a straight line through space.

The mathematics of such a large radius creates an environment in which the photons involved bounce off each other in such a way they group together. Such a particle moves faster than the speed of light (by about 1.3 times faster [4]) and interacts very little with the universe. Only direct hits with other photons can shake the group apart. Eventually, during a long trip across the universe, such collisions will occur. Therefore, the neutrino will eventually decay.

Possibly, partial hits occur and a few photons are knocked off the flying group. Such collisions are interpreted as flavor changes. [5]

Conclusion

The primary lesson learned in this chapter is that the forces of the universe depend upon collisions that occur in the middle of particles formed of photons. Another lesson is that the random action of the moving photons gives rise to these forces. The strongest effects of these forces are localized to some small space. When those forces are balanced, the weaker effects of the random motions are felt. Thus, strong forces rule over the center of atoms while weaker gravity affects particles far from that center.

Part 10

Super Matrix

This last part of the book is to tie up many loose ends about what has gone before and build a logical structure of the cosmos and its smallest elements. This is an attempt to answer questions such as, *Where do photons come from?* or *Why do they move in spirals?* An attempt is made to answer the broader question, *What is the physical universe?* Note, there is no attempt to answer the question, *What is the Theory of Everything (TOE)?* In fact, a goal here is to show that this is not a legitimate question.

The first task is to take on some cornerstones of contemporary physics and consider them from a different point of view. These would include Einstein's Theory of General Relativity, Expansion of the Universe, and the concept that entropy always increases. Presently, these concepts say very powerful things about the universe as a whole. Any reasonable theory should explain these or provide some deeper insight into them.

Then we move forward to present a kind of Theory of Everything. Unfortunately, we must pause a bit and consider how science generally approaches problems and how science attempts to understand real world issues. In a sense, it suggests that we cannot acquire a Theory of Everything. The concept of system analysis is presented. This is necessary to see how we must set limits on that which we study and the extent to which we can study.

With that under our belt, we take on some ideas of how the physical universe was created or perhaps where it may have come from.

The process begins with the idea that existence is simple but implementing that simple idea rapidly becomes complex. Thus, the idea must be massaged with a number of additional novel ideas to get the simple idea to work. The result is the helical pattern photons must follow.

All of this is woven together to show that the physical universe is a Forever Universe that is capable of never ending. However, logic enters the picture and reveals a few caveats that the universe must observe to be a perfect perpetual motion machine.

Chapter 35

General Relativity

When Newton released his *Mathamatica Princpea* to the world, he proposed that gravity was a force that acted at some distance on the centers of gravity producing objects. Any object that was made of matter could generate this force on any other object made of matter. He did not attempt to describe the source of this force other than suggesting that matter was its cause. Gravity was, an action at a distance. [1]

Newton reasoned that there is some property of matter that generated a force that pulled two massive objects together. Einstein took a somewhat different point of view. It followed Faraday's approach to the problem of force between charged particles. Faraday observed that a charged particle emanated a field of something that changed space around the charged particle. [2] That field changed the way other charged particles behaved in that field. His point was that two charged particles do not attract or repel each other but experience a force when sitting in one of these fields. That is, the force was generated by something local to the charged particle. The force was not caused by the distant particle even though the distant particle was responsible for the local environment at some distance. In essence, with gravity, Newton suggested action at a distance. With electromagnetism, Faraday suggested the action was caused locally. Einstein took Faraday's thought path.

Theory of General Relativity

Einstein thought that matter somehow changed the space around an object to cause those objects immersed in the changed space to appear to be moving or appear to have a force on it. [3] From this point of view, two matter objects did not pull themselves toward each other. Rather, they actually did nothing. From this perspective, that which the objects were immersed in, changed and any motion or force apparent in said object was an illusion. The observer cannot see the something that is changed around said object and perceives the object is moving or causes the force upon it.

This is somewhat like seeing a dry leaf sliding across the ground on a fall day. Picture yourself sitting in a wicker chair on a warm fall day. The air is for the most part still. You look at the ground and see a golden brown leaf that has been kicked around for a day or two. It suddenly moves as if it is alive. Your intellect tells you that it is a slight wind or gross motion of air around you where you sit. However, the leaf seems to have its own mind. It slides for ten inches and stops. Then it moves a bit to the left. Then it shuffles forward a few more inches. It seems possessed. The leaf seems to have a mind of its own. It is perhaps searching for a bit of food.

Such as what Einstein perceived of the behavior of matter in some kind of altered space. Einstein called his idea the Theory of General Relativity. An explanation of general relativity begins with a mental experiment. This mental experiment results in what Einstein called the Equivalence Principle.

The Equivalence Principle
First, visualize an elevator in free space. That means in outer space, away from all objects that cause gravity. On the bot-

tom of it, there is a rocket engine. The engine is on causing the elevator box to accelerate.

Second, visualize an elevator box sitting on the ground on earth.

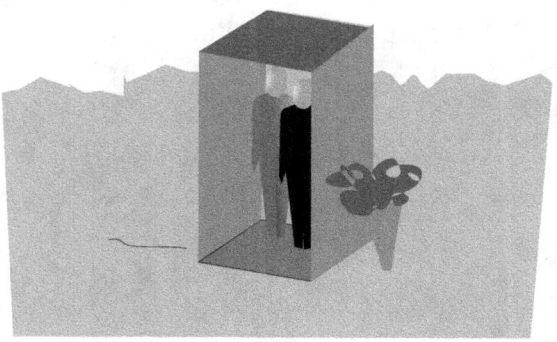

Both boxes are sealed and there are people inside each box.

Further, assume that the people cannot see outside of the boxes and they cannot hear anything outside of the boxes. We are privileged to see into the boxes to observe what happens during this experiment. Then, further assume that the rocket on the bottom of the elevator box in space is accelerating the box at the same rate a stone would fall near the elevator box resting on the ground on the earth. That is, the box in space is accelerating at the same rate gravity causes objects to accelerate downward on earth.

If these elements were true, the people in both boxes would feel a downward pressure the same. If each person were to drop a ball that hit the floor of such boxes, the balls would fall with the same speed. Each person could not determine if they were in the box in space or on earth. Einstein called this the Equivalence Principle. [4] The point of this is to indicate that acceleration and gravitation produce the same force on material objects to which they are subjected.

Warped Space

Now, we must add one more element to this picture. Assume that there is a small light bulb on the side of each elevator box. Assume the light is shining horizontally across each elevator box. The people inside each box observe the light as it travels across the space of the elevator. Consider the box in space first.

When the light leaves the light bulb, traveling as photons, the photons are no longer moving with the box. At the moment they escape the light bulb, they are moving with the box. However, as they move away from the light bulb, they are no longer accelerating. Therefore, the people in the space box will see the light beam curve downward as the light moves across the box. That is, there is nothing to accelerate the photons along the direction of motion of the box. Thus, the box accelerates away from the photons. However, from the point of view of the people in the box, the photons move away from them apparently causing the light to curve downward on its travel across the box.

Based on the Equivalence Principle, the people in the box on earth will see a light there curve down as well.

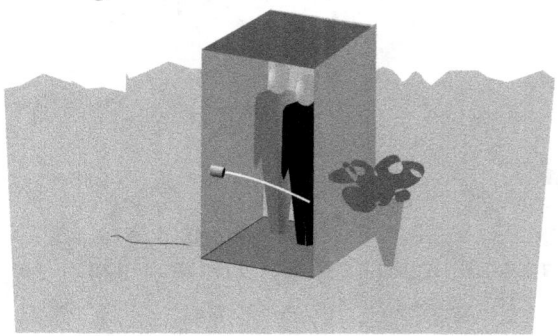

This can be shown to be true.

To pull all of this together, we must consider another law of physics. That is, light moves in straight lines through space. Applying this law to what we have just seen tells us what General Relativity is. Because the light in the box on earth is moving in a curve across the box, the space near the earth must be curved. [5] This allows the law that light moves in a straight line to remain accurate.

Another Look at Warped Space

Einstein's theory proposes that matter changes space that produces what we see as motion and force. Einstein proposed that matter warped the space around the matter object. That is, one object warped space around itself such that another object immersed in that space would appear to move. In reality, the object is not moving and is stationary in this space.

The space is changing just as the air around the leaf was
changing causing the leaf to appear alive.

This is a somewhat difficult idea to grasp so here is a bit more
explanation. Consider two rocks sitting in space.

Let's view these as if space is not warped. The rocks will just
sit there with nothing happening. The following is a graph
representing this non-moving state.

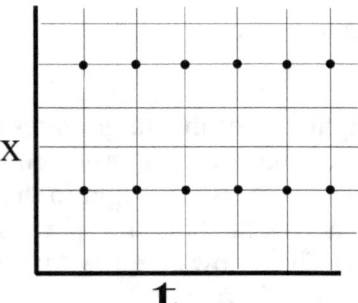

Study this graph closely. The vertical component of the
graph, X, represents the distance between the two rocks in the
previous picture. As we have decided that the rocks are in a
space that is not affected by the presence of matter, the rocks
do not move and X is always the same. The graph represents
that. The upper line of dots represent the rock far away and
the lower line of dots represent the rock near us. You can see

the dots extend to the right. The lower bar of the graph represents time. The dots extending to the right represent the change of time. The graph tells us that the rocks are always the same distance from each as time ticks on. That is, in space they are not moving.

When Einstein approached the problem of gravity, he approached it from the point of view that time is but another dimension like distance. Thus, he viewed the universe as something called space-time. The graph is a representation of space-time even thought it represents only one dimension of distance.

The next step is to realized that the matter in the rocks do affect or warp the space around them.

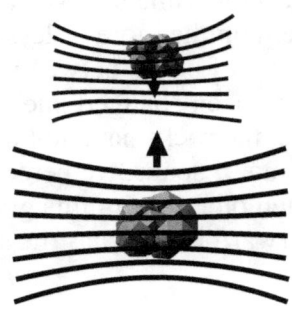

This is represented by the following graph.

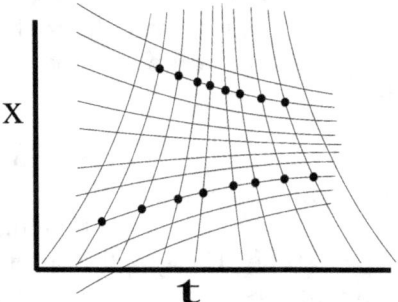

This graph represents a warped space-time. This is indicated by the lines representing X. Now those lines are curved, representing curved space. Note, from the perspective of the graph, the rocks are still the same distance apart. If you count the boxes between them on the more or less vertical lines, there is the same number of boxes between them. However, the size of each box has been changed due to the warping of space-time. In essence, the rocks have not moved. As we cannot perceive the warping of space, we perceive that the rocks are moving toward each other. From this graph, you can see that a rock moving forward in time has the space change around it.

If you understand this small discussion, you understand what Einstein's General Relativity is. Do not worry why space is curved. Do not attempt to understand what light is, what gravity is, or what space is. Right now, focus on the idea that light in a box on earth curves, as does a beam of light in a box accelerating in free space. All of the hand waving that occurs is about the intricate mathematics of drawing lines around massive objects to show where light would go when moving around objects made of matter. While working with this concept, the physicists are not concerned with the meaning of light, space, and so on. Outside of this they do. However, when discussing this, they are only concerned with the math of why those beams of light curve downward.

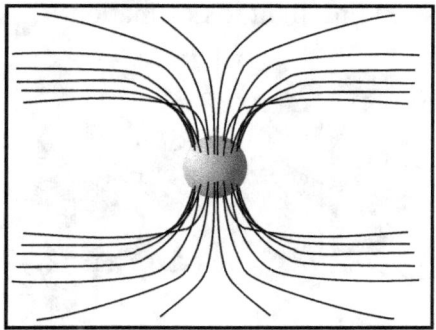

It gets exciting when this math is applied to black holes that seem to have infinite mass. Physicists get excited about this.

It's just about those little light beams bending in an elevator box. When the physicists are done with the math, they draw lines about anything that is made of matter and say that the line represents the effect of gravity. Nothing is being pulled. We are just sliding along these invisible lines.

One little Flaw

Hopefully, you got all this. Now, we can get on with pointing out one little flaw in this scenario. Light does not travel in straight lines. [6] Einstein's work was done without the logic of quantum mechanics, specifically, Quantum Electro-dynamics (QED).

In studying that, one learns that photons move in every pos-

sible direction when leaving a light source. Consider a light
source A that is sending light to destination B.

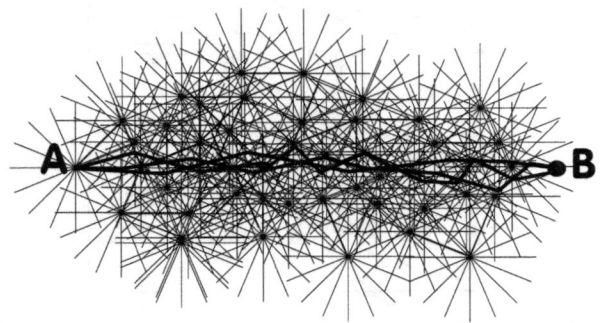

The dots in the above picture represent electrons. The lines
represent the paths of photons leaving the electrons or entering some other electron. Note there are many lines going in
many directions. When the light leaves A, it travels in all
directions. Particles around the light source A, specifically
electrons, absorb the photons and reemit them. Again, the
emission is in every possible direction. Thus, as the light
(photons) travels away from A, it is going almost everywhere. Now, if you look at everywhere light can travel and
measure its intensity, you will see that there is a very high
probability most of the light going to B will travel near a
straight line from A to B. That is, if we observe the light
arriving at point B from A, most of the light arriving at point
B travels this straight-line path. To repeat, the light leaving A
travels in every possible direction. At each point a photon is
absorbed and emitted, it again travels in every possible direction. This total randomness results in a situation in which a
majority of the light traveling from A to B, travels on a
straight line from A to B. This observation requires a change
in the definition of what a ray of light is.

A ray of light is a beam of photons that carry information
from point A to point B in a straight line. However, the photons that carry the information are constantly changing during
this trip. Chances are that the photons that are emitted from

point A never get close to point B. As soon as the photons leave the point A source, they are captured by whatever is nearby and that whatever, emits a photon. Chances are good that the emitted photon is not the same photon that was absorbed. Thus, the photons that apparently travel along the beam are constantly changing. [7]

So, part of this clarified definition is that a ray of light transmits some data from A to B along a straight line. However, many different photons cooperate in the transfer and many of them do not move directly from A to B. The photons leaving from point A never get to point B.

What does this have to do with General Relativity? Well, when a ray of light leaves the light bulb from the side of the elevator, the emitted photons are immediately absorbed by any particles nearby, usually electrons. Then they are re-emitted. Electrons are matter and are affected by acceleration and gravity. While a photon or photons participating in the ray of light traveling across the elevator are in the electron, they are thus subject to the same forces the electron is subject to. Thus, acceleration will affect how the ray moves across the elevator. Note the distinction here. Photons are not affected by gravity. However, light rays are. Photons moving through space are not the same as a ray of light moving through space. Up to now, both of these were considered the

same. The point is that gravity does not affect a photon of
light but does affect a ray of light.

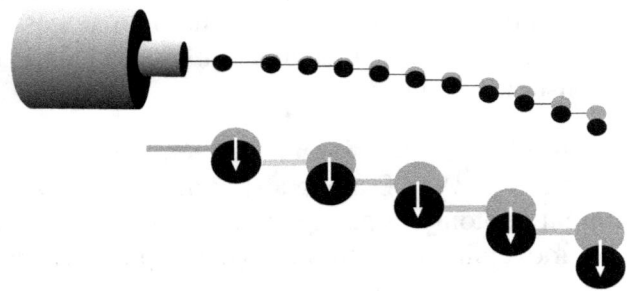

In the above picture, the straight lines represent the motion of
photons leaving a light bulb. The photons are absorbed by
electrons, shown here is dark disks. The electron absorbs a
photon and perhaps drops due to gravity. Then a photon is
emitted to carry the ray forward. The point here is that the
photon is not affected by the acceleration or gravity in its
travels from particle to particle; or electron to electron. This
conclusion is drawn from the material presented earlier in this
book. Specifically, photons only interact when colliding with
other photons.

Does this mean General Relativity is wrong? No, it does not.
It does suggest that a shift in point of view is necessary to see
what General Relativity represents. In the material presented
in this book, General Relativity maps out the density of the
background sea. In a sense, it is saying that space is the background sea. From that perspective, to say the background sea
is warped would be accurate. As we have seen, the presence
of particles (matter) introduces more photons into the sea.
Therefore, massive objects do warp space from the point of
view that the background sea is space.

Conclusion
Hopefully, we have clarified a great deal about General Relativity and space. Here, a point of view has been suggested

that space is defined as the Background Sea of Photons. That is the space we are accustomed to dealing with in our everyday life. This is logical. We can wave our hand through it as if it were the big nothing. When we suck the air out of a container, we suck all the air out. There is nothing there, which is to say, there is no matter there. However, the Background Sea of Photons is always there.

We have also clarified what a ray or beam of light is. It is a collection of photons passing data from one point to another. Collectively, the participating photons support Snell's Law, a ray of light moves from point A to point B in a straight line. [8] We also have a better understanding about subspace or the Super Matrix. It is the underlying structure supporting the universe. From the perspective of the beings that live within it, it is a homogeneous, invisible structure that supports the entire physical universe. It can only be detected by observing the motions of the particles it supports.

An additional conclusion can be drawn from what has just been presented. Per this discussion, gravity waves would be a systematic shift in the background sea. The term sea here is very appropriate. When an object of mass moves: the density of the background sea changes with it. Just as in a water wave, a part of the background sea will move and push another part of the sea out. This would appear as a wave in the background sea. As gravity is the result of differing densities of the background sea, this wave would appear as a gravity wave.

Chapter 36

Expansion of the Universe

The concept of an expanding universe hinges on two cornerstones of physics. One is Einstein's Theory of General Relativity. The other is the red shift of light coming from distant celestial objects. In this chapter, we take a close look at these subjects and attempt to understand how they imply expansion. To do this we spend some time attempting to gain understanding of both. We start by considering a history of the Doppler Shift and Einstein's General Relativity in regards to the expansion of the universe. We look at the work by Edwin Hubble and then analyze the two cornerstones.

We will find that the results accepted by the community are dubious. This claim can be made with a casual examination of the subject. Ostensibly, the initial conclusion of stellar red shifts was that the universe is expanding. The implication is that the fabric of the universe is expanding. This implication further implies that galaxies can move apart faster than light can travel. [1] The suggestion here is that matter is not moving apart at the speed of light but space itself is expanding faster than the speed of light. If an expanding universe of this nature were true, then the ruler used to measure this expansion would be expanding as well and render it useless as a measuring device. Ergo, if the universe were expanding, we could not observe it. [2] Proponents of expansion then say that the gravity within galaxies prevents them from expanding and the space between galaxies is expanding. [3] This would appear to be an Occum Band-Aid. That is, such an explanation seems questionable.

The conclusion of this author is that, while matter may be

spreading or even gathering in the universe, the universe is not expanding or contracting.

We begin with some history.

The Doppler Shift and Expansion

The use of the red shift of light coming from far stars to measure the speed of distant celestial objects has a long history. Christian Doppler explained the effect named after him in 1845. [4] He predicted that the phenomenon could be used to determine speed of stars relative to Earth's position. [5] The first Doppler red shift observation occurred in 1848 by French physicist Hippolyte Fizeau. In 1868, British astronomer William Huggins used it to determine the velocity of a star moving away from the Earth. [6] In 1887, Vogel and Scheiner discovered the annual Doppler effect, the yearly change in the Doppler shift of stars located near the ecliptic due to the orbital velocity of the Earth. [7] In 1901, Aristarkh Belopolsky verified optical redshift in the laboratory using a system of rotating mirrors. [8] In 1912, Vesto Slipher discovered that most galaxies had considerable red shifts. Subsequently, Edwin Hubble discovered an approximate relationship between the red shifts of galaxies and the distances to them called Hubble's Law. [9]

General Relativity and Expansion

The red shift was not the only device to support an expanding universe. General Relativity predicted it also. Willem de Sitter, a Dutch astronomer, the first that used Einstein's theory in 1917 to work out a mathematical description of the tendency for the spiral galaxies to have redshifts, as observed by Vesto Slipher. [10] It was an approximation however, as the hypothetical universe he used did not contain matter. Five years later Alexander Friedman discovered an error in Einstein's solutions and worked out his own solutions to the general relativity equations. Friedman described two possibilities for the universe that were consistent with general relativity. One was expanding, the other contracting. Both were realistic enough to contain matter. Friedman also showed that it was impos-

sible using theory alone to tell which description corresponded to the real universe. [11] Abbe Georges Lemaitre
was a Belgian priest and mathematician who claims to be the
first to derive the Hubble relationship using general relativity
equations in 1927. The claim is that he did it while isolated in
Belgium. [12] Lemaitre is considered the originator of the big
bang theory. [13]

This claim implies the theory, as opposed to Friedman's
work, predicted expansion. Lemaitre further proposed that, as
expansion was clearly present, the expansion originated from
a singular point, which came to be known as the Big Bang.
Believing Lemaitre's claim of isolation is difficult as he had
constant contact with Eddington and visited the United States
during the ten year time period Hubble was performing his
measurements. [14] Furthermore, in a paper presenting his
findings, a table was mentioned from which a Hubble like
relationship was developed. The paper was in French and
when translated to English, the reference of the use of this
table was not translated. [15] While considering the wor-
thiness of Lemaitre's statements or the truthfulness of those
that reported it are of interest, the item of real interest is to observe that the theoretical use of General Relativity probably
was not totally used to determine the expansion of the universe. Friedman had been over that path and found that the
theory could not establish real world events. [16]

Even Einstein could not establish this and was required to
review Hubble's work to conclude how the general theory of
relativity should be structured. [17]

One cannot overlook the fact that Lemaitre was responsible
for the Big Bang concept however. This concept has had an
enormous impact on contemporary physics.

Hubble
Hubble's law is the most powerful reason for the belief in the
expansion of the universe and the Big Bang.

In 1919, Edwin Hubble became the chief astronomer at the Mount Wilson Observatory. [18] At that time, astronomy was growing and the understanding of space was very limited. Human awareness had moved outside of our solar system and viewed the galaxy as the limits of the universe. Hubble studied this universe and observed a pulsating star. He realized that he could use the pulsations of this star to determine how far away it was from earth. He found it was outside of our galaxy. With this conclusion, he realized that there was a great deal of stellar material that had not been seen. [19]

Can you see what a significant discovery this is?

Now we understand that our planet is but a nano spec in the cosmos. However, then, we thought it was a nano spec in a galaxy. He, single handedly, revealed that our galaxy is a nano spec in the universe. He single handedly opened a door at the edge of our awareness, to reveal what was out there. He has not received significant recognition for this accomplishment. This is unfortunate.

Then he took another step that revolutionized our science even more. He observed that celestial bodies are speeding away from earth. He used Christian Doppler's, Doppler Effect. From this, he produced the Hubble equation. Hubble was a dedicated scientist and proceeded with his work carefully. It is worth noting that, although he produced this relationship, he did not suggest the universe was expanding. [20]

Doppler Effect Musings

We need to understand what the Doppler Effect is. We need to forget the cosmos for a moment and study this phenomenon. Today it is noticed mostly when we hear helicopters flying overhead. Most are familiar with the chop, chop sound of the helicopter blade. The blade is rotating over the body of the helicopter. The blade is also twisting as the whole blade rotates in a large circle. When the blade moves into the air as the helicopter moves forward, the blade must twist flat to this oncoming blast of air. When the blade moves

away from the oncoming blast of air, it must twist at a greater
angle to push more air downward. On that side of the large
circle of motion, the blade is moving with the onrush of air so
the blade must take a bigger bite of air to keep that side of the
helicopter upright. Thus, the blades are twisting very rapidly
to lift the helicopter and keep it level. This rapid twisting back
and forth creates the chop-chop sound characteristic of
helicopters.

This noise gives rise to a clear demonstration of the Doppler
Effect. This is apparent when the helicopter is flying toward
you and away from you as it flies overhead. As the machine
flies toward you, the chop-chop sound is fairly, quick. When
the machine goes over your head and moves away from you,
the rate of chop-chop decreases. This is due to the Doppler
Effect of the sound waves. When the helicopter is moving
toward you, the sound waves are colliding with your ears
quickly as the waves are generated by something moving
toward you.

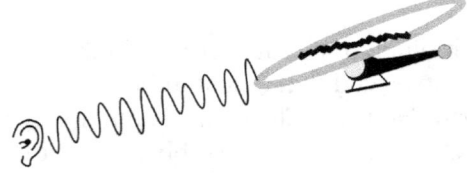

As the helicopter moves away, the sound waves are longer as
the thing generating them is moving away from you.

As mentioned, the speed of celestial bodies relative to earth
had already been measured using the Doppler Effect. If a star

were moving toward us, the light waves we see would be shorter.

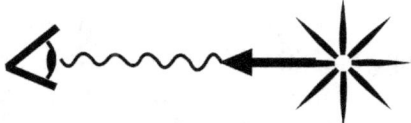

If a star were moving away from us, the light waves we see would appear longer.

Now, with a helicopter, the medium we detect it with is sound. Longer and shorter waves produce a different tone or sound. We do not hear light; we see light and perceive a color dependent on the wavelength of the light. The theory is that the color of light from stars could be studied to determine if they were moving toward or away, and how fast they were moving.

One of Hubble's first tasks was to locate nearby stars that gave off a characteristic color of light. Via a variety of different methods, he studied a particular kind of star in our galaxy that constantly produced that kind of light. [21] Then he searched for stars of those kinds in galaxies far away. In this way, he could know what kind of light the far star was emitting. Then, he could look at the light from that star and, using the Doppler Effect, determine if it were moving away from us or toward us. He could also determine the rate of motion.

Based on the Doppler shift, He discovered that far stars are moving away from us. He also discovered that the further a star was from us, the faster it was moving away from us. This led to the concept of an expanding universe. That is, if the rate stars were moving away from us depended on their distance from us, all the stars in the universe were moving away from each other. This suggests that the universe is expanding. Furthermore, it suggests that, if the universe is expanding, we could mentally reverse the expansion and trace backward to see where the expansion began. This leads to the idea of the Big Bang. That is, everything in the universe came from some small point long ago. Observing how fast the stars are moving now and how far away they are; gives us some ability to determine how long ago the universe was this small point and thus, how old the universe is. [22]

General Relativity Musings

In this section, we attempt to grasp the relevance of general relativity to the expansion of the universe. Some time will be spent doing this, as it is complicated.

The Theory of General Relativity predicted the existence of Black Holes. Likewise, the theory predicted the expansion or contraction of the universe. The goal in this chapter is to discern how the theory did this. Unfortunately we, the uninformed public, are spoon-fed the results the high power mathematicians arrive at. Then they give it to us in some language we cannot understand.

The goal in this section is to demonstrate how some equations can show that black holes can exist and how some equations can show that the universe is expanding or contracting. This is not an explanation of general relativity. The effort here is to connect a comment about equations with something we can understand. When the informed say, *This equation shows the universe is expanding,* the common person has no idea what this means. The goal here is to show how some equations can be interpreted to reveal such information.

The understanding of this is not necessary to understand the new age presented here. Rather, this is to show why present understanding should be left behind. Jump to "More Caveats," the next section, if you wish to avoid this tedium.

What Equations Do?

First, we discuss how mathematical equations are used. To understand this, we need to look at a simple example. We will consider a small equation we can understand and study it. Then we will use general relativity equations in a similar way.

Consider a weight on a spring.

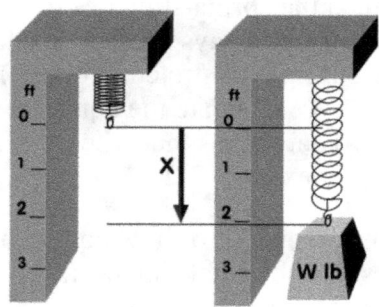

The structure on the left has a spring without a weight on it. The one on the right has a weight on it. We know with Hook's Law that the spring will extend with weight on it. Hook was a man that lived during Newton's time and developed this equation. The equation appears as follows:

$$X=KW$$

X=Distance spring extended.
K is Hook's spring constant.
W=Weight on spring.

K, Hook's constant, is a value that indicates how much the spring stretches when weight is applied. In general, the equation is used like this.

Some result = Some Equation [Parameters]

We have some equation. Then we have some parameters. We plug the parameters into the equation, evaluate the equation and we get some result. Here is how it looks with Hook's Law.

2 ft = Hook's Equation [1 ft/lb, 2 lbs]

The parameters put into Hook's equation is K that equals 1 ft/lb and W a weight that is 2 lbs. K is called a spring constant and gives how much the spring stretches for each pound of weight put on the spring. It says here, for each pound (lb) put on the spring, the spring will stretch a foot (ft). Once we have the parameters we evaluate the equation and get an answer which is 2 ft. That means that when we hang a weight of 2 lbs on the spring, it extends 2 ft.

This serves as an explanation of how equations can be used. One enters values into the right side of the equation to get the answer on the left side of the equation.

General Relativity Equations

We will not present Einstein's equations. The math involved is far beyond this book. However, we will consider them as equations and examine them in a similar way as we did Hook's Law. The method is to show input to the equations and then show possible outputs from the equations. In the texts that purport to explain these equations, the relationship between the inputs and outputs has been offered. For example, the general relativity equations are to show how light bends when near large masses that produce a large gravity field. We will view this as follows:

Amount Light Bent = GR Eq. [Mass]

The General Relativity Equation (GR Eq.) is an equation just as Hook's Law. The input to the equation is put into the brackets. The equation is evaluated with that input and produces an answer. Here the input is a value for a large mass. The output is the amount the light is bent.

We will then use this method to observe what the general relativity equation has to say about the space, matter, and the universe as a whole. To do this we use a variety of values for the inputs to the equations and see what outputs are produced. For example, we will use small and large values for mass. We will see that the equation shows that small mass has no effect on light, a large mass causes light to bend, and a huge mass will cause light to curl and essentially stop moving.

We begin with the most common example.

General Relativity and Bending Light
The following picture describes a mass and surrounding space.

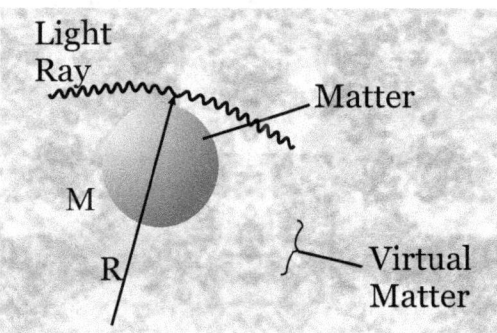

There is a lot of detail here. Right now, let's just consider mass M and a light ray passing near M. These are the input parameters to the equation. Notice the arrow labeled R. That is the radius of the light ray nearest mass M. R is an output from the equation. The radius of the light ray indicates how much the light beam is bent as it passes near M.

According to Einstein's Theory of General Relativity, the

light ray will bend near the mass as space is warped. We can use a general relativity equation to determine the radius.

R=GR Eq. [M]

If we plug in the mass of the earth for M, we can get an idea of how much the light ray curves downward as it passes across the elevator box. This is shown as the curvature (R) of the light ray in the picture just shown.

Space-Time and Momentum-Energy
The equations Einstein developed for general relativity described reality from the point of view of space-time and momentum-energy. The following picture is an attempt to show what the equations dealt with.

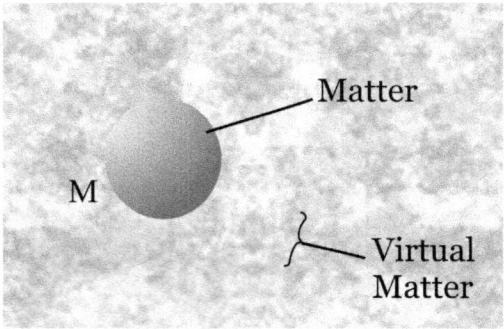

This picture shows one ball of matter. The grainy background is to represent the background sea of photons and the appearance of virtual particles. Note that virtual particles appear randomly in the background sea and suddenly melt back into the background sea. Although Einstein did not visualize space this way, he did attempt to describe space as containing energy and momentum.

The primary point is that Einstein designed the equations to represent reality. If he were to be successful, he would have described all of this with accuracy. Time has shown that his description did this very successfully.

This one point must be emphasized. The model had to match reality. This could not be done with theory alone. During Einstein's task of creating this theory, he had to constantly examine the real world and make his model match that. Theory does not exist alone.

General Relativity and Small Masses
Now, let's take up the example of a small mass with light passing nearby. The following picture depicts this. It depicts mass M and a light ray passing nearby.

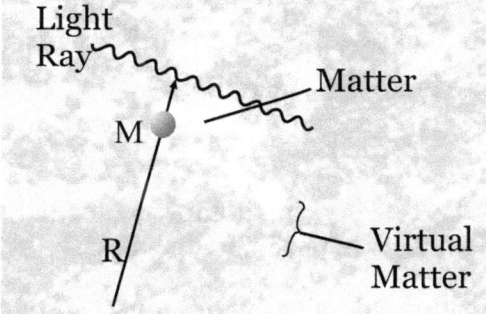

The mass is the input to the equation.

R=GR Eq. [M]

We are after output that indicates how much the light bends. The output again is the radius R of the light wave curvature. The radius output would be very large implying the light ray is not bending. We see that little changes.

General Relativity and Black Holes
Look what happens when we make M a very large mass.

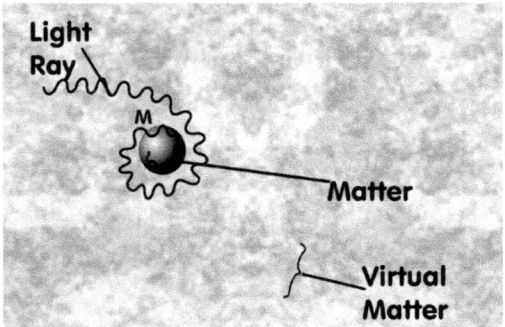

We put a large M into the equation.

R=GR Eq. [M]

Now, R will be very, very, small implying the light ray curls
into itself and remains trapped near the mass.

Trying combinations of values in the equation suggests that
very large masses attract everything in their vicinity and prevents them from going elsewhere. The celestial bodies pos-
sessing this character have come to be known as Black Holes.
This is derived from the idea that if light cannot escape them,
they must be black.

Relativity and the Cosmos
Now we have seen a process of using the equations to study
what can happen to space around bodies of matter. We did
this by trying out various values in the equations to see the effect of the equations. We see we have predicted real world
phenomena: the effect of small masses, the effect of matter on
light and the effect of very large masses, black holes.

In this section, we wish to expand this study to the entire universe so we can see how the cosmos interprets the equations.

As we have seen, we experiment with the equations by trying out various combinations of the parameters. We approach the task here in the same way. One problem is that we do not know how the universe is laid out. The task then, is to try combinations of universal structure. Here are some possibilities. One is that space is infinite and the amount of matter in it is infinite. Another possibility is that space is very large but finite and filled with matter. Yet another is that space is very large and matter occupies some small part of that space.

Right now, we are going to limit the possibilities. Our primary purpose is to illustrate how some equations thought to be representative of the universe could be used to study universal structure. Let's look at the possibilities presently selected, and see what we can see.

The first possibility is that space and matter is infinite. Should that be the case, we should be receiving light from an infinite number of stars. Then, according to Olbers' Paradox, the sky should be lit brightly because we would be receiving an infinite amount of light from all angles of the sky. [23] Therefore, it is unlikely that this is a real combination.

The second possibility could occur when the Super Matrix came into being and photons were placed into the matrix space evenly and then let go. This would not be conducive to a Big Bang but would provide for photons clustering into stars, galaxies and so on.

The third possibility would be a bit more striking. In this, the Super Matrix is mostly void of photons except at one unidentified spot. This scenario is somewhat like placing a small bottle of strong perfume in the corner of a gymnasium. The molecules of the perfume would diffuse throughout the large room. That would be somewhat like a Big Bang. The perfume would spread quickly. However, it is unlikely the perfume cloud would accelerate through the room but start quickly and slow as the room was filled. Perhaps the photon gas flashes out to fill the void. Then matter, forming slower, attempts to catch up and is pulled out by the virtual particles appearing in the photon gas that flashed outward.

We now consider each of these seeing how the equations might behave.

Infinite Space, Infinite Matter
Assume we have infinite space that is filled with an infinite amount of matter. The following picture illustrates this. The black dots represent matter in the space. The grainy texture represents the appearance of virtual particles.

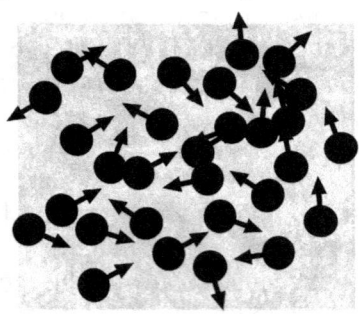

Even though the virtual particles appear in the sea and melt back into the sea very quickly, it happens often enough that the combination of all virtual events generates a gravity field similar to permanent matter. As the space is infinite and the number of celestial objects is infinite, everything is filled up and the force of gravity is equal everywhere. Thus, the celestial objects can meander in a variety of directions.

The equations that indicate this would appear as follows:

V_n = GR Eq. [P(0-n),M(0-n0),Space Energy Density, n]

P(0-n) is the position of each celestial object.
M(0-n) is the mass of each celestial object.

Space Energy Density is the density of the background sea.
'n' is the object we wish to calculate the velocity for.

The output is the velocity of each chosen object. Vn is a velocity for a particular object. If we try calculations for each object n, Vn would show that the objects are moving in random directions.

Matter Filled Finite Space
In the next possibility, assume we have a finite space filled to the edge with matter. Again, the black dots represent matter in the space and the grainy texture represents the appearance of virtual particles.

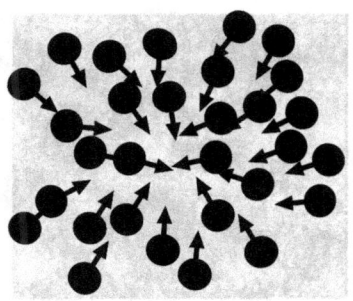

Again, the virtual events generate a gravity field similar to permanent matter. However, as the space is finite there are no virtual events outside the space filled by real matter. Therefore, there is nothing to balance the matter inside the space and the celestial objects. The result is that the objects all move toward the center of that gravity field.

The equations that indicate this would appear the same as just presented.

Vn = GR Eq. [P(0-n), M(0-n0), Space Energy Density, n]

P(0-n) is the position of each celestial object.
M(0-n) is the mass of each celestial object.

Space Energy Density is the density of the background sea.
'n' is the the object we wish to calculate the velocity for.

The output is the velocity of each chosen object. Vn is a velocity for a particular object. If we try calculations for each object n, Vn would show that the objects are moving toward the
middle of the space.

Fininte Space, Limited Matter
For the last possibility, assume we have a finite space of
which the center contains a limited amount of matter. Again,
the black dots represent matter in the space and the grainy
texture represents the appearance of virtual particles.

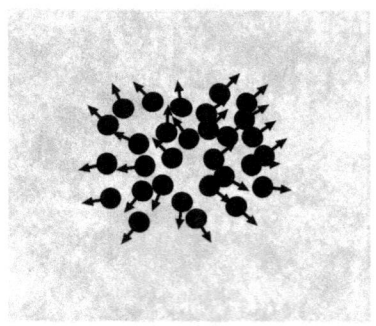

Again, the virtual events generate a gravity field similar to
permanent matter. However, as the matter in space is small
compared to the size of space, the virtual events outside the
real matter can cause a gravitational field. If the objects are
clustered close enough and the virtual events inside the group
is small compared to that outside the group, the objects all
move outward attracted by the gravity generated by the virtual
particles outside the field of matter.

The equations that indicate this would appear similar to those
just presented. The difference would be an indication of the
size of the background sea.

Vn = GR Eq. [P(0-n), M(0-n0), Space Energy Density, S, n]

P(0-n) is the position of each celestial object.
M(0-n) is the mass of each celestial object.
Space Energy Density is the density of the background sea.
S is the size of empty space relative to celestial objects.
'n' is the the object we wish to calculate the velocity for.

The output is the velocity of each chosen object. Vn is a velocity for a particular object. If we try calculations for each object n, Vn would show that the objects are moving away from the middle of the space.

Relativity and the Cosmos Conclusion
Using Einstein's equations of general relativity is similar to the application of Maxwell's Equations to electromagnetic fields in space. Maxwell's equations explain the properties of electromagnetic fields with great accuracy. However, to apply the equations to a space, one must know the boundary conditions of that space. Logic suggests that we need to know the boundary conditions of the space being studied to use Einstein's equations in a similar way. However, while applying general relativity equations, we cannot see to the edge of space, much less the fields of celestial mass. We cannot determine the boundry conditions and therefore cannot effectively apply Einstein's General Relativity equations to the problem. [24]

More Caveats
Over time as more and more people study the red shift phenomenon and our knowledge of the universe grows, the established reasons for the expansion of the universe are becoming weaker.

Cosmological Model
As discussions of red shift have matured, Doppler red shift has fallen from favor. The reasoning is that the effects of the long distance from the far star to earth distort the effects of the Doppler shift. The popular idea is that the light began long

ago from some distant star. During that time the space has
changed over which the light traveled. That is, space was
more compressed when the light began its travel. During that
time, the space between the star and earth expanded. This is
to have caused a shift in the light. [25] Another factor is that
the light coming here is formed near a star, which has a very
heavy gravity field. Due to that, the light wave is crushed. As
it leaves the gravity field, the light enters uncrushed space
causing it to get longer, thus red shifting. [26]

QED Redshift

This section presents a different interpretation of the red shift
phenomenon. The Hubble logic was done without understanding of Quantum Electrodynamics. This has an effect
here just as it did with Einstein's General Relativity presented
in the previous chapter. Hubble assumed that light travels in a
straight line. This assumption requires that a photon of light
leaves a far star and travels intact to our eye here on earth. If
that were the case, the Hubble Law would seem to be entirely
accurate.

The redefinition of a ray of light, from the earlier chapter, presents an alternate possibility. The following picture represents a star relatively close to earth. The light from the star
radiates in all directions. Light moves by being absorbed by
any material around the emanating object. In space, this is
often space dust. That can take on the form of hydrogen molecules floating in space. When the emanated light hits these
particles, the light is absorbed and reemitted in random directions. Some of this may be toward earth. In the following pic-

ture, the thick line connecting earth to a star represents a common path of light from the star to earth.

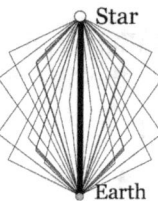

If an analysis is done of the light coming from the star, the straight path from the star to earth will be seen to be the most common path. Note that that path does not consist of the same photons. This is a statistical average of all the different photons carrying the data. Therefore, we observe that light moves in a straight line from the star to us on earth.

Consider a star much further away. Once again, the light is emanated in all directions from the star. However, note that less is emitted directly toward earth. Because the star is further away, much more light is emitted away from a direct path to earth. Thus, the line directly from the star to earth is no longer the most common path to earth. This is shown as a thin line connecting earth to a star.

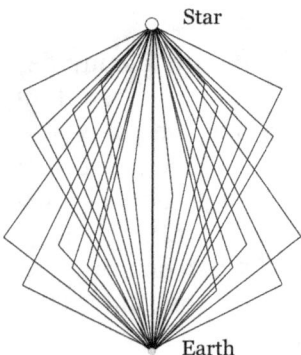

The surrounding paths now contribute more to the light that travels from the star to earth. As you can see, in this scenario, the light seen on earth seems to have traveled much further than that which would take a straight-line path. The primary point here is that light does not take one path to get to earth. Light takes different paths that require the light to travel a longer distance. Right now, this is not clear why this is important. We need to discuss interference of a split beam of light. To do this we leave the realm of space and look down at a puddle in a parking lot near the street.

The phenomena we need to understand appears when some oil floats on top of a mud puddle. We look at the oil slick and see a variety of colors. When oil floats on water, it creates two reflective surfaces. [27] The following picture depicts this.

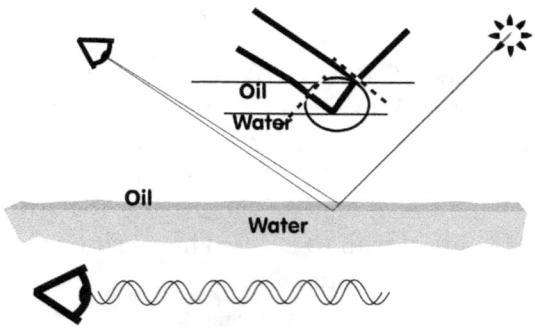

The image in the middle of the picture depicts light coming from the sun, hitting the oil, being split into two parts, and being observed by a human. The blown up image in the upper part of the picture shows the detail of the light being split at the two boundaries. The light beam in the circle is the distance the lower ray must travel to get to the eye. The two beams that are rejoined when entering the eye are shifted from each other just a bit. When they enter the eye together, the color is shifted from the original color of the light. This is depicted at the bottom of the picture.

The point of this little experiment is to show because a star is far away, the direct line between the star and earth is not the major contributor to the light we see. Some beams are taking a side trip and travel farther than a beam taking the direct trip. These multiple beams can interfere with each other giving rise to a different color. This can appear as a red shift in color. This shift is then a result of the distance of the star, not its motion.

Conclusion

There now seems to be several reasons for a red shift.

- Doppler red shift.
- QED red shift.
- General Relativity red shift.

There is not sufficient evidence to establish the heavens are moving apart. The initial reason to make the claim that everything is expanding was the Doppler red shift. Its use for that is in doubt in fact and by the community. [28] The community purports that the correct view is the one of General Relativity red shift establishing what is called a cosmological model. That however depends upon the observation of Doppler red shift. The observation is required to determine how to apply general relativity. Therefore, using some cosmological model based on general relativity does not make sense. Then, QED red shift suggests that the red shift has nothing to do with universal expansion or the velocity of the stars. The QED red shift does not imply that distant celestial objects are systematically moving away from earth based on their distance away. The conclusion here is that, unless otherwise demonstrated, stars are moving in a variety of directions relative to earth but there is not an ordered motion of stars away from earth. The conclusion would be that the universe is not expanding.

From the point of view of this author, the universe is finite. Possibly a number of photons were dumped somewhere in the middle of the Super Matrix. The strong perfume model would cause the photon background sea to expand outward. Matter would form in the middle of the background sea. The dark matter exterior to the matter filled area could create gravity outside the matter clump to pull it outward as in something that would appear as an expansion. Inertia would probably carry the clump farther than some balance point and then there would be a contraction. Inertia would carry that past some balance point inward. Then, the matter in the universe would oscillate back and forth like a weight on a spring. Each oscillation would be smaller and smaller until the matter came to a quiescent state.

In essence, there is no expansion of the universe. That does not preclude a lot of moving around. Space is not warped. However, the conclusion is that we just do not know. We don't have the data to make such a decision.

Chapter 37

Entropy

The universe is cooling off. Eventually everything will be ice and nothing will live. The stars will eventually burn out and the sky will be dark.

Today, this is the accepted belief of the future of the universe. The universe is so big however; this doom is far in the future. This concept is a result of entropy. We need to understand what it is.

However, why is this subject introduced along with general relativity and expansion of the universe? Essentially this part of the book is about how the Super Matrix theory supports the universe. It is based on helical photon motion, the Laws of Photon Interaction, and the Laws of Photon Transmission. As these laws are the cornerstone of the Super Matrix Theory, they are the basis of entropy as well. The information available from the traditional understanding of entropy only covers half of its behavior. There is an unknown half, which is very important. Thus, we study it here.

Entropy is part of the study of thermodynamics. Thermodynamics is the study of heat and energy. We can understand heat intuitively. A hot cup of coffee has heat in it. We know that if we let the coffee sit on the table for a while it will cool. Therefore, we understand that heat in one object will move to somewhere else that is cooler. What we may not understand is that the entropy in the coffee and surrounding environment has increased.

What is Entropy?

Essentially, entropy is the amount of work some system has
done that cannot be done again. [1] The primary issue is that
the process is irreversible. That is, the energy cannot move
back easily as it moved forward. However, the energy can be
moved back if some outside process intervenes. Then the entropy of both systems increases.

Let's use an example of an ice cube floating in a glass of water.

Consider the ice and water a closed system. That is, we view
the two as a unit and do not allow energy to enter or leave the
system consisting of the two.

The water has energy called Q1. The ice has energy called
Q2. The total energy is the sum of Q1 and Q2 or Q1 + Q2.

Some of the energy in Q1, the water, moves to the ice Q2.
Let's call this amount of energy QM.

The ice melts so now the energy in the water is Q1-QM and
the energy in the water that was the ice is Q2+QM. The total
energy then is Q1-QM + Q2-QM which is the same as
Q1+Q2.

The first point of this is to observe that QM energy has moved
from the water to the ice. The second point to observe is that
the amount of QM energy cannot be moved from the water
that was the ice cube to the water that wasn't the ice cube.
The entropy of the system has been increased by the amount
QM.

Note that the energy in the system has not changed. No heat
from outside the system has gotten into the original ice and
water system. In addition, no energy has left the original ice
and water system.

Entropy is a device that attempts to explain that energy can
move from one place to another and cannot go back.

Now, one could get something from outside the system and
cool part of the water to make it ice again. Likewise, some of
the water could be heated a bit so the system went back to its
original state. That would be an ice cube floating in a glass of
water. However, that would require that the entropy in that

outside system increase more than the entropy increase in the initial motion of energy from the water to the ice energy transfer. The point of this is that, one could get the ice back but entropy would increase even more.

Keep in mind that our goal here is to understand what entropy is. With that in mind, let's look at a simple analogy.

A Simple Analogy

Consider moving marbles from one cup to another cup. In this analogy, there is one rule. Marbles can only be moved from cups with more marbles to cups that have less marbles. Then, we have a definition. One unit of entropy is the act of moving one marble from one cup to another.

Let's try this out. We have two cups in front of us. Cup 1 has 8 marbles while Cup 2 has 4.

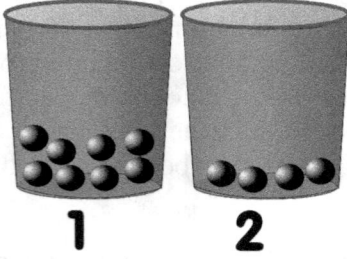

Move two marbles from Cup 1 to Cup 2.

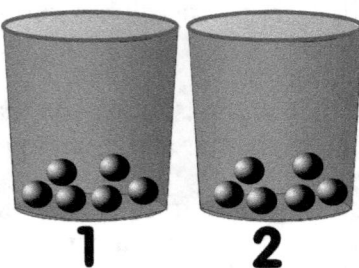

We can do this because Cup 1 has more marbles than Cup 2.

At the end of this sequence, both cups have 6. Entropy is now 2.

We cannot go back to the original state with 8 marbles in one cup and 4 in the other because we cannot violate our only rule. However, we can get back to the original state if we get help from outside of the initial two cups.

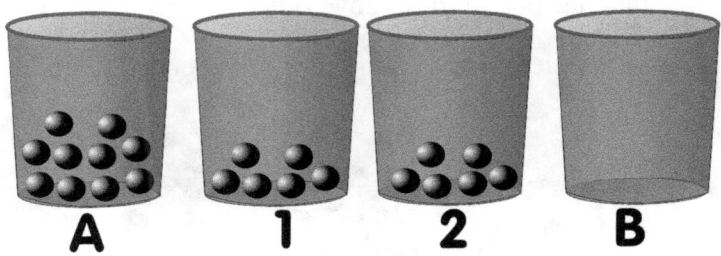

Put an empty cup to the right. Also, put a cup with 10 marbles in it to the left of the setup. Now, we can move 2 marbles from Cup 2 to the empty cup. We can also move 2 marbles from the left new cup to Cup1.

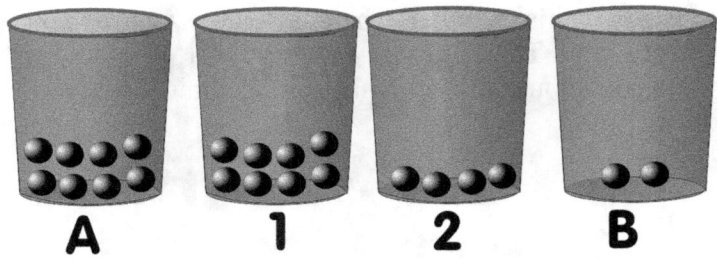

Cup1 and Cup 2 are back to their initial conditions. Note that we moved 4 marbles to do this. Therefore, the entropy is now 6.

Notice the overall effect of this. In the first transfer, the entropy was 2. In the second transfer, the transfer to return the middle cups to their original state, entropy increased to 6.

The Entropy of Work

Before we draw some conclusions about this, let's consider
another example. In this next example, we consider engines
around us that work for us. What work was accomplished
with the ice and water example? Well, the ice cooled the
water for us. We got a colder drink. Let's consider a small
machine that has potential to move objects.

In this example, let's use a piston in an open ended cylinder.

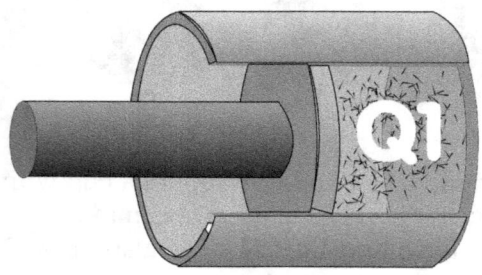

There is gas in this chamber. There is some pressure in the
chamber as well. The energy in this gas is Q1. The gas can
force the piston out of the chamber doing some work.

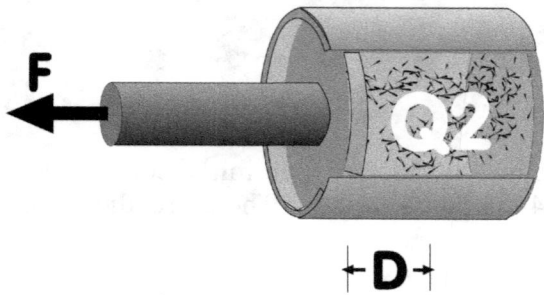

The rod attached to the piston can push on something doing
actual work for us. That work opposes the motion of the pis-

ton so the piston must apply force F to whatever it is pushing against. In addition, the piston moves some distance D. During this process, the work done is F X D. That is, force times the distance the force is applied, is the work (energy) performed. During this motion, the energy inside the chamber has dropped. After the motion, the energy inside the chamber is (Q1)-(F X D).

The energy of entropy is F X D.

Normally, when using heat to do work, the hot air is allowed to escape from the chamber through a hole in the side of the chamber. Then, the piston can be pushed back into the chamber, new air is allowed into the chamber, and it is heated to repeat the action of pushing the rod out of the cylinder with some force. The point is that to get the piston reset to do some more work again, entropy increases even more, though no work is done.

With each new stroke, entropy increases.

What have we learned? We can cool water with a transfer of energy. Furthermore, we can do some work when energy is transferred. The amount of the energy transferred increases entropy. That amount increases every time some energy moves. Entropy increases even if the energy is moved back to where it came from.

The important observation here is that an ice cube in water: cools the water. A piston moves so it can do some work. Likewise, a sun can burn to warm a planet. The real statement is that entropy always increases which means that eventually the sun will burn out and stop heating the earth. The statement, *The stars will eventually burn out and the sky will be dark,* implies there will not be any heat at a higher temperature to move to a lower temperature to do any useful work.

Now, we come to the reason all of this is discussed. The laws of photon interaction and transmission describe the motion of photons through the Super Matrix. These in turn, affect thermodynamics and entropy. These laws support the entropy we have studied. However, we have not touched one area yet. The laws cover photon motion while a photon can move from one node to another. We have not discussed the possibility where there is no node to jump to. What happens then? In normal space, this situation would not occur or would not occur very often. This is not true in a black hole. There, the photons are pressed together. When the universal clock ticks and all photons move, the photons here cannot move because so many occupy the same space. Things change.

The point of this is that entropy in this special case changes. Entropy here can decrease. As we have seen, energy transfers are thought to be irreversible. In this environment, they are reversible because the very essence of existence is now manipulated in a different way. Heat can appear to go backwards. The conclusion here is that in taking the universe as a closed system, entropy is conserved. That is, entropy increases outside black holes and decreases in black holes.

There is another interesting analogy. Chances are that the universe began with an empty space with a number of photons clustered in a small part of this space. This could have been an original black hole. It exploded as the photons spread from that point as a small bottle of strong perfume in a large room. They clustered and matter sprang into existence. As chaos ensued, gravity, via this new matter, formed stars and eventually formed black holes. They eventually decayed. This new material then moves out to build new stars. The point here is that black holes may be mimicking the original Big Bang.

Conclusion
The conclusion of this chapter is that entropy in the universe is conserved. In normal space, it increases. In black holes, it decreases.

Chapter 38

Systems Analysis

Why is systems analysis being presented this late in the book? Normally, these concepts are presented before the details of some subject are revealed. The reason was to spare you dear reader, the agony of going over something that seemed to have nothing to do with reality. The goal was to throw interesting things your direction with the hope they would entertain you. However, the task before us has changed and we need to clarify what we need to study to understand where all of this goes.

Briefly, the following outlines the process of systems analysis. The first action, in systems analysis, is to describe that which we are trying to understand. An assumption is made that, whatever is being analyzed has parts. Then that description is divided into parts. Each part consists of some description of that part. The description consists of prose, drawings, and models to describe each part. Lines are drawn between the parts to show how they relate. The lines also consist of descriptions to describe a relation. Often a line to a part represents an input or output to the part. An output line from one part will be an input line to another part. To continue the analysis, each of these parts is divided again to explain the function of the first level divisions. The process of describing these lower parts and connecting them with lines demonstrating relations is repeated. This continues until the whole is understood.

Understanding comes when the process produces parts small enough to be understood without further subdivision.

Systems analysis is introduced here to establish where to stop explaining the universe. Without some kind of limit, one could continuously ask the question, "What came before that?" Systems analysis helps establish a barrier or limit to how far our understanding can go or should go when studying a specific subject.

A Flashlight

To make this clear, consider how to understand how a flashlight works.

Per the process just described, a flashlight is something we hold in our hand that we can carry to shine light when there is no other light.

To pursue this analysis task we draw a box on a piece of paper and express in prose or pictures what the flashlight is.

The next task is to divide this into parts. Note that the precise way to divide it is not prescribed. The person dividing it selects parts that are conducive to describing the function of a flashlight. If perhaps this analysis were being done to enable someone to build a flashlight, the following description would be different. Our goal here is to understand what a flashlight is. Therefore, our analysis would be done with that purpose in mind. This first division could consist of the following parts or boxes. In each of these boxes, we can add pictures and an explanation of what each part is. We can also draw lines between the boxes to show how they relate.

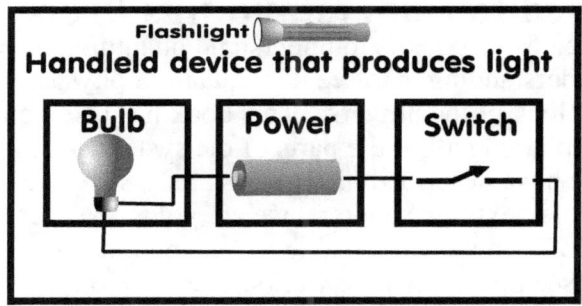

At this point, the analysis may be complete. The analysis here shows a light bulb, a battery, and a switch. If the audience this explanation is intended for understands what each of these is, the analysis need go no further.

However, more understanding may be needed. Each of these parts may be further subdivided. For example, the light bulb may be further subdivided as follows.

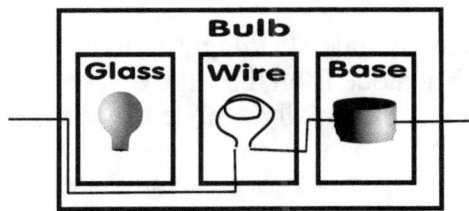

These three parts show that a bulb consists of a glass envelope, a wire, and a base. Each of these boxes contains a description and the boxes are connected to show how they relate. Note: the overall box has the two lines in it that still connect it to the larger parts.

Again, this division of parts can continue until a subdivided part is small enough to be understood.

We Need to Understand this Process

A purpose of this book is to understand quantum physics. Part of this understanding is to see how quantum physics is a building block of the universe. This book has taken on the task of explaining all of the parts of the system called universe. We have described many parts along the way. In a sense, in the next few chapters, we are going to describe the final parts. Then we will be able to reassemble the parts in our head and see how the whole works. The issue on the table, that has not been discussed, is where the analysis stops.

In essence, the analysis has followed this path. We began with universe. It was subdivided into matter. Matter was subdivided into molecules. Molecules were divided into atoms. Atoms were divided into subatomic particles. Subatomic particles were subdivided into photons. Photons are shown to be bits of data that exist within the nodes of the Super Matrix.

The nodes are to be our lowest level of analysis.

The implied task of understanding is to understand the physical universe. The node is on the edge of a description of the physical universe. Essentially, a node is not part of the physical universe but is a part we must understand in some way to rebuild all of these parts to grasp the universe as a whole. Thus, as in systems analysis, we describe the node with prose, drawings, and pictures. What it is: cannot be explained. However, an attempt is made to show how it affects the parts above it in the structure called physical universe.

Conclusion

The subject of systems analysis has been introduced to identify how much is to be explained and indicate what cannot be
explained. That is, this book intends to explain the parts of the
physical universe and stop at some lower level that cannot be
explained. That lower level is not part of the physical universe but that which supports it. While that which supports it
is not explained, its properties are revealed from the point of
view of how it affects the existence of the physical universe.
The result of this analysis is that we should have a good
understanding of what the physical universe is. Then, we also
have an understanding that there is a level to which we cannot
understand but we see that area as not part of the physical
universe.

Please do not assume that it cannot be understood. That is not
the intent of this book. The intent is that area becomes a subject for another area of study. The intent of this chapter is to
draw a line. One side of this line is to be explained by this
book. The other is unknown.

Chapter 39

The Spawning Universe

The forgoing chapter discussed the idea of subdividing a system into logical subparts and explaining what each is. In this chapter, the physical universe is the system we are considering. This is to introduce the idea that the smallest part of the physical universe is a node. Where does it come from? This book proposes a spawning universe that is not the physical universe. The spawning universe created the nodes that make up our physical universe. The goal now is to become comfortable with the idea of a spawning universe. In addition, this chapter examines some characteristics of the universe that was spawned. A key part of the existence of the physical universe is the algorithms in the Laws of Photon Transmission and Laws of Photon Interaction. There are some major constraints on those algorithms that, as of yet, have not been discussed. Those are taken up here. In essence, this chapter ties up many loose ends that have not been explained. The concept of the system is used to show that we can understand the universe by subdividing it down to the nodes of the Super Matrix. We need to understand how nodes affect the physical universe but do not need to know what a node actually is to understand how it affects the physical universe.

In a way, this chapter need be the only material presented about this subject. All other parts of this book are an attempt to prepare you for this presentation. This presentation is about the bootstrap from which all existence of the physical universe unfolds.

Possibilities

We begin by making it clear that this book does not purport to explain what the spawning universe is. Thoughts about what it could be are offered in an attempt to convince you that having a spawning universe is possible if not necessary. Perhaps we might move faster if we begin by explaining what it is not.

It is not a parallel universe.

Parallel quantum universes appear as a concept because those that support quantum thought cannot explain how a wave function can represent many states at the same time. For example, a cat in a box may be dead or alive depending on some radioactive element that shares the box with the cat. This represents two states: cat alive, cat dead. One would not know until opening the box and verifying which state exists. A solution for the quantum purist is to claim that there are two parallel universes; one where the cat is dead and one where it is alive. The thought here is that the parallel universe concept exists to explain the present state of knowledge of quantum physics. Other words for this concept are:

- Multiverse
- Megaverse
- Parallel Universes
- Many Worlds Interpretation

The spawning universe referred to in this book is not like these concepts. Furthermore, most present day concepts of other dimensions, planes of existence, or any other world place usually incorporate some kind of space and or time. From the author's point of view, the spawning universe has no space, time, energy or matter. This is very difficult for us to conceive of. And, for those that think we must measure it, the lack of these properties precludes measurement. It is perhaps a universe of pure thought. If that were the case, the photons created are objects of pure thought. Attempting to understand the range of our physical universe could suggest that it came from some "earlier" universe. Also, this spawning universe could well have spawned other universes. This idea is only raised to understand what is on the table before us. This book is only concerned with the idea that there is a spawning universe that manifested something we call photons in our, one and only, physical universe. We do not know what they are. We can only describe them from an operational point of view. That is, what is the mathematical behavior of a photon from an observer's point of view in our physical universe?

In an attempt to get some sense of comfort about this, we will consider a number of analogies of how the nodes can relate to each other to enable data photons to move through the Super Matrix. The purpose of going over these analogies is to drive home the point that the nodes do not exist in some time or space. Time and space is the result of how the nodes pass the data, not how they are structured.

Here are some possibilities. You could probably add to this list.

- Tiny computers.
- A bunch of people passing data.
- A 2D universe manifesting a 3D universe.
- A Universe with no dimensions.

Tiny Computers
This was introduced earlier. The Super Matrix could consist of very small computers linked together. Perhaps each computer is the size of a grain of sand whatever size means in this theoretical spawning universe. Each one is connected to six other computers and is designated as up, down, left, right, front, and back. As described, a photon consists of a set of data that is passed between these small computers. Now, the computers could exist in some physical universe we are not aware of. Or, they could exist in some kind of universe that is not physical but has some way to function like computers we are familiar with. Moreover, their relative organization need

not follow the rectangular pattern depicted before. They could be placed in a straight line.

Then, six connections between each enable data to be transferred from node to node to move as in up, down, left, right, back, and forth.

The links in six directions enable someone inside the spawned universe to perceive it as three-dimensional.

A Bunch of People Passing Data

Consider a bunch of people relaying messages to each other. The messages would be photon data sets. Each person has some kind of a direct link to six other people. They would be: Mr. Up, Mr. Down, Mr. Left, Mr. Right, Mr. Front, and Mr. Back. Then, each person has a number that clearly identifies where they are.

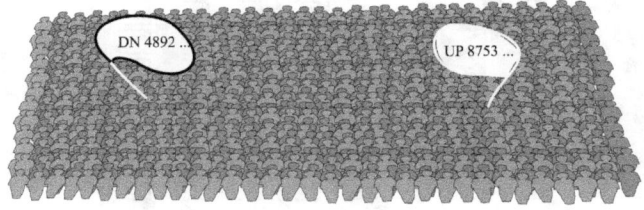

Somehow, each person can communicate with the six that are important to each. Maybe they yell in that universe. Maybe they flash some kind of lit up sign. Maybe they have a wire

running to that particular person. Then, each person may or
may not have a data set called photon. If a person does not
have a data set, the person does nothing. If a person gets a
data set, he makes a calculation using the Photon Rules of
Transmission to determine who it should be passed to. The
photon should be passed to one of the six. Should a person get
two photon data sets, the Photon Rules of Interaction are applied to determine where each photon data set goes.

Note that the organization of the people has nothing to do
with the three dimensional universe they hold in place. The
Super Matrix is supported because each person has those six
people they are to communicate with.

A Two Dimension Universe
The spawning universe could be two dimensional. This is
often called Flatland. The inhabitants of Flatland could have
set up little boxes. Inside each box is a Flatland brain. The
Flatlanders have trained bugs to communicate photon information. Each bug has information about a particular photon.
The bugs are trained to go to a particular box. The brain inside the box changes the data in the bug and tells it to go to
one of six boxes representing up, down, left, right, front and
back. The bug carries the information to the brain in that box
and life continues.

You may be concerned about the time required for a bug to
travel from box to box. Here, time in the spawning universe
does not matter. Whatever the time here, the spawned universe will perceive it as one tick. The "time" in the Flatland

universe could be long. Regardless of the "time" there, this time in the spawned universe would be one tick. That is, it would be very fast.

Note that Flatland does not extend itself to make a third dimension. The three dimensional universe is totally manufactured in Flatland. A being in Flatland would not jump into the three dimensional universe but would simply observe a totally different set of dimensions they have mentally constructed.

A Universe with No Dimensions
Imagine a universe with no dimensions. Can that be? Perhaps this would be a universe of pure thought. The beings of such a universe could exist by being aware of others in the universe that thinks. They could imagine a bit of data that would represent a photon. If they could identify others they could assign those with names of up, down, left, right, front and back. Then they could pass some data back and forth between themselves. If there were a lot of them, they could create a three dimensional universe without knowing what three dimensions are.

The following depicts some number of beings of thought that relate information to each other.

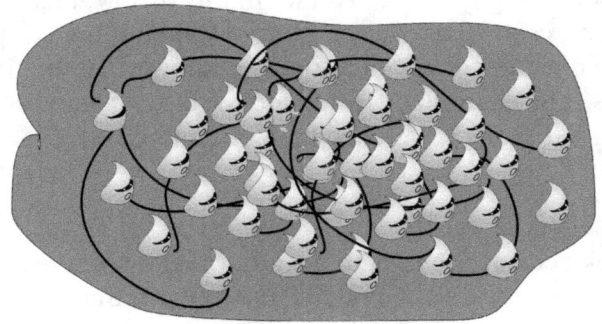

However, if they are accustomed to living in a universe that has no dimensions, they might just perceive themselves as being all at a single point. The ball in the following picture here represents that single point of no dimension. The lines surrounding the ball depict lines of communication between the components of those that occupy the single point universe.

We cannot conceive how many beings would perceive each other in such circumstances. However, if they can recognize each other and communicate to each other, it would not make a difference. They, as a group, could still create a Super Matrix in their own minds that would support a universe like ours.

What is the point?
The essence of this discussion is that the theory here explains the physical universe to the point where the descriptions end being about the physical universe. In this theory, the photon is an idea or a collection of data about its position in a Super Matrix. This book does not purport to know what the Super Matrix is. The examples offered are to serve to suggest possibilities that seem far fetched but communicate its complexity.

Just Accept It
This book accepts the idea that we have data photons that move from some kind of node to another in a Super Matrix. We will not look back and say, *What is a node?* There are a few things that should be noted here. First, the physical universe existing in the three dimensional array called the Super

Matrix exists due to the Photon Rules of Transmission and the Photon Rules of Interaction. They dictate that photons will only be passed to one of those six nodes attached to a particular node. So, at this point a decision has been made that the problem of what created photons or the Super Matrix is left behind.

More Loose Ends

When the Super Matrix was first conceptualized, the assumption was that the photon merely flew from node to node as it made its presence known in the universe. In the real world, we know that when a photon flies in any direction, it appears to travel the same distance in any direction during a similar amount of time. With what has been presented here, not all directions of motion are equal as the photon is moving through a rectangular array.

The following picture depicts the problem.

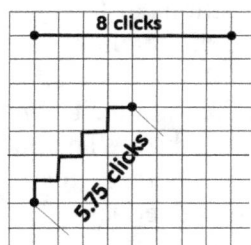

This illustrates two photons moving 8 clicks. In the rectangular array, the one moving horizontally moves 8 clicks through the space. The one moving at an angle only moves 5.75 clicks even though it made 8 transitions. This clearly does not represent reality. Even if we make the array click size smaller, we would not gain anything.

So, let's try a different scenario. Suppose the photon travels through space in some different pattern. Let's study this as if it were in a two dimensional space. Here we can try a photon

that moves in a circle. The heavy line in the following picture
is a path of a photon.

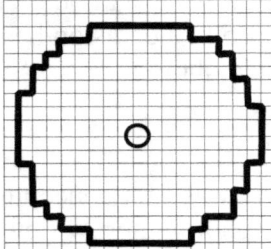

Each square represents one tick of time and one click of distance. Note that the photon still travels some jagged path.
However, we traditionally consider particles at their center of
mass. In this case, that would be the center of all the positions
the photon hits during this cycle of travel. Note that even with
this ragged path, the center of the photon would not move
much. The small circle in the picture above represents the
equal distance of all positions of the photon as it moves in this
jagged circle. If the size of the clicks (the distance traveled)
is half that depicted in the above picture, we would get even
less relative motion in the middle. The next picture shows
clicks half the size of that above. The path of the photon appears much smoother than before. In addition, the center of
motion of this path moves even less.

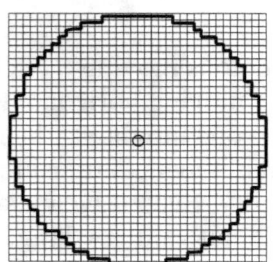

Thus, if the motion of the photon consisted of a circular pattern moving through the array, the motion of the center of the
composite motion would be much straighter. The following

pictures show motion of such circular paths in two different directions.

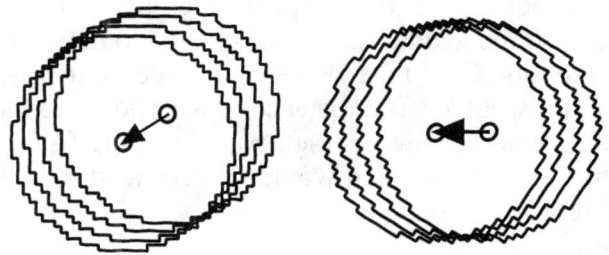

The picture on the left shows the motion of the center of the photon moving at an angle trough the array. It appears very straight. The motion displayed on the right shows the photon moving almost horizontal. It travels nearly the same distance as the center in the left picture. Now, if the click a photon moves is very small compared to the radius of the photon path, the direction of the center through a rectangular array would be very similar regardless of the direction of motion.

Let's expand this circular motion to the three dimensional world. We can also have the center move perpendicular to the plane of the circular path. The motion would appear as a helix.

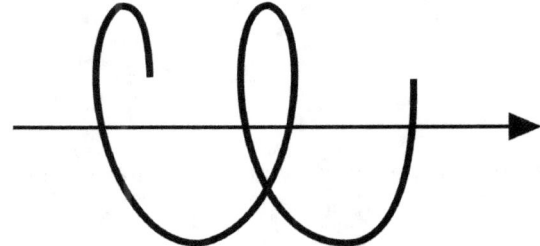

This logic is the reason earlier parts of this book treated the motion of a photon as the path of a helix.

There is still the problem of the center wobbling. With this design however, that wobble would not be noticed by the macro world. Then, there is another thought. Previously we have attributed the Heisenberg uncertainty principle to either the wave phenomena of the photon or the fact many photons whirl about so the average center of such a cloud would always be moving. Could the Heisenberg uncertainty principle be this center wobble? The distance of the photon to the axis of the helix would change as the photon is going from node to node. That is something that would be very worthwhile investigating.

One Final Note

Earlier in this book a promise was made to explain how photons could collide even thought they were points. Now we can do that. This chapter has explained that a photon is a piece of data that is transferred from one node to another node in the Super Matrix. Thus, it is clear that the photon appears as a point in this three dimensional array. It has no width but only a position. This is a very good definition for a point. But it occupies a location that can be occupied by another photon. Hence, when two photons attempt to move into a common node, a collision occurs.

When some area of the Super Matrix has many, many photons, many collisions will occur.

Conclusion

An observation has been made here that a photon is not part of the physical universe. It is merely some data that is transferred from node to node. The photons manifest particles by gathering together in groups, which is what we perceive to be the physical world. The transfer of the photons between nodes manifests space and time. We essentially have little information about what the Super Matrix is. We just understand that it supplies the nodes that enable the transfer of data from some point to another.

Chapter 40

A Forever Universe

The issues presented in the chapter titled "Expansion of the Universe" of this book are to demonstrate the universe is not expanding. Yet, the purpose is to show that the universe is not static either. The issue is that black holes decay and spread raw material throughout space. This material spawns stars, planets and possible life. Then, the stars eventually collapse and are sucked into black holes from which the cycle begins anew. We live in a forever universe.

In this scenario, the universe is filled with black holes that recycle matter and energy to be used again. The universe is constantly changing but never dying. The universe is a machine that will run forever.

The Big Bang

An observation here is that the big bang did not necessarily begin from a single point. The observation is that we do not have an answer to this question of what happened then. The thought of this author is that the Super Matrix sprang into existence from wherever it came from. It did not need to spread out because there was no space to spread out into. That is, the Super Matrix manifested space and time. It was populated with photons. These are pieces of data written into the nodes of the Super Matrix. Then they were allowed to move from node to node at the same time. Chaos ensued. Raw energy was released in an instant to smash and clash. The photons gathered following the rules of transmission and interaction. Matter materialized. Stars formed and here we are today.

There are a few consequences of this scenario. The first is about the size of the universe. Most think the universe is infinite. In the opinion of this author, the universe is finite. It is an artifact spawned from some other universe. It is large enough to appear infinite. The Super Matrix must consist of some finite number of nodes. Likewise, there must be some finite number of photons in the Super Matrix. The finiteness of the universe and the idea that black holes recycle matter and energy suggests another consequence. During the process of the cycle of construction and destruction in the universe, information is lost. Data is contained in some ordered configuration of particles and matter. Eventually, all of this stuff is sucked into a black hole and reset. Information can be maintained forever if there is some sentient force that copies the data from one set of particles to another. However, as there is a limit on how much matter is in the universe, the amount of information that can be saved is limited. This concept suggests that there is some period of time for which information is maintained. Past that point, it is deleted from existence.

For example, let's assume the period of information retention is a billion, billion years. Then, everything older than a billion, billion years no longer exists.

Apparently, the universe has been built to create the illusion it is symmetrical, logical, and forever. Logically, however, it cannot be infinite. Time must have a beginning and it must have an end. One can easily ask the question, what came before time. Alternatively, one can ask the question, what happens after time ends. These are not logical questions if one can see that time was created.

Whatever the universe is, it is logical. If it does not appear that way, it is because the one observing does not understand.

Conclusion
The conclusion here is that the universe is a perpetual motion machine. Life within it moves on continuously. That which supports life falls into black holes that crunch it and reset it so

new life can spring up and live. The universe, though large, is of some finite size. A consequence of this is that we have no way of detecting how old the universe is. Furthermore, information that emerges as the universe rolls on will eventually be lost unless there is a sentience that repetitively records it keeping it in the here and now. Even that task is limited.

Final Discussion

The purpose of this section is to put a finish on this book. To do this we consider a vision of this book by the author. He has a vision that this book describes the physical universe. This vision has several parts. One is that the material presented in this book is an extension of that which has gone before. That is, this book does not propose past expressions of information related to quantum physics was a mistake and this book reveals the truth of the universe. Rather, this book serves to acknowledge the effort of the past and add to it.

Another part of this vision is that the ideas in this book create another tier in the development of physics. The first tier, if you will, is classic physics. This is marked by viewing the universe as made up of tiny hard balls that obey Newton's Laws of Motion. The next tier is quantum physics. It is marked by the idea the universe consists of waves of energy and matter. In this system, matter waves are described mathematically as probability distributions.

This book purports to establish another tier, which is a statement that some universe spawned nodes of a Super Matrix that in turn maintain the existence of photons. In this vision, photons are pieces of data shared by the nodes. The interaction of photons: manifest time, space, matter, and energy creating the physical universe. No attempt is made to describe the spawning universe other than an attempt to logically describe the appearance of node interaction and photon interaction from the view of inside the physical universe.

Newton's Laws of Motion are a logical analogy of this effort.

His laws described the mathematical relationship between
bits of matter that produced gravity. His mathematics did not
attempt to explain the source of the gravity. During his time,
he was accused of foisting gremlins and spiritual entities on
the scientific community. Likewise, this book suggests we
can study photons and nodes mathematically without knowing what they are.

Hopefully, the first half of the book, which is about the history of quantum physics, is presented in a way that a non-
scientist reader can understand the concepts. The knowledge
presented in the first half of the book is used to introduce new
concepts in the second half of the book. The concept of helical photon motion can explain to the non-scientist why light
has the properties listed in that chapter. That in turn, presents
other mysteries elaborated on in the first half of the book. The
text then continues from that point forward to explain these
mysteries.

Another part of this view is that this book is but an introduction to the subject and is incomplete. Unfortunately, some
ideas in this book are in conflict with other ideas in this book.
This is an indication that the basic concepts have not been extended correctly. Unfortunately, this is not befitting the scientific method. Many conclusions have been offered without
some experiment to back them up. In addition, the developments lack the analytical depth rigorous science requires.
This book went far beyond the intention of an initial plan. The
first branch in the initial path forward was a desire to represent the people that advanced physics in the past. Reading a-
bout these people had a profound effect on the author. The
desire to share this was overwhelming. The second branch
was an attempt to explain everything with a single thread of
thought. The goal was to present a complete picture to the
reader. The breadth of this task is large. Filling in the gaps
with well-tested ideas was seen as impossible. So some
creative thinking was used to smooth over rough spots.

This book is standing as is with two hopes. One is that the

non-scientist reader senses the profound contributions those in the past offered. They offered part of their lives in the cause of advancing physics, often without hope of personal gain. The second is the hope that this material prompts experiments or some kind of work to prove or disprove strange ideas presented in this book

While this thinking may be Pollyanna-like, it has a place in science. Whatever we produce must be placed on a table for the community to judge. Even if it is wrong, an analysis of why it is wrong could bring up some truth that would be valuable. Hopefully, the ideas presented in this book are thought provoking and generate some further interest.

Then, consider the concepts from the first half of this book when they were new. Most had some element of truth and a lot wrong. However, the person that came next improved on the idea and science moved forward. Rutherford knew the structure he proposed for the atom was wrong immediately after it was released. Unfortunately, his first idea became an icon in our society. However, the next person, Niels Bohr or de Broglie fixed it and science marched on.

What is Next?
Part of the initial plan was to write software to simulate space, particles, photons, and so on. The goal was to present these in this book to back up the concepts. The task was overwhelming and the decision was made just to present the concepts. Now the plan is to continue with software models after the book is complete and some energy has been invested in selling it. Here is a software development plan.

Single Slit Experiment
This would be a program written to simulate the flight of photons through space. This beam would be directed at a slit to observe how the beam might slip through the slit and make the well-known diffusion pattern on a barrier screen.

Double Slit Experiment
This would be a program written to simulate the flight of photons through space. This beam would be directed at two slits to observe how the beam might slip through them and make the well known interference pattern on the barrier screen.

Particle Persistence
A program would be written to develop the interaction between photons and demonstrate how a particle maintains its persistence.

Particle Motion
A program would be written to develop the interaction between photons and demonstrate how a particle persists and moves through space.

Electromagnetic Force
A program would be written to model the existence of two electrons and demonstrate how photons generate an electric force.

Conclusion
As this is being written, the science community is realizing that there is no Higgs Boson. Furthermore, the people in CERN have discovered that neutrinos move faster than the speed of light. Why this is true has been presented in this book. Perhaps these events will trigger interest in the ideas of others about the structure of the universe. This book may be one that pops up to capture that interest.

A conclusion of this book is that the universe consists of a single particle called a photon of which the multitude in this universe behave as a gas in a universally large bottle that has no bounds. In this philosophy, photons are actually not a part of the physical universe but manifest space, time, matter and energy as they interact with each other. Finally, this should bring about an awareness that there is another universe separate from our own that deserves investigation from which photons were spawned.

Well, those are the plans. A web site has been organized to keep those interested informed. The project to date has been a daunting task. As this is being written, the web site is just a shell. It is a place to go to get more information should more information evolve. The name of the site is www.omenquest.com. The page on that site dedicated to this task is www.omenquest.com/supermatrix.htm.

Appendix A

Sine and Cosine Functions

The wave function is very critical in the description of quantum physics. It normally appears as a sine or cosine function. To have a good understanding of what is going on, the sine and cosine functions should be understood. For those readers not versed in trigonometry, we will go over the basics here.

One of the goals here is to explain the functions:

$$y = A \cos(\alpha)$$
$$y = A \sin(\alpha) \quad \alpha = \text{alpha}$$

The Greek symbol alpha is often used to represent angles. We will also attempt to explain the following function that is used to represent water waves as well as electromagnetic waves and matter waves.

$$y = A \sin\left(2\pi\frac{x}{\lambda} - 2\pi\frac{t}{T}\right)$$

We Begin with an Angle

Here is an angle of forty five degrees. Most people will understand the meaning of this. However, this simple picture does not go into the detail we need about angles and their use.

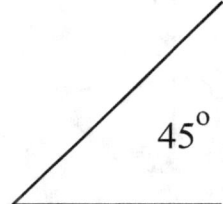

Consider a wheel.

The wheel rotates. We want to describe this rotation in a mathematical way. Describing the motion as a change of angle is one way to do this. Let's begin by assuming the wheel is spinning. How would we describe the motion? One way to describe the motion is by saying it rotates. That is, one could say it rotated 20 times. If one were to ask how fast it is moving, one could say it is rotating 20 times a second. Here the basic unit of measure is one rotation. Another way to do that is with angles. Above we saw an angle of forty-five degrees.

Let's apply that method to the wheel. Here is a picture of a
wheel that rotated 45 degrees.

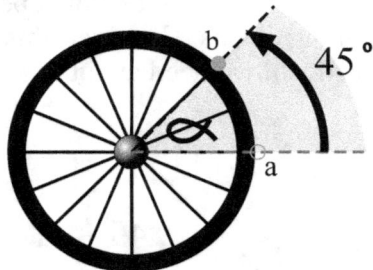

When we begin this experiment, a spot on the wheel is at a.
After the wheel is rotated, the spot is at b. This is based on the
idea that there are 360 degrees in a full circle or rotation of
motion. If we consider this from a rotation point of view, the
wheel rotated 1/8 of a rotation. Consider partial rotations in
these various positions.

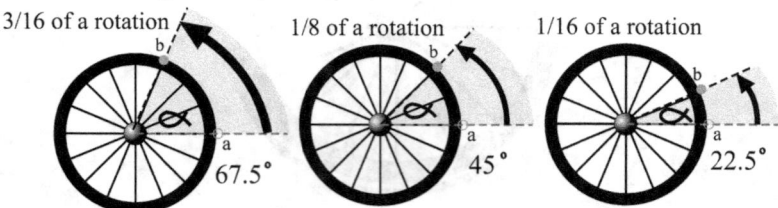

This shows how we can describe small motions of wheel rotation. We need one little more step to see how this is applied
to the task. Before, we spoke of the wheeling turning twenty
or more times. This can be represented in degrees. As one
complete rotation of the wheel is 360 degrees, 20 rotations
would be equivalent to 7200 degrees of rotation. The major
point here is that degrees of motion are not limited to 360 degrees or less. One could even say something is rotating at ten
degrees per second. Then, if it were a wheel, we would be
describing a continuous motion of a spinning wheel. Often

omega is a Greek letter used to represent how fast something is rotating. That would be written as:

$\omega = 10^{0}$ per sec ω = omega

At this point, we have seen that some amount of rotation can be described as some number of rotations or some number of degrees. There is another way to describe rotation. That is with radians. Realize that as a wheel rotates, one point on the edge of the wheel moves some distance in a curved line. That distance might be one-half the way around the curvature of the wheel or many times around the wheel. Radians are defined by dividing that distance traveled by the radius of the wheel.

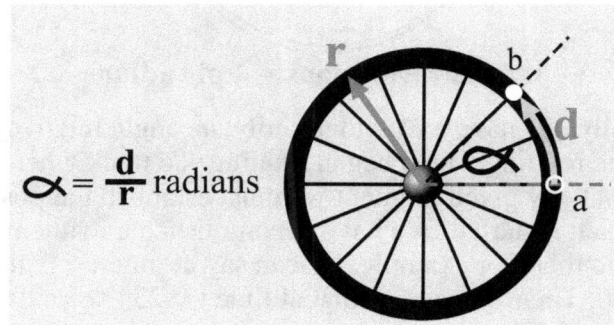

$$\alpha = \frac{d}{r} \text{ radians}$$

If we assume the radius of the wheel is 1, then

One revolution = 360 degrees = 2 pi radians.

This is because the distance around a circle = radius X 2 pi. In algebraic equations, the angular position or alpha is usually thought of in terms of radians.

We have covered two very important points. One is that rotation can be described in terms of radians. Another is that angles are not limited to small values such as 45 degrees (1/8 pi radians) or 90 degrees (1/4 pi radians). Angular motion or change in position can appear as follows:

alpha = 0.34 radians
alpha = 1 radian
alpha = 4 pi radians

The last would describe a wheel that has rotated twice.

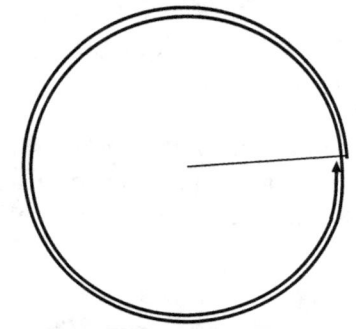

∝ = 2 revolutions = 4 pi radians

Essentially, we have a way to describe an angle relative to
where the rotating wheel began rotating. To track where the
wheel is at any given moment we must establish the position
from which it started. We have accomplished a mathematical
way to do this. For example, we can say at time t = 0, the angle was 0. Then we can say that at time t = 234 seconds, the
wheel is at 1475 radians.

Sine and Cosine

Our next task is to introduce the concepts of sine and cosine.
In a sense, sine and cosine is a way to represent angles as
with rotation, angles and radians. However, the sine and cosine do not depend upon a rotation of one side of the angle.
Sine and cosine are called trigonometric relations. They are
usually referred to as sin (pronounced "sign") and cos (pronounced "ko-sign"). There are other trigonometric relations:
tangent, cotangent, secant and cosecant. We will only present
sin and cos.

The following picture shows how this device is implemented.

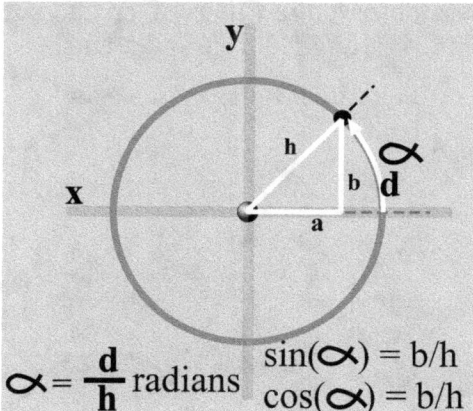

$$\alpha = \frac{d}{h} \text{ radians} \qquad \begin{array}{l} \sin(\alpha) = b/h \\ \cos(\alpha) = b/h \end{array}$$

This picture shows an x and y set of axis superimposed over the wheel. The dot is shown to be some angle alpha, from the x-axis. The line b is the distance from the x-axis to the dot. The line a, is the distance from the center of the wheel and axis to the dot along the x-axis. The letter h represents the hypotenuse and is the distance from the center of it all to the dot.

The angle in radians as shown is d/h. The sin is defined as a ratio of two sides of the triangle. Sin is the side opposite from the angle divided by the length of the hypotenuse. The cos is defined as the length of the side adjacent to the angle divided by the length of the hypotenuse. Thus, the sine of alpha is b/h while the cos of alpha is a/h.

As we have observed earlier, the rotating line or hypotenuse can move around more than one rotation. Thus, the hypotenuse will spin through all four quadrants of the space defined

by the axis. Let's observe the values of sin and cos as the
hypotenuse swings through all four quadrants.

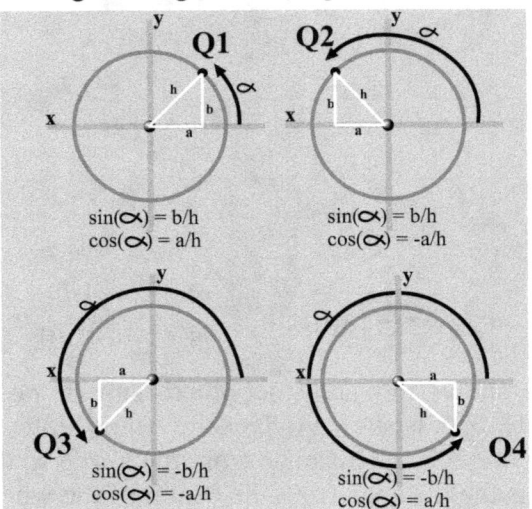

Note the sin and cos values in the first quadrant (Q1). Both
are positive. In the second quadrant (Q2) sin is plus and cos
is negative. In quadrant three (Q3), both are negative. In the
fourth quadrant, the sin is negative.

The next task we face is to see how the angle and the dimension labeled b in these pictures is plotted graphically.

Straight Line Angular Motion Plot
An important point to notice is that alpha increases as the
wheel spins around the center. That is, the value of alpha increases as time progresses. The value of the sin and cos of alpha bounce between -1 and 1. This is an important property of
the sin and cos values. Let's look at this closer.

During the first rotation of the wheel, alpha will go from 0 to
2pi. The sin of alpha will go from 0 to 1 to 0 to -1 and back to
0. The cos of alpha will go from 1 to 0 to -1 to 0 and back to 1.

If we were to graph the circular motion of a point on the perimeter of the wheel on a straight line, the graph would appear as follows.

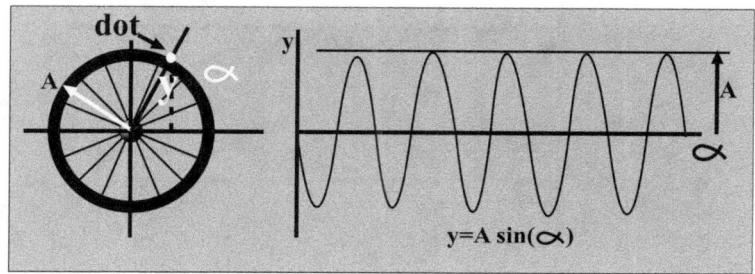

And in a similar fashion for the cos.

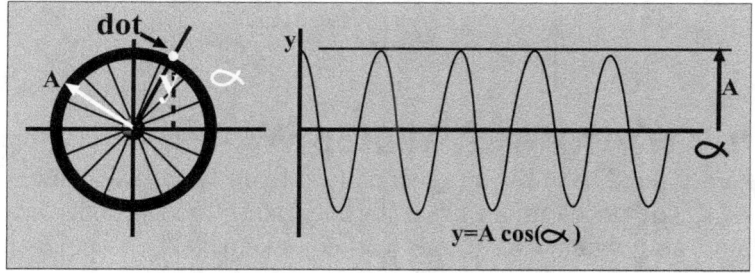

Note the horizontal line in the graph labeled alpha, is constantly increasing. Then, note the value for y or the distance of the dot from the horizontal axis is going up and down as alpha increases.

How Does Amplitude Fit In?

Note that because the values of sin and cos range from 1 to -1, y in the equation y=A cos(alpha), will always be less than A and greater than -A. The following three graphs illustrate the fact that the value of A determines the height of the associated curve.

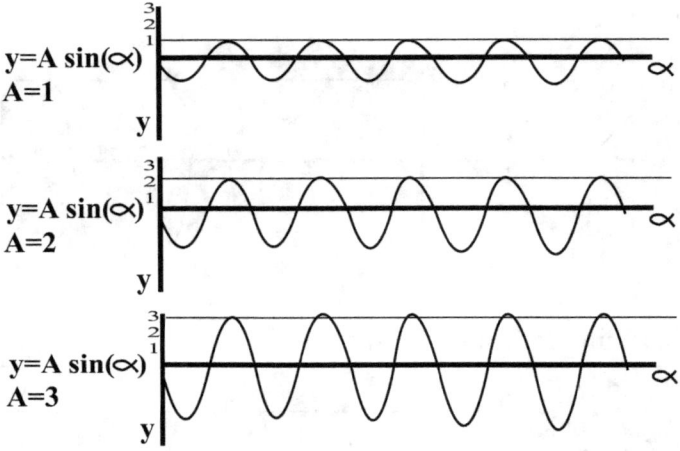

This Works with Water Waves

Now we need to make a big transition from wheels to water waves. The problem is that, for some, using the same math for wheels and water waves does not make sense. The upper part of the picture below shows that the math represents a dot moving on a spinning wheel. The distance y on the wheel is displayed graphically to the right of the wheel as the vertical axis. Alpha, the angular position of the dot, is plotted as a horizontal line in the graph. The graph represents the up and down motion of the dot as it rotates around the wheel.

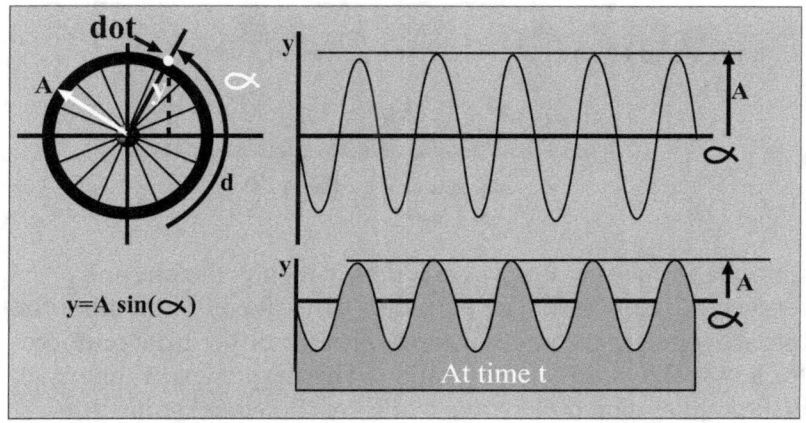

$$y = A \sin(\alpha)$$

At time t

The same function represents the shape of a water wave shown at the bottom of the picture. There is no resemblance between water waves and wheels. Yet, the same mathematical equation describes some part of each in a similar way.

Alpha or angular position in a water wave does not make sense to our mind's eye. The math needs to be changed a bit so we can use the same equation for both. We need to call the distance the dot travels as it moves around the wheel, x. Then alpha can be replaced with a relation using x. The following shows this relation.

$$\alpha = 2\pi \frac{x}{\text{circumference of wheel}}$$

So we get,

$$y = A \cos\left(2\pi \frac{x}{\lambda}\right) \quad \lambda \text{ is wavelength or circumference of wheel}$$

Since the graphic representation of the dot on the wheel is so much like a water wave, characteristics of the equation are named from the point of view it is a wave.

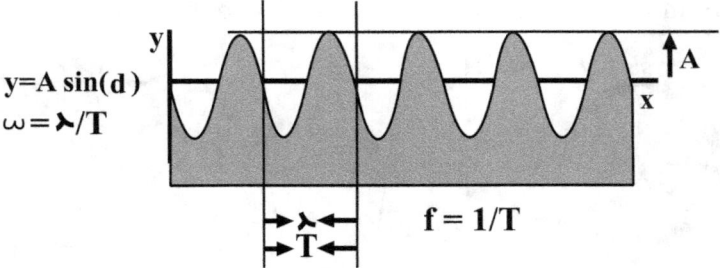

Lambda is equal to one wave length which is the circum-
ference of the wheel. The period of the wave is T, the time for
one wavelength to move distance lambda or the time required
for a wheel to spin one revolution. The frequency of the wave
is the rate at which the wave cycles or the wheel spins. It is
the reciprocal of the period or 1/T.

Here is yet another factor. As you are aware, a water wave is
not only undulating up and down, it is moving transversely.
Then, if x is fixed, alpha can be expressed as some function of
time:

$$\alpha = 2\pi \frac{t}{T}$$

If we incorporate this into our water wave equation, it appears
as follows:

$$y = A \sin\left(2\pi\frac{x}{\lambda} - 2\pi\frac{t}{T}\right)$$

The time factor is subtracted from the distance factor. This is
somewhat arbitrary. This is the wave function normally presented when discussing wave functions.

Appendix B

Special Relativity

Einstein and associates applied themselves to the problem of the speed of light and the meaning of its velocity in the real universe. A primary source of these ideas came from Maxwell's theory of electromagnetism and lack of evidence of ether. [34] These factors lead to Einstein's popular postulates.

The Postulates

Postulate one; *The laws by which the states of physical systems undergo change are not affected, whether these changes of state be referred to the one or the other of two systems in uniform translatory motion relative to each other.*

Postulate two; ... *light is always propagated in empty space with a definite velocity [speed] c which is independent of the state of motion of the emitting body.*

To continue we need to discuss this a bit, so we have some kind of familiarity with it.

Postulate One

Let us get a handle on Postulate one. You should read Einstein's statement repeatedly to get the gist of it. Here is the author's gist, *The things an observer observes, does not depend upon the observer's constant speed relative to any other*

observer's speed. As is often done in this book, let us look at
this with a picture.

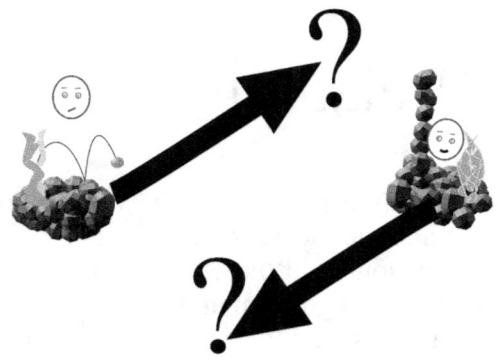

This picture shows Jack and Joe. Each is on a rock flying
through free space. The flames represent chemical reactions.
One is watching a ball bounce and the other has balanced
some stones. Postulate one indicates that both Jack and Joe's
view is not dependent upon what speed each is moving. The
big arrows indicate that one rock is moving at some speed
past the other. The question marks indicate that both Jack and
Joe do not know who is moving past the other. Both perceive
that the other could be moving or stationary.

Postulate Two
Now, let us discuss the constant speed of light. Assume John
is sitting on a rock in the middle of space. His rock has a jet
pack on it so he can do some traveling. He also holds a dollar
bill in front of his face and holds a stopwatch. He waits.
Eventually, a photon comes along that travels the length of
his bill. He starts the stopwatch when the photon crosses the
far end of the bill. Then he stops the watch when it crosses his

end of the bill. He records this time as = 0.00000000056 seconds.

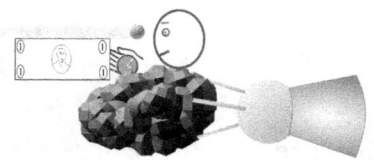

He turns on his rocket engine and lets it roar for ten hours. He must hang on tight to his rock as he accelerates.

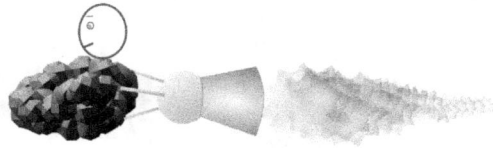

He knows he is moving forward in space much faster than he was before. He remembers he lit his engine for ten hours. He gets out his stopwatch and dollar bill and waits. Eventually a photon flies along the edge of his dollar bill. Again, he measures the time for the photon to fly along the edge of the bill to be 0.00000000056 seconds.

He concludes that, "Yep, Einstein was right. The speed of light is constant."

Conclusions of the Postulates

Einstein and company, utilizing these postulates, eventually
emerged with the concepts of Special Relativity. We need to
understand a bit about these to make progress.

Time Dilation

Perhaps the most noteworthy feature of Special Relativity is
the change in time when one object moves faster relative to
another object. This is time dilation. Without getting into the
math, this phenomenon appears as follows. Let us introduce
Jack and Joe again.

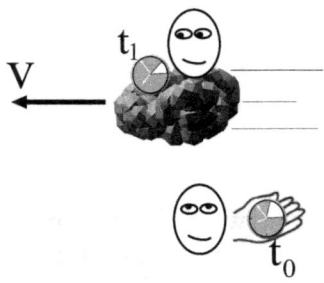

Jack is sitting on a rock that is flying through space at great
speed. Joe is just floating in space holding a clock. From our
point of view, Joe is not moving. Jack happens to fly over
Joe's head the moment we look at both of them. We have
enough time to see the clocks they are holding. In the white
area on Joe's clock, the one not moving, we see it has moved
almost a quarter of the dial. We can see that during the same
time, the time change on Jack's clock on the rock is about half
of that. We know the clocks are the same. This picture shows
that the clock moving rapidly runs slower than one that is not
moving in our non-moving reference frame.

Note that normal speeds will not produce this effect. The
speed around us such as cars and airplanes is not enough for
this to become noticeable even though the effect is present.

Now, here is a warning. Many people read about this change in time. Then they refuse to read more about it. They say, *This is just too complicated.* It is not complicated. When one rock goes faster than another, time slows down on the one moving faster. True, this does not make sense. Nevertheless, it is not complicated. It is simply not a normal observation.

Lorentz-FitzGerald Length Contraction

The sizes of objects also change at relatively high speeds. The following duplicates Jack and Joe's experiment using rulers instead. Joe is sitting on a motionless rock holding a ruler. Jack, on the moving rock, is holding a ruler the same size as Joes. Jack's ruler appears shorter because he is moving quickly past Joe. This is a consequence of Special Relativity.

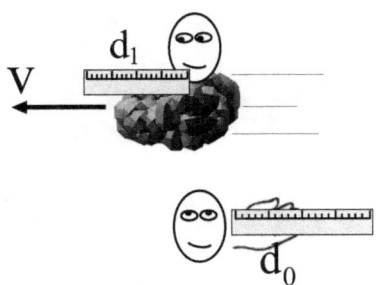

The lengths of the rulers are d sub 1 for the ruler moving at high velocity and d sub 0 for the ruler not moving. We will use this phenomenon later so we need to get a mathematical relationship between the velocity and the change in size. The equation is as follows.

$$d_1 = d_0 \sqrt{1 - \frac{v^2}{c^2}}$$

This equation indicates that if we multiply the length at rest by the square root of 1 minus v squared divided by c squared; we will get the precise length of the ruler moving by at a high speed. Jack, who is not moving, observes this length. Joe, on the rock, does not see the change in size, as he is moving with the ruler.

If this is the first you have heard of this phenomena, you will probably find it incredible. Right now, there is no accepted explanation. Those that work with it simply accept it. This theory is probably one of the most proven theories in physics.

Appendix C

New Age Special Relativity

To get a better understanding of the definitions used for time and space, let us see how these relate to concepts from special relativity. Two important concepts are time dilation and Lorentz-FitzGerald contraction. They appear as:

Time Dilation

$$MT_n = \frac{MT_m}{\sqrt{1 - \frac{v^2}{c^2}}}$$

This equation shows the relation between measured time in a non-moving (MT sub n) reference frame to a reference frame moving very fast (MT sub m). The time in the moving reference frame is slower than that measured in a non-moving reference frame.

Lorentz-FitzGerald Contraction

$$D_m = D_n \sqrt{1 - \frac{v^2}{c^2}}$$

This equation shows the relation between distances in a non-moving reference frame (D sub n) and a reference frame moving very fast (D sub m). Distances in a moving reference frame are shorter than in a non-moving frame.

The following uses the concepts that time and distance are the result of jumps of photons from node to node in the Super Matrix. These minta form all of our real world measurements of time and distance. Particles consist of the photons whirling about a common point. We use the concept that a particle consists of composite single photons moving about a common point in space in a square pattern as introduced in the chapter, "Introduction to Particles."

The goal here is to show that a study of this structure can produce the same result as Einstein and others in special relativity.

These experiments are performed in a normal Euclidian space. That is, the primary reference frame has the traditional x-axis and y-axis and the dots or photons are moving at a constant speed in this reference frame. Note that the dots or photons, as observed from the moving reference frame, remain moving at that constant speed relative to the non-moving frame.

Time Dilation

The purpose of this section is to explain why time slows down on an object that moves faster than some other object.

To clarify this statement, consider two platforms. One platform is stationary. One platform is moving. Let's assume a spring pendulum clock is on each platform. Here is the image of a spring pendulum clock.

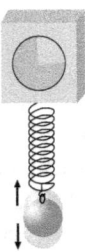

A spring pendulum clock consists of a spring with a weight on it. The spring is attached to a clicker in the box. When the ball is at the bottom position, the weight on the clicker causes the spring to be jerked up just a bit. This gives the ball some energy each bounce to keep the weight bouncing. Each jerk advances the hands on the clock. This provides us with a measure of time. In this appendix, this time is referred to as measured time and has the acronym MT.

The pendulum clocks are identical. When both clocks are on a non-moving platform, both show the same amount of time passing. Mary (M for moving) is on one platform and Neil (N for non-moving) is on the other platform. Mary is on the one moving. Neil is on the non-moving platform.

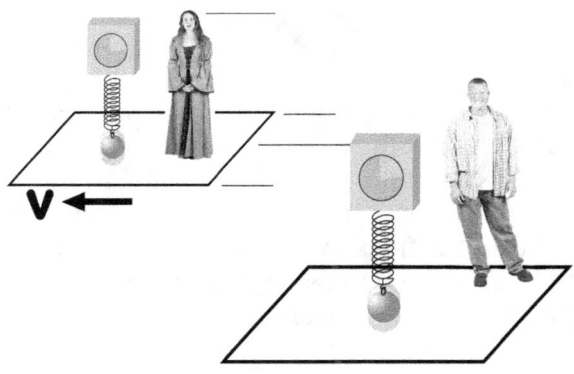

Mary's platform is moving past Neil's platform at some speed. Mary's pendulum clock will bounce fewer times than Neil's pendulum clock. This should clarify the statement made above.

Our goal is to explain this. Recollect the concept of the universe being made of a single particle called photons. The photons cluster to form particles that further cluster to form objects in this universe. Thus, the clocks, people, and platforms consist of photons. This section purports that the motion of the photons in moving and non-moving objects account for the discrepancy in time between Mary and Neil's platforms.

A Closer Look at Time

We need to look at how we measure time closely. In a pendulum clock, a pendulum bouncing up and down measures time. Time is the number of bounces of the pendulum from A to B.

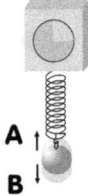

The concept of time being a count is very important. To use a pendulum clock, count the number of bounces of the pendulum from A to B. What does this mean relative to one clock flying past another clock? This means that if Neil's pendulum bounces 10,000 transitions from A to B, Mary's pendulum clock could bounce 8660 transitions from her A to B. This would imply that Mary is traveling at half the speed of light past Neil.

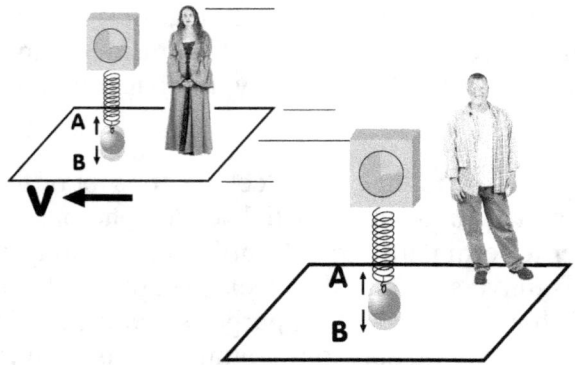

Here is a quick explanation. The particles that make up the pendulums consist of photons moving about. The photons are moving about each other at the speed of light. When the pendulum clock is not moving, all the photons move some specific distance per bounce. When the pendulum clock and its platform move, the photons in the clock must move some additional distance per bounce. This is to support motion of the pendulum bouncing and the motion of the entire clock through space. As a result, the pendulum by Mary bounces slower from the point of view of Neil, which is not moving.

There are some consequences to this.

Mary does not see this change in time. Since she is moving with her pendulum clock, that is her true measure of time. Neil's pendulum clock in her moving reference frame moves slower for the photons in his clock would seem to move farther than is necessary for the bouncing pendulum. The photons in Neil's clock must move, relative to Mary, to support the existence of the pendulum but also move to support the motion relative to Mary. Remember, this is the state of things in Mary's reference frame.

This is still confusing and we intend to clear it up. However, to do that, we need to look at all of this in even more detail.

Observing Different Times

We need to clarify how we measure time. In the real world, time is measured by some repetitive series of events. Perhaps the first repetitive event used was the motion of the sun through the sky. Here we have selected the motion of a pendulum bouncing up and down. During this analysis, we are going to measure the time of motion of the particles moving within the pendulum. This creates another problem for we are looking at the time required for photons to move within a pendulum, the time required for photons to move within a moving pendulum, how long it takes a pendulum to move from A to B, and then we count the number of bounces the pendulum makes to specify the time we measure. As all of these events are referred to as time, it might be confusing. Let's describe each.

The following picture depicts each of these.

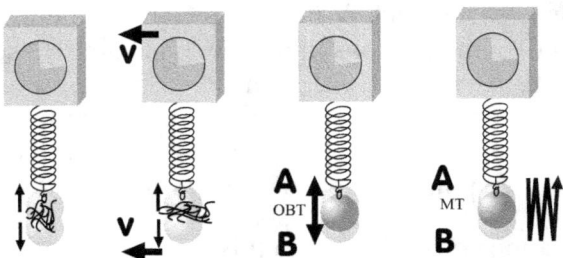

The first image to the left shows the photons moving about in the pendulum. Each photon moves some time during each bounce of the pendulum. The second image from the left shows the clock moving. The photons in the pendulum must move, not only to support the existence of the pendulum, but must move to the left also to enable the whole clock to move to the left. Therefore, the photons must move farther during each bounce. The third image represents the time for one bounce of the pendulum (One Bounce Time, OBT). The rightmost image represents the total bounce time of the pendulum (Total Bounce Time, TBT). The time we read from the clock is the number of bounces of the pendulum (Measured Time, MT). This last time is the one we read from the clock. Each bounce of the pendulum moves the hands of the clock a bit.

It is important to realize the time we use in the real world is the number of bounces the pendulum makes. However to determine that, we need to measure the time the pendulum bounces from A to B (OBT) once. Moreover, we need to measure the total bounce time (TBT) as the clock continues to bounce. Then we can calculate the number of bounces or measured time (MT). Measured time would then be total bounce time divided by one bounce time.

$$MT = \frac{TBT}{OBT}$$

We will use this relationship in a moment.

Thus, our next goal is to determine one bounce time (OBT) and a total bounce time (TBT).

Even More Detail
We need more detail. Therefore, our next task is to construct special clocks that have a similar structure as pendulum clocks but consist of only the necessary components. The goal is to show that a photon in a pendulum clock must move further, when the clock is moving. Because the clock photons must move further, more time is taken to move from A to B. Hence, time goes slower on an object moving past another object.

Here is a clock made with one photon bouncing up and down. The clock consists of two mirrors. One at A and the other at B. This clock is not moving. The clock keeps time by bouncing up and down as it reflects off the mirrors. That is, the measure of time consists of the number of bounces of this photon.

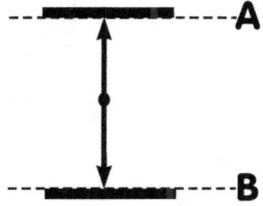

Here is a clock made with one photon bouncing up and down
and is moving to the left.

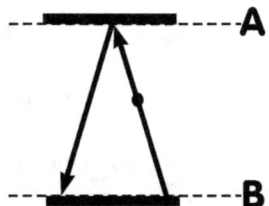

You can see that when the clock is moving, the photon must
travel further to support the clock and the motion of the clock.
Realize that the mirrors are the same distance apart. This is
true for the moving and non-moving clocks.

The key to understanding this is to realize that observers are
not aware of this extra motion. Because the object consists of
many photons that are darting all over the place, this extra
motion is not seen. If we look at the motion of one photon, as
this special clock allows, we can see this extra motion and
calculate with precision what is going on.

Consider the following picture. This picture demonstrates the
critical components in the motion of a photon in a clock.

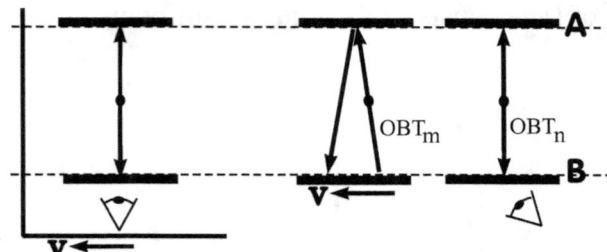

The image on the right represents a photon clock at rest. The
photon is bouncing up and down from A to B. The time for
one motion up or down is one bounce time (OBT sub n).
During this time it travels the distance D sub n. (the n is for
non-moving)

The image in the middle represents a photon clock moving to the left at velocity v. The photon is bouncing up and down from A to B. The time for one motion up or down is one bounce time (OBT sub m, for moving). During this time it travels the distance D from A to B. Note that in the real world, only the distance D perpendicular from A to B is used to measure time.

The image on the left represents the clock from the perspective of a reference frame moving at velocity v with the clock. In this reference frame, the appearance should appear exactly as the photon motion in the reference frame of the non-moving clock.

The First Postulate
It is important to note here that these drawings were made with Einstein's postulate one of special relativity in mind. The stationary observer, under the image on the right, must see the same clock as the observer moving with the clock at velocity v. That is what postulate one explicitly states. Other structures of this picture were formed using that postulate as well. All three images must have the height of D to accomplish the demands of the postulate. The clock box in the middle must show the photon moving at an angle from the perspective of the non-moving reference frame to enable the clock box moving to the left to have the appearance shown in the moving reference frame.

To repeat, this image appears as it does to obey postulate one of Einstein's special relativity.

This appears as a mystery for two reasons.

1. Observers today are not aware that physical particles consist of a mass of whirling photons.

2. Observers cannot see the extra motion photons must travel because of the mass of them whirling around each other.

Therefore, the time of that extra motion is ignored when calculating the measured time from the perspective of the non-
moving reference frame.

Now the Details

If we consider the path of one photon in each clock, one moving and one not moving, we can determine the times involved
with some accuracy. We proceed with this in mind.

1. In the moving platform, we see the photon moves from A to
B.

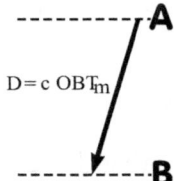

It moves at speed c. The time to move this distance is OBT
sub m. Therefore, the distance is c times OBT sub m.

2. The photon moves horizontally with the velocity v during
the same time the photon moves down at the angle. This is
OBT sub m. Therefore, the distance D sub l is v times OBT
sub m.

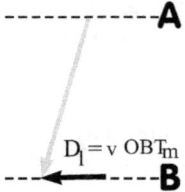

3. The photon in the non-moving platform moves from A to B in time OBT sub n. The distance is c times OBT sub n.

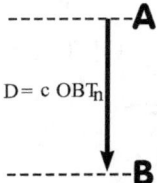

Relating the Distances

To establish the relation of moving time and non-moving time, we need to find a relatonsip between these distances.

These three distances form a right triangle and appears as follows.

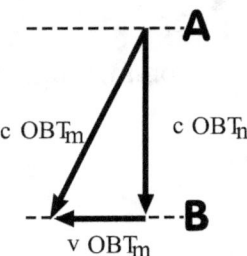

Recollect what the Pythagorean theorem is. In a right triangle, the relationship between the sides of the triangle is as follows.

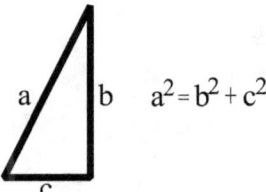

If we apply the Pythagorean Theorem, we get a relation between these three distances indicated above.

$$c^2 \ OBT_m^2 = v^2 \ OBT_m^2 + c^2 \ OBT_n^2$$

If we solve this equation for OBT sub m we get:

$$OBT_m = \frac{OBT_n}{\sqrt{1 - \frac{v^2}{c^2}}}$$

Recollect that when we observe the photon go from A to B on the moving platform we measure the time as OBT sub m. The time for the photon to go from A to B on the non-moving platform is OBT sub n.

Now we need the relation we established before between measured time and total bounce time.

$$MT = \frac{TBT}{OBT}$$

Apply this to moving and non-moving times.

$$MT_m = \frac{TBT}{OBT_m}$$

and

$$MT_n = \frac{TBT}{OBT_n}$$

Solve each of these relations for OBT and substitute for OBT in the equation for OBT sub m. We get the following.

$$MT_n = \frac{MT_m}{\sqrt{1 - \frac{v^2}{c^2}}}$$

Simplify and we get the relationship between Mary's time and Neil's time.

$$MT_n = \frac{MT_m}{\sqrt{1 - \frac{v^2}{c^2}}}$$

This is the time dilation equation from Einstein's special theory of relativity. Now we see a mathematical relationship between the time on Mary's platform relative to Neil's platform.

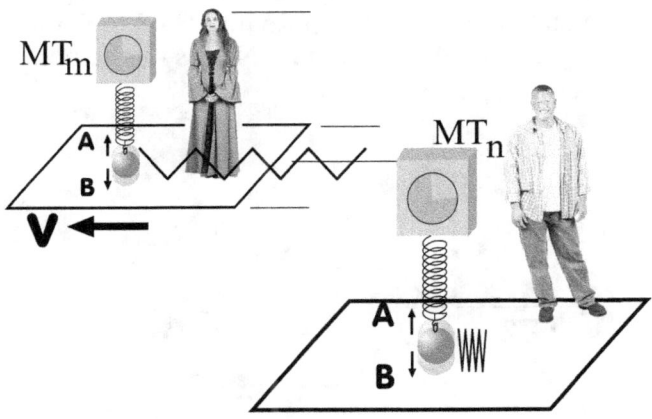

Bear in mind this represents the time on each platform relative to Neil's non-moving platform. In this non-moving frame of reference, Mary's pendulum must travel farther as indicated by the jagged line near the ball on her platform. Neil's ball only goes up and down. Because Mary's platform and clock are going faster than Neil's is, her time is the measured time MT sub m and is related to Neil's measured time MT sub n by the equation above.

This is the essence of Einstein's special relativity.

Lorenz-FitzGerald Contraction

The purpose of this section is to explain why objects moving faster than other objects are shorter.

Once again, to clarify this statement, consider two platforms. One platform is not moving. One platform is moving. Let's assume a bar of gold is on each platform. The gold bars are identical. When both gold bars are on a non-moving platform, both are the same length. Again, Mary (M for moving) is on one platform and Neil (N for non-moving) is on the other platform. Mary is on the one moving. Neil is on the non-moving platform.

Mary's platform is moving past Neil's platform at some speed. Mary's gold bar will be shorter than Neil's bar.

In the imaginary clock established in the previous section, we used a photon bouncing up and down as a clock. That was effective to observe the details of time. Here we wish to observe the details of an object. Time still plays an important part of this analysis but we need to somehow look at an object moving and not moving. To do this we will create an object that consists of a set of photons that form a box around a bouncing photon that keeps time. Here is a box holding a horizontal photon clock.

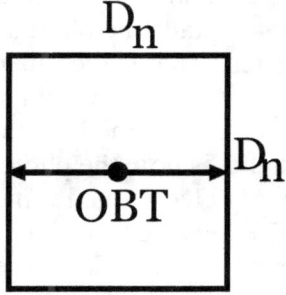

The clock is D sub n high and wide. The black dot represents the photon moving side-to-side. One motion from side to side is one bounce time (OBT).

The following picture is to clarify our task. The box clock on the right is not at motion with respect to the page. This is referred to as frame n. An identical box clock on the left is moving to the left with velocity v.

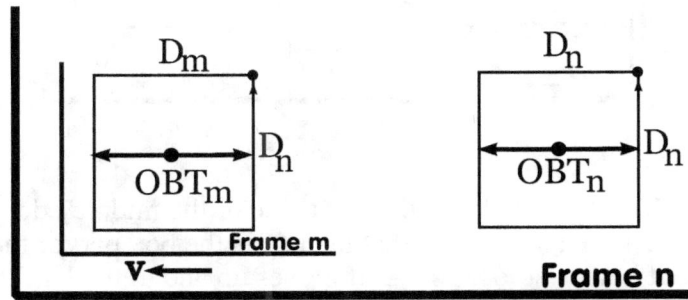

Note that the left box is shown in reference frame m moving to the left at velocity v. The box appears as it would to an observer moving with that reference frame. Our task is to determine D sub m relative to reference frame n, the frame attached to the page.

Motion Detail in Frame n

Here is a similar picture but with more detail. In this picture, an image is inserted that details the photon motion of the left box as it moves to the left. That is the image in the middle. That is how the photons in the box would appear in the non-moving reference frame to have the box move to the left at velocity v. Moreover, that is how the photons would need to move so the observer depicted in the leftmost image would see the box normal. The moving box and the non-moving box are depicted in reference frame n. The large gray arrows at the top represent time as calculated from the previous section.

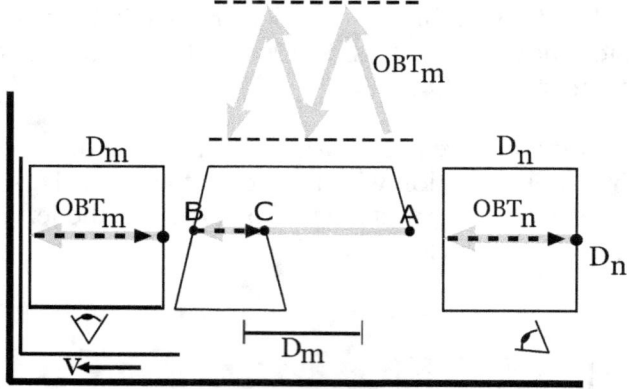

To clarify a bit, the dot in the right side of the middle odd image represents a photon that manifests the box part of the clock. It begins at location A. It moves up and to the left, then to the left, then down to the left, to the right a bit, and finally up and to the left a bit to make one trip around the Cycle of Existence at C. It essentially represents a particle of one photon moving at one-fourth the speed of light to the left.

During this cycle, the clock photon, the photon that only goes back and forth horizontally, begins its travel from the dot pictured on the right at A, goes back and forth and ends at location C. During one trip of the box photon, the clock photon travels back and forth four times. The total distance traveled is 4 D sub n.

The strategy here is to calculate various parameters for the non-moving box and the moving box then compare them to establish a relationship between the width of each box. There is a problem with the middle box in that its motion in the non-moving frame is very odd. We do not have a ruler that can measure the width of that box. To measure the width of both boxes we use the speed that a photon travels across each box and the time of that travel. The distance is related by the following equation.

$$v = \frac{d}{t}$$

With this in mind, we gather some parameters for the non-moving box or the rightmost image.

    Velocity = c
    Time   = MT sub n
    Distance = 4 D sub n

The velocity of all photons is of course the speed of light c.

Getting the time the photon moves is a bit tricky. From what has gone before, we know the travel time for one trip of the photon horizontally is OBT sub n. That is one bounce time horizontally. However, from what has gone before we know that time is measured in the number of transitions a clock particle makes instead of how long the clock particle takes to travel. Therefore, the measured time here is MT sub n.

The distance the photon moves back and forth in the non-moving box clock during one cycle of existence is D sub n.

Now let's get similar parameters for the middle box.
    Velocity = c
    Time   = MT sub m
    Distance = 4 D sub m

Again, velocity is c.

We get the time of motion of the photon bouncing horizontally from the clock lines at the top of the picture. That
clock photon moves up and down twice during one Cycle of
Existance. The time for motion in one direction is OBT sub
m. So the time for the motion of the horizontal moving photon
is 4 OBT sub m. However, as before, measured time is the
number of transitions of the clock particle. This measured
time here is MT sub m.

The width of the box in the middle is difficult to determine as
there is no starting point or end point to measure. The line at
the bottom of the picture labeled D sum m represents the
width of the middle box. This is an unknown value. We have
enough data to calculate this dimension.

Using the relationship for velocity, we know that

$$v = \frac{d}{t}$$

So,

$$V_m = \frac{D_m}{MT_m}$$

and

$$V_n = \frac{D_n}{MT_n}$$

As

$$V_m = V_n = c$$

we have

$$\frac{D_m}{MT_m} = \frac{D_n}{MT_n}$$

Recollect the equation for MT sub n, the equation for time dilation.

$$MT_n = \frac{MT_m}{\sqrt{1 - \frac{v^2}{c^2}}}$$

Substitute that for MT sub n in the previous equation and we get the following.

$$\frac{D_m}{MT_m} = \frac{\frac{D_n}{MT_m}}{\sqrt{1 - \frac{v^2}{c^2}}}$$

Simplify and we get:

$$D_m = D_n\sqrt{1 - \frac{v^2}{c^2}}$$

Now we see a mathematical relationship between the length or distance on Mary's platform relative to Neil's platform.

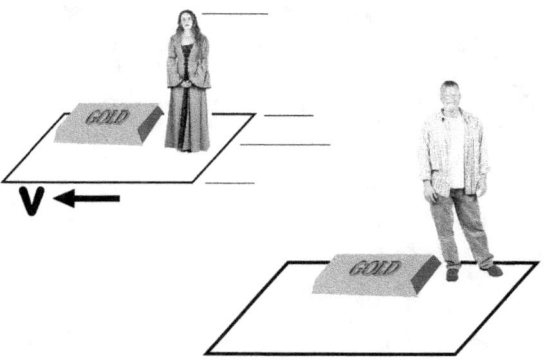

Bear in mind this represents the length on each platform relative to Neil's non-moving platform. In this non-moving frame of reference, Mary's length appears shorter because the photons in the bar of gold appear to move slower and cannot travel as far as the photons in Neil's bar of gold. The equation above shows the relation of the lengths of the two bars relative to Neil's frame of reference.

This is the Lorentz-FitzGerald contraction equation from Einstein's special relativity.

Appendix D

Kinetic Energy

The main part of this book has attempted to show that energy is the area bounded by a moving photon. This appendix is to develop a mathematical expression of the kinetic energy of a moving particle. We will carefully follow the steps outlined in the main part of this book. Note that the kinetic energy of a moving particle is different from the rest energy of a particle. Therefore, the energy of a non-moving particle is subtracted from that of a moving particle to capture the energy of a particle's motion.

Begin with a non-moving particle.

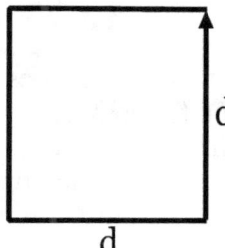

At the lower right corner, just as the constituent photon be-

gins its upper motion, a force from outside the particle causes
the upward motion to veer to the left.

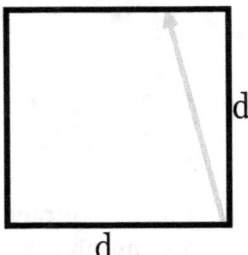

This change in the Cycle of Existence causes the particle to
move to the left at velocity v.

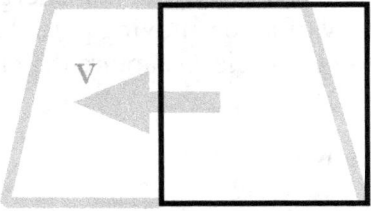

As we have discussed, this pattern does not make a lot of
sense in the non-moving frame of reference. We know,
however, that the height of the particle seen from the non-
moving frame is d and the horizontal distance is somewhat

less than that. The horizontal distance is given by Lorentz-FitzGerald contraction. This is depicted here,

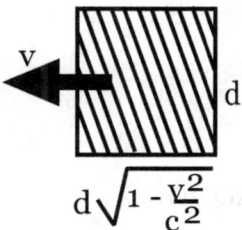

We want to calculate the area of this moving particle. The area is height times width, as is observed from a stationary reference system,

$$A = d^2 \sqrt{1 - \frac{v^2}{c^2}}$$

Per the discussion in the main part of this book, we convert the distance vectors in the equation for area to time vectors. That is, the d is replaced with t. We have seen that energy is,

$$E = \frac{1}{A}$$

Then, substituting for A, we get.

$$E = \frac{1}{t^2 \sqrt{1 - \frac{v^2}{c^2}}}$$

We can expand the fraction formed by the radical in the denominator with the Taylor-Binomial series,

$$\frac{1}{\sqrt{1 - \frac{v^2}{c^2}}} = 1 + \frac{1}{2}\frac{v^2}{c^2} + \frac{3}{8}\frac{v^4}{c^4} + \frac{5}{16}\frac{v^6}{c^6} + \cdots$$

Plugging this into the energy equation, we get,

$$E = \frac{1}{t^2}\left(1 + \frac{1}{2}\frac{v^2}{c^2} + \frac{3}{8}\frac{v^4}{c^4} + \frac{5}{16}\frac{v^6}{c^6} + \cdots\right)$$

Since,

$$c = \frac{d}{t} \quad \text{or} \quad t = \frac{d}{c}$$

we replace t with d/c and get,

$$E = \frac{c^2}{d^2}\left(1 + \frac{1}{2}\frac{v^2}{c^2} + \frac{3}{8}\frac{v^4}{c^4} + \frac{5}{16}\frac{v^6}{c^6} + \cdots\right)$$

The result is,

$$E = \frac{c^2}{d^2} + \frac{1}{2}\frac{v^2}{d^2} + \frac{3}{8}\frac{v^4}{d^2c^2} + \frac{5}{16}\frac{v^6}{d^2c^4} + \cdots$$

The first term is the non-moving mass of the particle we wish to ignore. Therefore, that is subtracted from this equation. The terms after the second with c squared and above are ignored as they are very small and only critical at very high speeds.

We are left with,

$$E = \frac{1}{d^2}\frac{1}{2}v^2$$

As,

$$m = \frac{1}{d^2}$$

The final equation for energy is,

$$E = \frac{1}{2}mv^2$$

Note that this relationship is calculated with the assumption that energy is defined as the area bounded by a moving photon.

Appendix E

Potential Energy

The development in this appendix is similar to the development of kinetic energy in the appendix devoted to that subject. As indicated there, the main part of this book has attempted to show that energy is the area bounded by a moving photon. This appendix is to develop a mathematical expression of potential energy. We will carefully follow the steps outlined in the main part of this book. Note that the potential energy of a particle is different from the rest energy of a particle. The approach here is to show that the area of motion of a photon in a particle as the particle moves some distance, changes as the particle moves against some force during the transition between those two positions. The changes of area during transition are manifested as potential energy. Thus, the method here is to calculate the change of area the photon cuts out of space as the particle changes position in a field of force.

The traditional formula for potential energy is

$$E_{pe} = FD$$

E sub pe is potential energy, F is force, and D is the distance the potential is applied through.

Here is a traditional way of demonstrating potential energy. A particle of matter begins at some fixed point and moves away from that point against some force. Here the force is gravity. The particle is moved against the force with a constant velocity some distance D. The following is a mental experiment doing this and calculating associated energies.

Begin with a particle at rest.

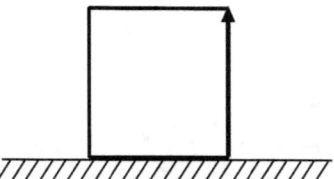

When the photon is at the lower left corner, it angles up a bit as the particle begins its motion upward.

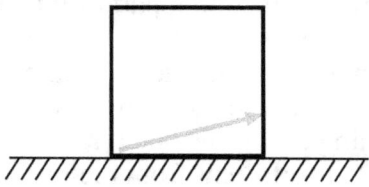

Then, the particle's Cycle of Existence is upward.

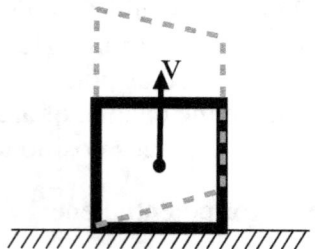

The particle was at rest and is now moving at velocity v. The distance it moves is d sub v. The time for this motion is t sub v. Thus, v is given by the following,

$$v = \frac{d_v}{t_v}$$

This appears as an acceleration of the particle upward. Acceleration is the change in velocity divided by change in time. The initial velocity was zero. The final velocity is v. The assumption is that the acceleration was constant. That is during one Cycle of Existence the velocity was zero. During the next cycle, it was v. Therefore, the total change in velocity is 1/2 of v during time t sub v. Then, the acceleration is,

$$a = \frac{1}{2} \frac{v}{t_v}$$

Next, consider the change of area of the particle as it moves upward. The width does not change in the non-moving frame of reference. However, the height of the particle does due to its motion upward. As demonstrated elsewhere, the change is the Lorentz-Fitzgerald contraction. In the non-moving frame of reference the area is depicted with the following,

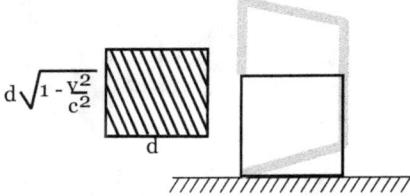

The particle is moving upward so the continious motion would appear as follows,

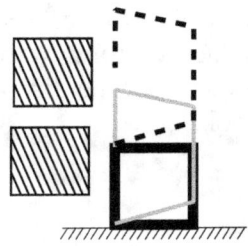

The Cycle of Existence is repeated constantly until the par-

ticle travels upward the distance D above the table top. This
is depicted a follows,

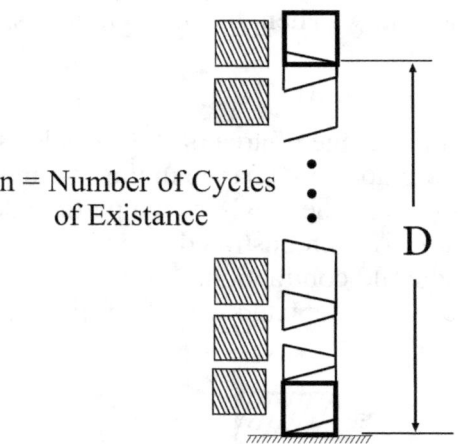

n = Number of Cycles
of Existance

D

The area of a single Cycle of Existence is the height times the
width of the moving particle which is,

$$A = d^2 \sqrt{1 - \frac{v^2}{c^2}}$$

The energy is the reciprocal of this or E=1/A. Note that we
also reinterpret the distance vector as a time vector to calculate the area. This is done per the discussion of energy in
the main part of this book.

$$E = \frac{1}{t^2 \sqrt{1 - \frac{v^2}{c^2}}}$$

Translate the reciprocal of the radical to its Taylor - Binomial expansion.

$$\frac{1}{\sqrt{1-\frac{v^2}{c^2}}} = 1 + \frac{1}{2}\frac{v^2}{c^2} + \frac{3}{8}\frac{v^4}{c^4} + \frac{5}{16}\frac{v^6}{c^6} + \cdots$$

Substitute this in the equation for E,

$$E = \frac{1}{t^2}\left(1 + \frac{1}{2}\frac{v^2}{c^2} + \frac{3}{8}\frac{v^4}{c^4} + \frac{5}{16}\frac{v^6}{c^6} + \cdots\right)$$

Using the following,

$$c = \frac{d}{t} \quad \text{or} \quad t = \frac{d}{c}$$

replace t in the equation for E,

$$E = \frac{c^2}{d^2}\left(1 + \frac{1}{2}\frac{v^2}{c^2} + \frac{3}{8}\frac{v^4}{c^4} + \frac{5}{16}\frac{v^6}{c^6} + \cdots\right)$$

Multiply the contents of the parenthesis by the fraction to get,

$$E = \frac{c^2}{d^2} + \frac{1}{2}\frac{v^2}{d^2} + \frac{3}{8}\frac{v^4}{d^2c^2} + \frac{5}{16}\frac{v^6}{d^2c^4} + \cdots$$

The first term in the equation is the rest mass energy. As we are after the potential energy less the rest mass energy, that term is subtracted out. Those terms after the second that contain c square or greater are very small and can be ignored. Thus, those are subtracted out as well. We are left with the following for the equation of energy.

$$E = \frac{1}{d^2}\frac{1}{2}v^2$$

As,

$$v = \frac{d_v}{t_v}$$

we can separate the v squared into parts giving,

$$E = \frac{1}{d^2} \frac{1}{2} \frac{v}{t_v} d_v$$

Now E here is the energy for one Cycle of Existence. There are n of these as the particle rises up from the table. Thus, we add n of them together to get E sub pe, the total potential energy in this experiment. The large squiggly line with the small n represents the summation of all n Cycles of Existance.

$$E_{pe} = \sum_n E = \frac{1}{d^2} \frac{1}{2} \frac{v}{t_v} \, d_v \, n$$

Considering the following equations,

$$m = \frac{1}{d^2} \, , \quad a = \frac{1}{2} \frac{v}{t_v} \, , \quad D = d_v n \, ,$$

the energy formula becomes,

$$E_{pe} = m a D$$

As,

$$F = ma$$

(from Newton's laws of motion) the equation becomes,

$$E_{pe} = FD$$

Thus, we get the standard equation for potential energy.

To clarify our purpose here: the goal was to show that energy is the area cut out of space by a moving photon. In potential energy, the energy is that area change when a particle moves from one point to another against a force.

References

Part 1: The Ancients

Introduction

None

Chapter 1: Common Ancient Beliefs

None

Chapter 2: The Ancients

1. Surendranath Dasgupta. (1975). A history of Indian philosophy, Vol. I. Motilal Banarsidass. Delhi, ISBN 81-208-0412-8 .

en.wikipedia.org. Vaisheshika. Retrieved June 2, 2011 from http://en.wikipedia.org/wiki/Vaisheshika.

S Radhakrishnan. (2006). Indian philosophy. Vol. II. Oxford University Press, New Delhi, ISBN 0-19-563820-4

2. A. A. Long (ed.). The Cambridge companion to early Greek philosophy. Pg 185.

3. Philip Wheelwright (ed.). (1966). Leucippus, in the presocratics. The Odyssey Press. Pg 177.

4. en.wikipedia.org. Leucippus. Retrieved June 2, 2011 from

http://en.wikipedia.org/wiki/Leucippus.

5. Jonathan Barnes. (1982). The presocratic philosophers. Routledge Revised Edition.

6. Russell Bertrand. (1972). A history of western philosophy. Simon & Schuster. Pg 64–65.

7. en.wikipedia.org. Democritus. Retrieved June 2, 2011 from http://en.wikipedia.org/wiki/democritus

8. Russell, Bertrand (1972). A History of Western Philosophy, Simon & Schuster, Pg 64-65

9. Pamela Gossin. (2002). Encyclopedia of Literature and Science,

10. en.wikibooks.org. General Chemistry/ Atomic Structure/ History of Atomic Structure. Retrieved October 8, 2011 from http://en.wikibooks.org/wiki/General_Chemistry/Atomic_Structure/History_of_Atomic_ Structure

11. universetoday.com. Democritus model. Retrieved October 8, 2011 from http://www.universetoday.com/60137/democritus-model/

12. philosophybasics.com. Empedocles. Retrieved October 8, 2011 from http://www.philosophybasics.com/philosophers_empedocles.html

13. en.wikipedia.org. Lucretius. Retrieved June 6, 2011 from http://en.wikipedia.org/wiki/lucretius

14. The concept of the four classical elements appeared in other lands and spoken about by other philosophers before Empedocles' time. For example, the Vaisheshika Hindu School had a concept of fire, earth, water, and air. See reference 1. In a text written between the 18th and 16th centuries

BC the four classical elements are presented. The reference to this as follows.

Francesca Rochberg. (2002). A consideration of Babylonian astronomy within the historiography of science. Studies in History and Philosophy of Science 33 (4): Pg 661–684. doi:10.1016/S0039-3681(02)00022-5.

15. gap-system.org. Light through the ages: Ancient Greece to Maxwell. Retrieved October 8, 2011 from http://www.gap-system.org/~history/HistTopics/ Light_1.html

16. en.wikipedia.org. Epicurus. Retrieved October 8, 2011 from http://en.wikipedia.org/ wiki/Epicurus

17. Kenneth Cole McLeisch. (1999). Aristotle: The Great Philosophers. Routledge, Pg 5, ISBN 0-415-92392-1

18. Russell Bertrand. (1972). A History of Western Philosophy. Simon & Schuster.

19. Peter Green. (1991). Alexander of Macedon. University of California Press, Ltd. Oxford, England. Library of Congress Cataloging-in-Publication Data. Pg 58-59

20. en.wikipedia.com. Aristotle. Retrieved June 2, 2011 from http://en.wikipedia.org/wiki/Aristotle

21. scienceworld.wolfram.com. Ether. Retrieved June 5, 2011 from http://scienceworld.wolfram.com/physics/ Ether.html

22. Jonathan Barnes. (1995). Life and Work. The Cambridge Companion to Aristotle. Pg 9

23. http://en.wikipedia.org. Lucretius. Retrieved October 8, 2011 from http://en.wikipedia.org/wiki/Lucretius

24. en.wikipedia.org. Buddhist atomism. Retrieved July 11,

2011 from http://en.wikipedia.org/wiki/Buddhist_atomism

25. Bradley Steffens (2006). Ibn al-Haytham: First Scientist, Morgan Reynolds Publishing, ISBN 1599350246

26. David C Lindberg. (1976). al-Kindi to Kepler, Theories of Vision. University of Chicago Press, Chicago. ISBN 0-226-48234-0, OCLC 185636643 1676198 185636643. Pg 60–7

27. britannica.com. Ibn al-Haytham. Retrieved 2008-08-06 http://www.britannica.com/EBchecked/topic/738111/Ibn-al- Haytham

28. www-history.mcs.st-andrews.ac.uk. Abu Ali al-Hasan ibn al-Haytham. Retrieved September 20 from http://www-history.mcs.st-andrews.ac.uk/ Biographies/Al-Haytham.html

29. sciencemag.org. Alhazen. Retrieved September16, 2008 from http://www.sciencemag.org/cgi/content/full/297/5582/773

30. Sadegh Sajjadi. Alhazen. Great Islamic Encyclopedia, Volume 1, Article No. 1917.

Part 2: The Renaissance

Introduction

1. en.wikipedia.org. Renaissance. Retrieved October 11, 2011 from http://en.wikipedia.org/wiki/Renaissance

Chapter 3: The Renaissance Men

1. en.wikipedia.org. Renaissance. Retrieved October 11, 2011 from http://en.wikipedia.org/wiki/Renaissance

2. en.wikipedia.org. Descartes. Retrieved October 11, 2011 from http://en.wikipedia.org/wiki/Ren%C3%A9_Descartes

renedescartes.com. René Descartes. Retrieved October 11, 2011 from http://renedescartes.com/

3. plato.stanford.edu. Pierre Gassendi. Retrieved October 11, 2011 from http://plato.stanford.edu/entries/gassendi/

4. en.wikipedia.org. Robert Hooke. Retrieved October 11, 2011 from http://en.wikipedia.org/wiki/Robert_Hooke

roberthooke.com. Robert Hooke. Retrieved October 11, 2011 from http://roberthooke.com/

5. en.wikipedia.org. Isaac Newton. Retrieved October 11, 2011 from http://en.wikipedia.org/wiki/Isaac_Newton

britannica.com. Sir Isaac Newton. Retrieved October 11, 2011 from http://www.britannica.com/EBchecked/topic/413189/Sir-Isaac-Newton

6. en.wikipedia.org. Francesco Maria Grimaldi. Retrieved October 11, 2011 from http://en.wikipedia.org/wiki/Francesco_Maria_Grimaldi

7. en.wikipedia.org. Christiaan Huygens. Retrieved October 11, 2011 from http://en.wikipedia.org/wiki/Christiaan_Huygens

britannica.com. Christiaan-Huygens. Retrieved October 11, 2011 from http://www.britannica.com/EBchecked/topic/277775/Christiaan-Huygens

8. en.wikipedia.org. Thomas Young Retrieved October 11, 2011 from http://en.wikipedia.org/wiki/Thomas_Young_%28scientist%29

britannica.com. Thomas Young. Retrieved October 11, 2011 from http://www.britannica.com/EBchecked/topic/653983/Thomas-Young

9. britannica.com. Augustin Jean Fresnel. Retrieved October 11, 2011 from http://www.britannica.com/EBchecked/topic/219805/Augustin-Jean-Fresnel

en.wikipedia.org. Augustin-Jean Fresnel. Retrieved October 11, 2011 from http://en.wikipedia.org/wiki/Augustin-Jean_Fresnel

10. en.wikipedia.org. Léon Foucault. Retrieved October 11, 2011 from http://en.wikipedia.org/ wiki/L%C3%A9on_Foucault

encyclopedia.com. Jean Bernard Leon Foucault. Retrieved October 11, 2011 from http://www.encyclopedia.com/topic/Jean_Bernard_Leon_Foucault.aspx

Chapter 4: The Meaning of Light

1. britannica.com. Huygens' principle. Retrieved October 12, 2011 from http://www.britannica.com/EBchecked/topic/277804/Huygens-principle

2. en.wikipedia.org. Corpuscular theory of light. Retrieved October 13, 2011 from http://en.wikipedia.org/wiki/Corpuscular_theory_of_light

3. en.wikipedia.org. Diffraction. Retrieved October 13, 2011 from http://en.wikipedia.org/wiki/Diffraction

4. en.wikipedia.org. Diffraction formalism Retrieved October13, 2011 from http://en.wikipedia.org/wiki/Diffraction_ formalism

5. sci.tech-archive.net. Theories of light 2: ether theories. Retrieved October 13, 2011 from http://sci.tech-archive.net/Archive/ sci.physics/2005-10/msg01264.html

6. en.wikipedia.org/wiki. Double-slit experiment. Retrieved October 13, 2011 from http://en.wikipedia.org/wiki/Double-slit_experiment

7. en.wikipedia.org. Huygens-Fresnel principle. Retrieved October 12, 2011 from http://en.wikipedia.org/wiki/Huygens%E2%80%93Fresnel_principle

Chapter 5: Wave Theory

None

Chapter 6: Polarization

1. vision.berkeley.edu. Waves & Superposition. Retrieved October 12, 2011 from http://vision.berkeley.edu/roordalab/VS203BWebsite/LectureNotes/Waves&Superposition 2011.pdf

2. polarization.com. The Discovery. Retrieved October 12, 2011 from http://www.polarization.com/history/history.html

3. en.wikipedia.org. Huygens-Fresnel principle. Retrieved October 12, 2011 from http://en.wikipedia.org/wiki/Huygens%E2%80%93Fresnel_principle

Chapter 7: Wave Functions 101

None

Part 3: The Scientific Revolution

Introduction

None

Chapter 8: Electromagnetic Personalities

1. en.wikipedia.org. Alessandro Volta. Retrieved October 28,

2011 from http://en.wikipedia.org/wiki/Alessandro_ Volta

answers.com. Alessandro Volta. Retrieved October 28, 2011 from http://www.answers.com/topic/alessandro-volta

2. en.wikipedia.org. Charles Augustin de Coulomb. Retrieved October 28, 2011 from http://en.wikipedia.org/ wiki/Charles- Augustin_de_Coulomb

answers.com. Charles Augustin de Coulomb. Retrieved October 28, 2011 from http://www.answers.com/topic/ charles-augustin-de-coulomb

3. en.wikipedia.org. Michael Faraday. Retrieved October 28, 2011 from http://en.wikipedia.org/wiki/Michael_Faraday

answers.com. Michael Faraday. Retrieved October 28, 2011 from http://www.answers.com/topic/michael-faraday

4. en.wikipedia.org. James Clerk Maxwell. Retrieved October 28, 2011 from http://en.wikipedia.org/wiki/ James_Clerk_Maxwell

5. en.wikipedia.org. Michael Faraday. Retrieved October 28, 2011 from http://en.wikipedia.org/wiki/Michael_ Faraday #cite_note-32

6. en.wikipedia.org. Michael Faraday. Retrieved October 28, 2011 from http://en.wikipedia.org/wiki/Michael_Faraday #cite_note-33

Chapter 9: Electromagnetic Theory

1. answers.com. Luigi Galvani. Retrieved October 28, 2011 from http://www.answers.com/topic/luigi-galvani

2. inventors.about.com. History of The Battery. Retrieved

October 28, 2011 from http://inventors.about.com/od/
bstartinventions/a/History-Of-The-Battery.htm

3. ias.ac.in. Resonance. Retrieved October 28, 2011 from
http://www.ias.ac.in/ resonance/Mar2002/pdf/Mar2002p35-
45.pdf

4. rare-earth-magnets.com. History of magnets. Retrieved
October 28, 2011 from http://www.rare-earth-magnets.com/
t-history-of-magnets.aspx

5. bart.tcc.virginia.edu. Chapter1 sec3. Retrieved October 28,
2011 from http://bart.tcc.virginia.edu/book/chap1/chapter1
sec3.htm

6. en.citizendium.org. Michael Faraday. Retrieved October
28, 2011 from http://en.citizendium.org/wiki/Michael_
Faraday

7. physicsclassroom.com. Estatics. Retrieved October 28,
2011 from http://www.physicsclassroom.com/class/estatics/
u8l3b.cfm

8. en.wikipedia.org. Electromagnetic theory. Retrieved October 28, 2011 from http://en.wikipedia.org/wiki/
Electromagnetic_theory

9. en.wikipedia.org. A Dynamical Theory of the
Electromagnetic Field. Retrieved October 28, 2011 from
http://en.wikipedia.org/wiki/A_Dynamical_Theory
_of_the_Electromagnetic_Field

10. en.wikipedia.org. Wave Retrieved October 28, 2011
fromhttp://en.wikipedia.org/wiki/Wave

Chapter 10: Electromagnetic Personalities

1. en.wikipedia.org. Antoine Lavoisier Retrieved October 28,
2011 from http://en.wikipedia.org/wiki/Antoine_Lavoisier

answers.com. Antoine Lavoisier, Retrieved October 28, 2011 from http://www.answers.com/topic/antoine-lavoisier

2. en.wikipedia.org. Dmitri Mendeleev. Retrieved October 28, 2011 from http://en.wikipedia.org/wiki/Dmitri_Mendeleev

aip.org. Periodic Table. Retrieved October 28, 2011 fromhttp://www.aip.org/history/curie/periodic.htm

3. en.wikipedia.org. John Dalton. Retrieved October 28, 2011 from http://en.wikipedia.org/wiki/John_Dalton

answers.com. John Dalton. Retrieved October 28, 2011 from http://www.answers.com/topic/john-dalton

4. en.wikipedia.org. Jean Baptiste Perrin. Retrieved October 28, 2011 from http://en.wikipedia.org/wiki/Jean_Baptiste_Perrin

nobelprize.org. Jean Baptiste Perrin - Biography. Retrieved October 28, 2011 from http://www.nobelprize.org/nobel_prizes/physics/laureates/1926/perrin-bio.html

5. en.wikipedia.org. Amedeo Avogadro Retrieved October 28, 2011 from http://chemistry.about.com/od/famouschemists/a/avogadro.htm

chemistry.about.com. Avogadro. Retrieved October 28, 2011 from http://chemistry.about.com/od/famouschemists/a/avogadro.htm

Chapter 11: Molecules and Atoms

1. novelguide.com. Antoine Laurent Lavoisier Retreived October 7, 2011 from http://www.novelguide.com/a/discover/ewb_09/ewb_09_03756.html

2. iun.edu. Dalton's Atomic Theory. Retreived October 7, 2011 from http://www.iun.edu/~cpanhd/C101webnotes/composition/dalton.html

3. dl.clackamas.edu Dalton's Atomic Theory. Retreived October 7, 2011 from http://dl.clackamas.edu/ch104-04/dalton%27s.htm

4. bulldog.u-net.com. Avogadro. Retrieved October 7, 2011 from http://www.bulldog.u-net.com/ avogadro/avoga.html

5. pbs.org. Dimitri Mendeleev. Retrieved October 7, 2011 from http://www.pbs.org/wnet/ hawking/cosmostar/html/cstars_mendel.html

6. Physical Science, Holt Rinehart & Winston (January 2004), page 302 ISBN 0-03-073168-2

7. davidparker.com. Nuclear twins: the discovery of the proton and neutron. Retrieved October 6, 2011 from http://www.davidparker.com/janine/twins.html

8. James Clerk Maxwell. (1873). Molecules. Nature, Pg 437-441

9. answers.com. Jean Baptiste Perrin. Retrieved October 28, 2011 from http://www.answers.com/topic/jean-baptiste-perrin

Chapter 12: Atomic Men

1. en.wikipedia.org. J._J._Thomson. Retrieved October 7, 2011 from http://en.wikipedia.org/wiki/J._J._Thomson

nobelprize.org. J.J. Thomson - Biography. Retrieved October 7, 2011 from http://www.nobelprize.org/nobel_prizes/physics/laureates/1906/thomson-bio.html

2. Nobelprize.org. Ernest Rutherford - Biography. Retrieved

October 7, 2011 from ttp://www.nobelprize.org/nobel_prizes/
chemistry/laureates/1908/rutherford-bio.html

newworldencyclopedia.org. Ernest_Rutherford. Retrieved
October 7, 2011 from ttp://www.newworldencyclopedia.org/
entry/Ernest_Rutherford

3. Encyclopędia Britannica Online. Henry Gwyn Jeffreys
Moseley. Retrieved October 7, 2011 from http://www.
britannica.com/EBchecked/topic/393528/ Henry-Gwyn-
Jeffreys-Moseley>.

Encyclopedia.com. Complete Dictionary of Scientific
Biography, Moseley, Henry Gwyn Jeffreys. Retrieved October 7, 2011 from http://www.encyclopedia.com

4. nobelprize.org. James Chadwick - biography. Retrieved
October 7, 2011 from http://www.nobelprize.org/nobel_
prizes/ physics/laureates/1935/chadwick-bio.html

New World Encyclopedia.org. James Chadwick. Retrieved
October 7, 2011from http://www.newworldencyclopedia.org/
entry/James_Chadwick?oldid=678675

Chapter 13: Atomic Theory

1. aip.org. A look inside the atom. Retrieved October 6, 2011
from http://www.aip.org/history/electron/jjhome.htm

2. J.J. Thomson. (1897). Cathode rays. The Electrician. Pg
39, 104

3. aip.org. Three experiments, one big idea. Retrieved October 6, 2011 from http://www.aip.org/history/
electron/jj1897.htm

4. galileo.phys.virginia.edu. Rutherford_Scattering.
Retreived October 6, 2011 from http://galileo.phys.virginia.

edu/ classes/252/Rutherford_Scattering/
Rutherford_Scattering.html

5. en.wikipedia.org. Geiger-Marsden experiment. Retrieved
October 6, 2011 from http://en.wikipedia.org/wiki/
Geiger%e2%80%93Marsden_experiment

6. hyperphysics.phy-astr.gsu.edu. Rutherford Scattering.
Retrieved October 6, 2011 from http://hyperphysics.phy-
astr.gsu.edu/hbase/rutsca.html

7. en.wikipedia.org. Rutherford model. Retrieved October 6,
2011 from http://en.wikipedia.org/wiki/Rutherford_model

8. http://www.britannica.com. Henry Gwyn Jeffreys
Moseley. Retrieved October 6, 2011 from http://www.
britannica. com/EBchecked/ topic/393528/Henry-Gwyn-
Jeffreys-Moseley

9. http://www.sas.upenn.edu. Moseley. Retrieved October 6,
2011 from http://www.sas.upenn.edu/~mabruder/
moseleypage.html

10. en.wikipedia.org. Henry Moseley. Retrieved October 6,
2011 from http://en.wikipedia.org/wiki/Henry_Moseley

11. http://www.britannica.com. Moseleys X-ray studies.
Retrieved October 6, 2011 from http://www.britannica.com/
EBchecked/ topic/41549/atom/48359/Moseleys-X-ray-
studies

12. britannica.com. Proton. Retrieved October 6, 2011 from
http://www.britannica.com/EBchecked/topic/480330/proton?
anchor=ref206857

13. davidparker.com. Nuclear twins: the discovery of the proton and neutron. Retrieved October 6, 2011 from http://
www.davidparker.com/janine/twins.html

14. helium.com. James Chadwick and his discovery of the neutron. Retrieved October 6, 2011 from http://www.helium.com/items/ 222794-james-chadwick-and-his-discovery-of-the-neutron

15. aip.org. A second generation of Curies. Retrieved October 6, 2011 from http://www.aip.org/history/curie/2ndgen1.htm

16. Chadwick, James. Possible existence of a neutron. Nature 129 (3252): 312. Bibcode 1932Natur.129Q.312C. doi:10.1038/129312a0.

Part 4: The Second Scientific Revolution

Introduction

None

Chapter 14: Men of the Second Revolution

1. en.wikipedia.org. Albert Abraham Michelson. Retrieved October 4, 2011 from http://en.wikipedia.org/wiki/Albert_Abraham_Michelson

nobelprize.org . Albert A. Michelson - Biography. Retrieved October 4, 2011 from http://www.nobelprize.org/nobel_prizes/physics/laureates/1907/michelson-bio.html

2. en.wikipedia.org. Edward Morley Retrieved October 4, 2011 from http://en.wikipedia.org/wiki/Edward_Morley

www.britannica.com. Edward Williams Morley. Retrieved October 4, 2011 from http://www.britannica.com/EBchecked/ topic/392445/Edward-Williams-Morley.

3. en.wikipedia.org. Max Planck. Retrieved October 4, 2011 from http://en.wikipedia.org/wiki/Max_Planck

nobelprize.org. Max Planck - Biography. Retrieved October 4, 2011 from http://www.nobelprize.org/ nobel_prizes/ physics/ laureates/1918/planck-bio.html

4. en.wikipedia.org. Albert_Einstein Retrieved October 4, 2011 from http://en.wikipedia.org/wiki/Albert_Einstein

nobelprize.org. Albert Einstein - Biography. Retrieved October 4, 2011 from http://www.nobelprize.org/nobel_ prizes/physics/laureates/1921/einstein-bio.html

5. en.wikipedia.org. Niels Bohr. Retrieved October 4, 2011 from http://en.wikipedia.org/wiki/Niels_Bohr

nobelprize.org. Niels Bohr - Biography. Retrieved October 4, 2011 from http://www.nobelprize.org/nobel_prizes/ physics/ laureates/1922/bohr-bio.html

6. en.wikipedia.org. Geoffrey Ingram Taylor. Retrieved October 4, 2011 from http://en.wikipedia.org/wiki/ Geoffrey_Ingram_Taylor

britannica.com. Sir Geoffrey Ingram Taylor. Retrieved October 4, 2011 from http://www.britannica.com/ EBchecked/ topic/ 584826/Sir-Geoffrey-Ingram-Taylor

Chapter 15: The Second Revolution

1. aip.org. Their apparatus. Retrieved October 4, 2011 from http://www.aip.org/history/einstein/ae20.htm

2. en.wikipedia.org. Luminiferous aether. Retrieved October 4, 2011 from http://en.wikipedia.org/wiki/ Luminiferous_aether

3. encyclopedia.com. Max_Planck. Retrieved October 5, 2011 from http://www.encyclopedia.com/topic/Max_ Planck.aspx

4. abyss.uoregon.edu. Planck's constant. Retrieved October 4, 2011 from http:// abyss.uoregon.edu/%7Ejs/ 21st_century_science/ lectures/lec12.html

5. newworldencyclopedia.org. Photoelectric effect. Retrieved October 4, 2011 from http://www.newworld encyclopedia.org/ entry/Photoelectric_effect

6. en.wikibooks.org. Chemical Sciences: A Manual for CSIR-UGC National Eligibility Test for Lectureship and JRF/Photoelectric effect. Retrieved October 4, 2011 from http://en.wikibooks.org/wiki/Chemical_Sciences: _A_Manual_for_CSIR-UGC_National_Eligibility_ Test_ for_Lectureship_and_JRF/Photoelectric_effect photoelectric effect

7. spiff.rit.edu. Einstein and the photoelectric effect. Retrieved October 4, 2011 from http://spiff.rit.edu/classes/ phys314/ lectures/photoe/photoe.html

8. www.britannica.com. Robert Andrews Millikan. Retrieved October 4, 2011 from http://www.britannica. com/EBchecked/ topic/382902/Robert-Andrews-Millikan

9. undsci.berkeley.edu. A science prototype: Rutherford and the atom. Retrieved October 4, 2011 from http:// undsci.berkeley.edu/lessons/pdfs/rutherford.pdf

10. britannica.com. Spectral line series. Retrieved October 4, 2011 from http://www.britannica.com/EBchecked/ topic/558836/spectral-line-series

11. rwc.uc.edu. Bohr model of the atom. Retrieved October 4, 2011 from http://www.rwc.uc.edu/koehler/ biophys/6a.html

Part 5: A Paradigm Shift

Introduction

None

Chapter 16: The Paradigm Men

1. en.wikipedia.org. Louis de Broglie. Retrieved September 30, 2011 from http://en.wikipedia.org/wiki/ Louis_de_Broglie

Nobelprize.org. Louis de Broglie - biography. Retrieved September 30, 2011 from http://www.nobelprize.org/ nobel_prizes/physics/ laureates/1929/broglie-bio.html

2. en.wikipedia.org. Max Born. Retrieved October 4, 2011 from http://en.wikipedia.org/wiki/Max_Born

nobelprize.org. Max Born - Biography. Retrieved October 4, 2011 from http://www.nobelprize.org/ nobel_prizes/ physics/laureates/1954/born-bio.html
3. en.wikipedia.org. Werner Heisenberg. Retrieved September 30, 2011 from http://en.wikipedia.org/wiki/ Werner_Heisenberg

Nobelprize.org. Werner Heisenberg - Biography. Retrieved September 30, 2011 from http://www.nobelprize.org/ nobel_prizes/ physics/ laureates/1932/heisenberg-bio.html

4. en.wikipedia.org Erwin Schrödinger. Retrieved September 30, 2011 from http://en.wikipedia.org/wiki/ Erwin_Schr%C3%B6dinger

nobelprize.org. Erwin Schrödinger - Biography. Retrieved September 30, 2011 from http://www.nobelprize. org/nobel_prizes/ physics/ laureates/1933/schrodinger-bio.html

Chapter 17: A Paradigm Shift

1. britannica.com. Lester Halbert Germer. Retrieved

September 30, 2011 from http://www.britannica.com/
EBchecked/ topic/231764/Lester-Halbert-Germer

2. encyclopedia.com. Stern, Otto. Retrieved 30 Sep. 2011
from http://www.encyclopedia.com/topic/Otto_Stern.aspx

Chapter 18: PSI

1. en.wikipedia.org. Probability amplitude. Retrieved October 2, 2011 from http://en.wikipedia.org/wiki/
Probability_Amplitude

Chapter 19: Boundary Conditions

None

Chapter 20: More Quantum Weirdness

1. www.britannica.com. Sir Geoffrey Ingram Taylor.
Retrieved September 30, 2011 from http://www.britannica.
com/EBchecked/topic/ 584826/Sir-Geoffrey-Ingram- Taylor

2. Jönsson Clauss (1974). Electron diffraction at multiple
slits. American Journal of Physics, 4:4–11.

3. reference.com. Slit experiment. Retrieved September 30,
2011 from http://www.reference.com/browse/Slit+
experiment

4. plato.stanford.edu. The Uncertainty Principle. Retrieved
October 1, 2011 from http://plato.stanford.edu/entries/qt-
uncertainty/

5. J. von Neumann (1955). Mathematical Foundations of
Quantum Mechanics. Princeton University Press.

6. Schrödinger E; Born, M. (1935). "Discussion of probability relations between separated systems". Mathematical

Proceedings of the Cambridge Philosophical Society 31 (4): 555–563. doi:10.1017/S0305004100013554.

7. curious.astro.cornell.edu. Does quantum entanglement imply faster than light communication? Retrieved October 2, 2011 from http://curious.astro.cornell.edu/question. php?number=612

8. Blaylock, Guy (2010). The EPR paradox, Bell's inequality, and the question of locality. American Journal of Physics 78 (1): 111–120. arXiv:0902.3827. Bibcode 2010AmJPh..78..111B. doi:10.1119/1.3243279.

9. Simon, D.R. (1994). On the power of quantum computation. Foundations of Computer Science, 1994 Proceedings., 35th Annual Symposium on: 116–123.

10, science.jrank.org. Virtual particles. Retrieved October 2, 2011 from http://science.jrank.org/pages/7195/Virtual-Particles.html

Chapter 21: What is Quantum Physics?

1. Arthur Beiser. (1963). Concepts of modern physics. McGraw Hill Book Company, Pg 60

2. inerton.wikidot.com. Quantum mechanics and de Broglie's concept. Retrieved September 30, 2011 from http://inerton.wikidot.com/ quantum-mechanics-and-de-broglie-s-concept

3. Nobelprize.org. Erwin Schrödinger – Biography. Retrieved September 30, 2011 from http://www.nobelprize. org/nobel_prizes/physics/ laureates/1933/schrodinger-bio.html

4. hawking.org.uk. Does God Play Dice? Retrieved September 30, 2011 from http://www.hawking.org.uk/ index.php/lectures/64

5. The Illusion of Reality. BBC Atom Part 3 of 3 Southern Star Entertainment UK Pic MMVII A video presentation. 27:00. Retrieved June 8, 2011 from http://www.youtube.com/watch?v=bF5-jTlolMk

Part 6: We are Beings of Light

Introduction

None

Chapter 22: What is Light?

1. Richard P. Feynman. (2006). QED. Princeton University Press Pg 14

2. http://en.wikipedia.org. Electromagnetic wave equation. Retrieved Sept 27, 2011 from http://en.wikipedia.org/wiki/Electromagnetic_wave_equation

3. www.physicsclassroom.com Polorization. Retrieved Sept. 27, 2011. from http://www.physicsclassroom. com/Class/light/ U12L1e.cfm

4. onderwijs1.amc.nl. Light: diffraction. Retrieved Sept. 27, 2011. from http://onderwijs1.amc.nl/medfysica/doc/LightDiffraction.htm

5. onderwijs1.amc.nl. Light: diffraction. Retrieved Sept. 27, 2011. from http://onderwijs1.amc.nl/medfysica/doc/LightDiffraction.htm

6. http:// www.electron.rmutphysics.com/. Electromagnetic waves. Retrieved Sept. 27, 2011 from http://www.electron.rmutphysics.com/ physics/charud/scibook/Physics-for-Scientists-and- Engineers-Serway-Beichne%206 edr-4/34%20-%20 Electromagnetic% 20Waves.pdf

7. en.citizendium.org. Quantization of the electromagnetic

field. Retrieved Sept. 27, 2011 from http://en.citizendium. org/wiki/ Quantization_of_the_ electromagnetic_field

8. www.ece.msstate.edu. Electromagnetic waves. Retrieved Sept. 27, 2011 from http://www.ece.msstate.edu/~donooe/ ece3324notes10.pdf

9. www.iwu.edu Deflection by e-field. Retrieved Sept. 27, 2011 from http://www.iwu. edu/~gspaldin/ DeflectionByE-field.pdf

10. en.wikipedia.org Electromagnetic radiation Retrieved Sept. 27, 2011 from http://en.wikipedia.org/wiki/ Electromagnetic_radiation

11. worsleyschool.net. Polorized light. Retrieved September 28, 2011 from http://www.worsleyschool.net/science/files/ polarized/ light.html

12. commons.wikimedia.org. Circular polarization. Retrieved September 28, 2011 from http://commons.wikimedia. org/wiki/ File:Circular.Polarization.Circularly.Polarized. Light_ Circular.Polarizer_Creating.Left.Handed.Helix. View_el.svg

13. jirkacech.com. Single slit experiment. Retrieved September 28, 2011 from http://www.jirkacech.com/ public/Thesis/node7.html

14. micro.magnet.fsu.edu. Double Slit Experiment Retrieved September 28, 2011 from http://micro.magnet.fsu.edu/ primer/ java/interference/doubleslit/

15. juliantrubin.com. The double slit experiment. Retrieved September 28, 2011 from http://www. juliantrubin.com/ bigten/youngdoubleslit.html

Chapter 23: All is Light

1. http://math.ucr.edu. Is the speed of light constant? Retrieved September 27, 2011 from http://math.ucr.edu/ home/baez/physics/ Relativity/SpeedOfLight/speed_of_ light.html

2. en.wikipedia.org. Euclidean space. Retrieved September 27, 2011 from http://en.wikipedia.org/wiki/Euclidean_space

3. Richard J. Trudeau (1987).The non-euclidean revolution. Birkhauser. ISBN-12: 978-8176-4782-7

4. en.wikisource.org. On the Non-Euclidean Interpretation of the Theory of Relativity Retrieved September 27, 2011 from http://en. wikisource.org/wiki/On_the_Non-Euclidean_ Interpretation_of_the_Theory_of_Relativity

5. Arthur Beiser. (1963). Concepts of modern physics. McGraw Hill Book Company, Pg 13

6. Arthur Beiser. (1963). Concepts of modern physics. McGraw Hill Book Company, Pg 14

7. www2.slac.stanford.edu. Special relativity. Retrieved September 27, 2011 from http://www2.slac.stanford.edu/ vvc/theory/ relativity.html

8. en.wikisource.org. On the Non-Euclidean Interpretation of the Theory of Relativity Retrieved September 27, 2011 from http://en.wikisource.org/wiki/On_the_Non-Euclidean_ Interpretation_of_the_Theory_of_Relativity

9. Arthur Beiser. (1963). Concepts of modern physics. McGraw Hill Book Company, Pg 33

10. nsrconline.org. Measurment. Retrieved November 6,

2011 from http://www.nsrconline.org/ curriculum_resources/
Com_overview.html

11. unc.edu. English Customary Weights and Measures
Retrieved September 27, 2011 from http://www.unc.
edu/~rowlett/units/custom.html

12. oscience.info. Standards of Measurement: Length, Mass
and Time. Retrieved September 27, 2011 from http://
oscience.info/ physics/ standards-of- measurement- length-
mass-and-time/

13. history.mcs.st-andrews.ac.uk. A history of time: Classical
time Retrieved September 27, 2011 from http:// www-
history.mcs.st-andrews.ac.uk/HistTopics/ Time_1.html

14. oscience.info. Standards of Measurement: Length, Mass
and Time. Retrieved September 27, 2011 from http://
oscience.info/physics/standards-of-measurement-length-
mass-and-time/

15. Arthur Beiser. (1963). Concepts of modern physics.
McGraw Hill Book Company, Pg 326

Part 7: Subquantum Physics 101

Introduction

None

Chapter24: Introduction to Particles

1. plato.stanford.edu. Ancient atomism. Retrieved September
29, 2011 from http://plato.stanford.edu/ archives/fall2008/
entries/atomism-ancient/

2. azonano.com. Researchers observes unexpectedly small
proton radius in a precision experiment. Retrieved September

29, 2011 from http://www.azonano.com/
news.aspx?newsID=18428

3. E. Rutherford. (1911). The scattering of alpha and beta
particles by matter and the structure of the atom. Philosophical Magazine, Series 6, vol. 21

4. images-of-elements.com. Particle zoo. Retrieved
September 29, 2011 from http://images-of-elements.
com/particle-zoo/

5. en.wikipedia.org. Standard model. Retrieved September
29, 2011 from http://en.wikipedia.org/wiki/Standard_Model

6. en.wikipedia.org. Standard model. Retrieved September
29, 2011 from http://en.wikipedia.org/wiki/Standard_Model

7. hyperphysics.phy-astr.gsu.edu. Fundamental force concepts. Retrieved September 29, 2011 from http://
hyperphysics.phy-astr.gsu.edu/ hbase/forces/funfor.html

8. quora.com. What does it mean to say a wave-function
collapses? Retrieved September 29, 2011 from http://
www.quora.com/ What-does-it-mean-to-say-a-wave-
function-collapses

Chapter 25: Introduction to the Background Sea

1. science.jrank.org. Virtual Particles. Retrieved September
29, 2011 from http://science.jrank.org/pages/7195/ Virtual-
Particles.html#ixzz1ZKXBvSin

2. Barnett, R. Michael, Henry Mühry, and Helen R.
Quinn. The Charm of Strange Quarks. New York: Springer-
Verlag, 2000.

3. Arthur Beiser. (1963). Concepts of modern physics.
McGraw Hill Book Company, Pg 324

4. Arthur Beiser. (1963). Concepts of modern physics.
McGraw Hill Book Company, Pg 324

Chapter 26: Quantum Weirdness Explained

1. Leonid A. Bendersky and Frank W. Gayle. (2001).
Electron diffraction using transmission electron microscopy.
Journal of Research of the National Institute of Standards and
Technology, 106 Pg 997–1012.

2. quora.com. What does it mean to say a wave-function
collapses? Retrieved September 29, 2011 from http://
www.quora.com/ What-does-it-mean- to-say-a-wave-
function-collapses

Chapter 27: Introduction to the Super Matrix

1. en.wikipedia.org. Bootstrap model. Retrieved May 8, 2011
from http://en.wikipedia.org/wiki/Bootstrap_model

Part 8: Mass, Energy, Space, and Time

Introduction
None

Chapter 28: Particlescope

None

Chapter 29: Space and Time

1. www.britannica.com. Pauli exclusion principle.
Retrieved September 28, 2011 from http://www.britannica.
com/ EBchecked/ topic/447124/Pauli-exclusion-principle

Chapter 30: Mass

None

Chapter 31: Energy

None

Part 9: Universal Gas

Introduction

1. en.wikipedia.org. Statistical_mechanics. Retrieved September 28, 2011 from http://en.wikipedia.org/ wiki/ Statistical_mechanics

Chapter 32: Particle Dynamics

None

Chapter 33: Energy Levels, Orbits, and Bonding

1. www.britannica.com. Pauli exclusion principle. Retrieved September 28, 2011 from http://www.britannica.com/ EBchecked/topic/447124/Pauli-exclusion-principle

Chapter 34: The Four Forces

1. sldnt.slac.stanford.edu. Standard model of particle physics. Retrieved September 28, 2011 from http://www. sldnt.slac. stanford.edu/alr/standard_model.htm

2. whillyard.com. Standard model. Retrieved September 28, 2011 from http://www.whillyard.com/ science-pages/ forces-bosons.html

3. en.wikipedia.org. Weak interaction. Retrieved September 28, 2011 from http://en.wikipedia.org/wiki/Weak_force

4. Arthur Beiser. (1963). Concepts of modern physics.

McGraw Hill Book Company, Pg 297

5. scienceagogo.com. Experimental results hint at neutrino flavor change. Retrieved September 28, 2011 from http://www.scienceagogo.com/news/ 20110516001121 data_trunc_sys.shtml

Part 10: Super matrix

Introduction

None

Chapter 35: General Relativity

1. en.wikipedia.org. Action at a distance. Retrieved June 19, 2011 from http://en.wikipedia.org/wiki/Action_at_a_distance_ %28physics%29

perimeterinstitute.ca. What is warped spacetime? Retrieved June 19, 2011 from http:// www. perimeterinstitute.ca/ Outreach/Explore_Our_Universe/What_is_Warped_Spacetime?/

2. en.wikipedia.org. Lines of force. Retrieved June 19, 2011 from http://en.wikipedia.org/wiki/Lines_of_force

3. perimeterinstitute.ca. What is warped spacetime? Retrieved June 19, 2011 from http://www.Perimeterinstitute. ca/ Outreach/Explore_Our_Universe/What_is_Warped_Spacetime?/

4. en.wikipedia.org. Lines of force. Retrieved June 19, 2011 from http://en.wikipedia.org/wiki/ Equivalence _principle/

5. /abyss.uoregon.edu. Relativity. Retrieved Sept 25, 2011 from http://abyss.uoregon.edu/~js/ast122/lectures/ lec20.html

6. Feynman, Richard. (2006). QED. Princton University

Press. ISPN-13; 978-0-691-12575-6

7. Feynman, Richard. (2006). QED. Princton University
Press. Pg 38. ISPN-13; 978-0-691-12575-6

Chapter 36: Expansion of the Universe

1. en.wikipedia.org Expansion of space. Retrieved June 9,
2011 from http://en.wikipedia.org/wiki/Expansion_ of_
space/

2. Alan B. Whiting (2004). The expansion of space: Free
particle motion and the cosmological redshift. ArXiv preprint.
arXiv:astro-ph/0404095. Bibcode 2004Obs ...124..174W.

EF Bunn & DW Hogg (2008). The kinematic origin of the
cosmological redshift. ArXiv preprint. arXiv:0808.1081.
Bibcode 2009AmJPh..77..688B. doi:10.1119/1.3129103. Yu.
V. Baryshev (2008). "expanding Space: The root of
conceptual problems of the cosmological physics. Practical
Cosmology 2: 20–30. arXiv:0810.0153. Bibcode
2008pc2..conf...20B.

JA Peacock (2008). A diatribe on expanding space. ArXiv
preprint. arXiv:0809.4573. Bibcode 2008arXiv0809.4573P.

3. en.wikipedia.org. Galaxy. Retrieved June 9, 2011 from
http://en.wikipedia.org/wiki/Galaxy

4. Alec Eden. (1985). The search for Christian Doppler,
Springer-Verlag, Wien. Contains a facsimile edition with an
English translation.

5. Alec Eden. (1985). The search for Christian Doppler,
Springer-Verlag, Wien. Contains a facsimile edition with an
English translation.

6. Huggins, William. (1868). Further Observations on the

Spectra of Some of the Stars and Nebulae. Philosophical Transactions of the Royal Society of London 158: 529–564. Bibcode 1868RSPT..158..529H.

7. The Columbia Electronic Encyclopedia. (2007). Doppler shift of stars. Columbia University Press

8. Reference: Bélopolsky, A. (1901). On an apparatus for the laboratory demonstration of the Doppler-Fizeau Principle. Astrophysical Journal 13: 15

9. Reference: Hubble, E. (1929). A relation between distance and radial velocity among extra-galactic nebulae. Proceedings of the National Academy of Science, 15:168-173

10. www.britannica.com. Vesto Slipher. Retrieved June 10, 2011 from http://www.britannica.com/EBchecked/topic/546850/Willem-de-Sitter

11. Friedman, A. (1999). "On the Curvature of Space". General Relativity and Gravitation 31 (12): 1991–2000. Bibcode 1999GReGr..31.1991F.doi:10.1023/A:1026751225741. English translation

12. en.wikipedia.org Georges Lema-AEtre. Retrieved June 10, 2011 from http://en.wikipedia.org/wiki/ Georges_Lema%C3%AEtre

13. en.wikipedia.org Georges Lema-AEtre. Retrieved June 10, 2011 from http://en.wikipedia.org/wiki/Georges_Lema%C3%AEtre

14. en.wikipedia.org. Georges Lema-AEtre. Retrieved June 10, 2011 from http://en.wikipedia.org/wiki/Georges_ Lema%C3%AEtre

15. blogs.discovermagazine.com. Update Lemaitre vs. Hubble. Retrieved June 10, 2011 from http://blogs.discover

magazine.com/cosmicvariance/ 2007/11/25/update-lemaitre-vs-hubble/

blogs.forbes.com. Why Hubble's law wasn't really Hubbles. Retrieved June 10, 2011 from http://blogs.forbes.com/ johnfarrell/ 2011/06/15/why-hubbles-law- wasnt-really-hubbles/ ?partner=contextstory

16. www.suite101.com. Predicting the expanding universe. Retrieved June 16, 2011 from http://www.suite101.com/ content/predicting-the-expanding-universe-a27557

17. en.wikipedia.org. Cosmological constant. Retrieved June 18, 2011 from http://en.wikipedia.org/wiki/ Cosmological_constant

18. en.wikipedia.org. Edwin Hubble. Retrieved June 18, 2011 from http://en.wikipedia.org/wiki/Edwin_Hubble

19. en.wikipedia.org. Edwin Hubble. Retrieved June 18, 2011 from http://en.wikipedia.org/ wiki/Edwin_Hubble

20. en.wikipedia.org. Expansion of space. Retrieved June 9, 2011 from http://en.wikipedia.org/wiki/Expansion_ of_space

Sten Odenwald, Rick Fienberg (February 1993). Galaxy redshifts reconsidered. Astronomy Cafe. May 29, 2011.

21. en.wikipedia.org. Hubble's law. Retrieved June 17, 2011from http://en.wikipedia.org/wiki/Hubble%27s_law

zebu.uoregon.edu. Milky Way. Retrieved June 18, 2011from http://zebu.uoregon.edu/~soper/MilkyWay/ cepheid.html

22. en.wikipedia.org. Age of the universe. Retrieved June 18, 2011from http://en.wikipedia.org/wiki/Age_of_ the_universe

23. en.wikipedia.org. Hubble's Law. Retrieved June 17, 2011 from http://en.wikipedia.org/wiki/Hubble%27s_law

24. This is an observation by the author.

25. cecelia.physics.indiana.edu. Redshift. Retreived May 20, 2011 from http://cecelia.physics.indiana.edu/life/redshift.html

26. cecelia.physics.indiana.edu. Redshift. Retreived May 20, 2011 from http://cecelia.physics.indiana.edu/life/redshift.html

27. en.wikipedia.org. Thin-film interference. Retrieved June 10, 2011 from http://en.wikipedia.org/wiki/ Thin-film_interference

28. utc.edu. Redshift essay. Retrieved June 10, 2011from http://www.utc.edu/Faculty/LingJun-Wang/Redshift Essay.pdf

cecelia.physics.indiana.edu. Redshift. Retrieved June 10, 2011 from http://cecelia.physics.indiana.edu/life/redshift.html

Chapter 37: Entropy

1. en.wikipedia.org. Entropy. Retrieved October 15, 2011 from http://en.wikipedia.org/wiki/Entropy

Chapter 38: Systems Analysis

None

Chapter 39: The Spawning Universe

None

Chapter 40: A Forever Universe

None

Please visit

www.omenquest.com

www.ingramcontent.com/pod-product-compliance
Lightning Source LLC
Chambersburg PA
CBHW072023190526
45166CB00015B/5